THE WORLD AND ALL THE THINGS UPON IT

THE WORLD AND ALL THE THINGS UPON IT

NATIVE HAWAIIAN GEOGRAPHIES OF EXPLORATION

DAVID A. CHANG

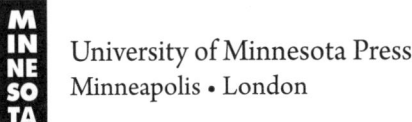

University of Minnesota Press
Minneapolis • London

Portions of chapter 5 were published as "Borderlands in a World at Sea: Concow Indians, Native Hawaiians, and South Chinese in Indigenous, Global, and National Space, 1860s–1880s," *Journal of American History* 98, no. 2 (September 2011): 379–98. Chapter 7 was previously published as "We Will Be Comparable to the Indian Peoples: Recognizing Likeness between Hawaiians and American Indians, 1834–1923," *American Quarterly* 67, no. 3 (September 2015): 859–86.

Copyright 2016 by David A. Chang

All rights reserved. No part of this publication may be reproduced, stored in a retrieval system, or transmitted, in any form or by any means, electronic, mechanical, photocopying, recording, or otherwise, without the prior written permission of the publisher.

Published by the University of Minnesota Press
111 Third Avenue South, Suite 290
Minneapolis, MN 55401-2520
http://www.upress.umn.edu

Printed in the United States of America on acid-free paper

The University of Minnesota is an equal-opportunity educator and employer.

21 20 19 18 17   10 9 8 7 6 5 4 3 2

Library of Congress Cataloging-in-Publication Data
Chang, David A., author.
The world and all the things upon it : native Hawaiian geographies of exploration / David A. Chang.
Minneapolis : University of Minnesota Press, [2016] | Includes bibliographical references and index.
Identifiers: LCCN 2015050876| ISBN 978-0-8166-9941-4 (hc) | ISBN 978-0-8166-9942-1 (pb)
Subjects: LCSH: Discoveries in geography—American. | Explorers—Hawaii. | Hawaiians—Travel. | Geographical perception—Hawaii. | Hawaiians—Historiography. | Hawaii—History. | Hawaii—Social conditions.
Classification: LCC G222 .C53 2016 | DDC 910.89/994—dc23
LC record available at http://lccn.loc.gov/2015050876

CONTENTS

Introduction Making Native Hawaiian Global Geographies — vii

1 Looking Out from Hawai'i's Shore — 1
The Exploration of the World Is the Inheritance of Native Hawaiians

2 Paddling Out to See — 25
Direct Exploration by Kānaka in the Late Eighteenth Century

3 A New Religion from Kahiki — 79
Christianity, Textuality, and Exploration, 1820–1832

4 The World and All the Things upon It — 103
Geography Education and Textbooks in Hawai'i, 1831–1878

5 Hawaiian Indians and Black Kanakas — 157
Racial Trajectories of Diasporic Kanaka Laborers

6 Bone of Our Bone — 195
The Geography of Sacred Power, 1850s–1870s

7 "We Will Be Comparable to the Indian Peoples" — 227
Recognizing Likeness between Kānaka and American Indians, 1832–1895

Epilogue Genealogies of the Present in Occupied Hawai'i — 249

Acknowledgments — 259
Notes — 263
Index — 305

INTRODUCTION

MAKING NATIVE HAWAIIAN GLOBAL GEOGRAPHIES

What if we were to understand indigenous people as the active agents of global exploration, rather than the passive objects of that exploration? What if, instead of conceiving of global exploration as an activity just of European men such as Christopher Columbus or James Cook or Ferdinand Magellan, we thought of it as an activity of the people they "discovered"? What could such a new perspective on the project of global exploration reveal about the meaning of geographical understanding and its place in struggles over power in the context of colonialism?

The World and All the Things upon It addresses these questions by tracing how Kānaka Maoli (meaning indigenous Hawaiian people) in the nineteenth century explored the outside world, generated their own understandings of it, and placed themselves strategically in the understandings of global geography they created.[1] This book looks at travel, spirituality, print culture, sexuality, gender, labor, education, and race to shed light on how constructions of global geography became a site through which Hawaiians as well as their would-be colonizers understood and contested imperialism, colonialism, and nationalism.

When Native Hawaiian people paddled their canoes out to explore the ships that Captain James Cook had anchored off Kauaʻi in January 1778, they began a new stage in the ongoing Hawaiian exploration of the world, a stage that is far larger in scope but much less understood by scholars than the European exploration of the Hawaiian archipelago. In one of the culminating stages of a tremendous age of exploration of the Pacific, their ancestors had first peopled the islands around 300 CE. Kānaka Maoli maintained contact via long-distance canoe travel with the islands far to the south until about 1400 CE, when such voyages came to an end. Four centuries later, the exploration of Cook's vessels and his men opened a second great age of exploration by Kānaka. This book traces how indigenous Hawaiian people explored the outside world and how they constructed Kanaka Maoli ways of understanding the globe. To trace how Hawaiians placed themselves in the ideas of global geography they were creating, this book takes a long-range view stretching from the period before Cook's arrival and into the second decade of the twentieth century.

A long-range view is essential to resist two fallacies: first, that it was Captain Cook in 1778 that introduced Kānaka to the idea that a world beyond their shores existed, and second, that annexation to the United States in 1898 brought an end to Kanaka people, Kanaka resistance, or Hawaiian history. Kanaka Maoli thinking about the nature of the world started long before Cook arrived at Kauaʻi, and thus this book opens by exploring what Kānaka knew about places far from their islands long before Cook sailed into nā kai ʻewalu (the eight seas, as Kānaka often term the waters surrounding their islands). Similarly, the illegal occupation of Hawaiʻi did not bring an end to independent Kanaka thought about the world and their place in it. Therefore, this book does not conclude until the early decades of the twentieth century, revealing that ideas about the nature of the globe remained at the center of Kanaka Maoli resistance to American colonialism. In fact, this book is itself an intervention into the politics of Hawaiian global geography. It takes part in the centuries-long process that it describes: a process of emphasizing that Hawaiians can and should understand their world from their own Hawaiian perspectives, not those of their colonizers, and that Hawaiʻi is best understood as a land deeply rooted in the Pacific sea of islands, not merely a peripheral dependency of some other power (notably the United States).

As generations of critical scholars in history, geography, literature, and cultural studies have now emphasized, the classic narrative of European discovery celebrates an untroubled quest for knowledge spurred by the Enlightenment. This story of discovery masks the fact that the act of exploring and the creation of the ideas and practices we commonly call European global geography served the process of European imperial expansion.[2] Despite these critiques, that narrative of discovery has rarely been met head-on by accounts of how non-European peoples in the age of high imperialism also generated their own global geographies. Similarly, scholars have emphasized that Western ideas about differences between a supposedly civilized West and an exoticized and allegedly savage non-West depended (and depend still) on works of geographical description. Rarely, however, have historians asked how non-Western people imagined and even forged their own geographies of their colonizers and the broader world.[3] To the extent that scholars have investigated these questions, many have generally emphasized "first encounters" between non-Europeans and Westerners rather than the longer process in which non-Europeans explored, generated, and engaged global geographies.[4] Thus we have a growing scholarly literature on "Pacific worlds" that traces flows of commodities, capital, bodies, diseases, and Western power, but we know precious little about how Pacific Islanders understood the changing world of the "age of discovery" and the age of empires it served.[5]

This study takes up that task. It emphasizes, moreover, that there is no better way to understand the process and meaning of global exploration than by looking out from the shores of a place, such as Hawai'i, that was allegedly the object, and not the agent, of exploration. This is not, however, a case study chosen for analytic purposes only: as a Native Hawaiian, I am committed to the study of Hawai'i and Kānaka Maoli and consider them to be worthy of study in their own right. I believe that demonstrating how Kānaka in the past actively engaged with the outside world in the pursuit of Hawaiians' own interests can speak to Kānaka in the present who are engaged in resisting American political, economic, and cultural colonialism and forms of neoliberal globalization that would reduce Hawai'i to the role of a military and recreational hinterland in the service of North American, European, and Asian powers.

Yet even for readers not directly invested in the study or the future of Hawai'i, the archipelago offers a number of advantages for understanding the history and the politics of thinking about global geography. First, whereas scholars must struggle to recover indigenous voices in most of the United States and much of the world, indigenous Hawaiian perspectives are richly documented early in the period after Western incursion: by the late 1830s, most Kanaka adults were literate. In the following decades, they generated tens of thousands of texts (from letters to chants, from newspaper articles to books, and from songs to broadsides) that allow us to access indigenous perspectives. Second, Hawai'i occupied a central place both economically and strategically in the nineteenth-century maritime world. Indeed, it was for this reason that colonial powers coveted control of the islands. Thus, understanding

Kanaka Maoli perspectives helps us better understand the process that built the globalizing world of capitalism and colonial powers in the nineteenth century.

Nineteenth-century Kānaka were hardly isolated: they actively engaged the world around them and the foreigners among them. Already in the late eighteenth century, Hawaiian women and men were embarking on American and European ships and soon could be found working as sailors and whalers in the Pacific and the Atlantic and as laborers in the U.S. fur trade. Others flocked to California in the 1848 gold rush. Meanwhile, Americans, Asians, and Europeans flocked to nineteenth-century Hawaiʻi, and Native Hawaiians were in close contact with these foreigners, many making families with them. In all of these travels and in the making of these relations with other peoples and places, as well as in the information presented in the Hawaiian schools, the Hawaiian-language press, and the foreign-language press, Hawaiians demonstrated an active process of geographical exploration.

This process of exploration is strikingly apparent in Hawaiian-language historical sources and strikingly absent from English-language historical accounts. In source after source—Hawaiian-language newspapers, Hawaiian Kingdom records, the Hawaiian-language letters of Kanaka Maoli missionaries to other Pacific Islands, school essays, and more—Hawaiian initiative and engagement are obvious. And yet this global engagement is almost altogether missing as a theme in the historical literature on Kānaka Maoli. For many of these works, which include the most influential works on the history of Hawaiʻi, this absence results from the biases of their American authors. Noenoe Silva has forcefully and persuasively pointed out the perverse colonialism of a historical literature on nineteenth-century Hawaiʻi written almost exclusively by people who cannot read the language of Hawaiʻi, the language in which Kānaka Maoli documented their ideas on tens of thousands of pages from the 1820s to the 1920s and beyond.[6] Colonialist bias explains the failure by American historians to learn and use this language. To borrow the vivid cartographic vocabulary used by Kanaka gender studies scholar Lisa Kahaleole Hall, we must "map" the colonial "strategies of erasure" that render Kānaka and their resistance invisible, and we must think through "sites of resistance" to those erasures.[7] For the nineteenth century, the project of mapping erasure and situating resistance that Hall names requires the use of the Hawaiian language. Without reading these texts, historians cannot see the inescapable evidence in them of how intensely Kānaka Maoli thought about the nature of the world and their place in it.

This is one of the principle reasons that when this book includes quotes from Hawaiian texts, the Hawaiian-language original is given first, before the English translation: to insist on the primacy of Hawaiian sources for understanding the Hawaiian past. Following the practice that is now common in Hawaiian studies, this creates interruptions—even eruptions—in the text. It is my hope that, like the volcanic eruptions that are still forming the Hawaiian archipelago as Pele creates new earth and new islands, these linguistic eruptions are constructive. Their purposes are multiple. Whereas translations create a sense of comfort for the reader, these eruptions serve to jar the reader from too easily assuming that she or he understands people and perspectives from the Hawaiian past. These eruptions forcefully insert Hawaiian people and perspectives into the text. They allow readers of Hawaiian to consider alternate translations. Most of all, quotations in Hawaiian serve as reminders to us that scholars must use Hawaiian-language sources to study Hawai'i in periods when Hawaiian was the dominant language in the archipelago, before being almost totally wiped out by the educational and cultural hegemony of U.S. colonial power. The particular history of Hawai'i and of studies of Hawai'i therefore compel these eruptions, which are intended to encourage readers and scholars to attend to Hawaiian-language sources, particularly in the study of intellectual history.

Because Native Hawaiian intellectual history has received very little attention from writers and scholars, the fact that Kānaka Maoli were deeply engaged with ideas about the world beyond the shores of Hawai'i is largely missing from our accounts of the Hawaiian past. This is true both in writing on Hawaiian history that is explicitly sympathetic to Native Hawaiian concerns and in writing that is not. For understandable reasons, much of the historical writing on Hawai'i is largely focused on loss: the loss of Hawaiian lands, the loss of a sovereign Hawaiian kingdom, and the loss (sometimes more perceived than real) of Hawaiian culture.[8] Unfortunately, this emphasis on lost land, lost power, and lost culture has too often obscured the ways that Kānaka have maintained an ongoing intellectual engagement with the present, the future, and the outside world—even while ever safeguarding what we call "tradition" today.

In recent decades, Kanaka intellectuals have led the way in demonstrating that the history of the lāhui (nation) and the history of its ideas must be studied together, with a full understanding that the Hawaiian nation has always been a diverse, complex, and changing intellectual and cultural world. Other scholars have followed their lead, and the extraordinary literature in Hawaiian studies now points to the central place of the history of ideas (that is, intellectual and cultural history) in the history of the Hawaiian nation. Much of this work centers on the crucial nineteenth century, when Kānaka unified, consolidated, and defended a Hawaiian nation-state—an effort that depended on extensive and intensive Kanaka intellectual labor. My work is deeply indebted to scholars (especially but not exclusively Kānaka) who have engaged in tracing the contours and exploring the depths of Native Hawaiian intellectual history. Many of these scholars would not necessarily consider themselves to be historians, let alone intellectual historians. Yet as pathbreaking works by Lilikalā Kameʻeleihiwa, Jocelyn Linnekin, Sally Engle Merry, Jonathan Kamakawiwoʻole Osorio, Noenoe Silva, John Charlot, J. Kēhaulani Kauanui, Noelani Arista, Hokulani Aikau, Kamanamaikalani Beamer, and others have demonstrated, Kanaka Maoli intellectual and cultural history is inextricably intertwined with the history of formal, governmental politics in Hawaiʻi. While ideas are important to politics in all places, the literature in Hawaiian studies has been particularly adept at centering the history of ideas in the history of contests over power. This is true whether one looks at politics in resistance to American colonialism (as Silva demonstrates in her exploration of the deployment of story, hula, and a sovereign newspaper apparatus), in the functioning of the state (as Osorio reveals in his meticulous tracing of struggles between Haole [Westerners] and Kānaka over ideas and meanings in kingdom politics), in the initiatives of political leaders (as Beamer emphasizes in his study of aliʻi and their proactive engagement with foreign ideas in the defense of Hawaiʻi), or in the other ways that Hawaiian studies scholars have understood politics.[9] In this history of Hawaiian ideas, Kānaka have been engaged with the future as well as the past.

Similarly, in their thinking about the nature of the world of which they are a part, Kānaka have been engaged with places far away as well as places close to home. *The World and All the Things upon It* draws its title from a work of global geography (titled "Ka Honua Nei a me na Mea a Pau Maluna Iho" in Hawaiian) that the Hawaiian intellectual J. H.

Kānepuʻu published in 1877 in serialized form in a Hawaiian-language newspaper. This book borrows that title to signal that it emphasizes that indigenous geography means looking out at the global world from indigenous perspectives as well as looking closely at homelands. The tremendous strength of Kanaka Maoli geographers has been to demonstrate the relationship of moʻokūʻauhau (genealogy) that ties Kānaka and ʻāina (land) together. From that relationship comes a profound knowledge of and sense of connection to land and sea, wind and rain, uplands and lowlands. Kānaka sing these connections in mele (songs), narrate them in moʻolelo (stories), and render them with astonishing specificity in the knowledge of wahi pana (often translated as "storied places," because renowned sites are tied to narratives). Katrina-Ann R. Kapāʻanaokalāokeola Nākoa Oliveira rightly underscores that Hawaiian places are "ancestral places" in more ways than one.[10] Kānaka today know they move through places the kūpuna (ancestors) lived and knew and that the islands and the Kānaka are literally kin. The first Kanaka, Hāloa, was the child of Wākea, who fathered the islands, and the grandchild of the female god Papahānaumoku, "Papa-who-births-islands." Scholars of indigenous geography elsewhere have noted similarly deep ties to the land and have emphasized that indigenous geography is crucial to understanding the dispossession of indigenous people and working for ecological restoration and social justice. As Kanaka geographer Kali Fermantez puts it, "re-placement" by "reconnecting to Hawaiian ways of knowing and being that are rooted in space" can be a powerful means to counter the "dis-placements" Hawaiians face in their own homeland.[11] This project proceeds from these studies and depends on them as the foundation for a departure in a different direction: tracing how Kānaka have looked outward from Hawaiʻi to construct Kanaka global geographies. As Kanaka geographer Carlos Andrade has argued, what we need is not just a geography of Hawaiʻi, but a Hawaiian geography—that is, geography understood through Hawaiian systems of knowledge.[12] This book emphasizes that a Hawaiian geography must be a *global* geography as well as a geography of the islands. This project is a crucial counterpart to the more established focus of indigenous geography and is equally urgent intellectually and politically, given that the engagements of Kānaka and other indigenous people extend around the globe.

This book emphasizes that the history of Hawaiian ideas and of Hawaiian exploration must not be limited to the words and actions of chiefs and monarchs. All Kānaka were engaged in the process of exploration, but to date we have learned the most about the travels of the most powerful and most privileged. This book therefore explores the activities of less prominent people and the texts they might have read or been taught from. *The World and All the Things upon It* is a social and cultural history that aims to shed light on means and results of Hawaiian global exploration that general readers might find are less expected (such as missionization by Hawaiians) or that scholars have not studied in detail (such as Hawaiian laborers' engagement with the nineteenth-century American racial system or Hawaiians' writings about American Indians). Because of that focus, this book leaves much out, including the royal tours, foreign travels, and diplomatic missions of ali'i and mō'ī (chiefs and monarchs). It does not even recount Davida Kalākaua's tour of the world (the first circumnavigation of the globe by a reigning monarch of any nation) or his effort to build a confederacy of Pacific nations. These travels and initiatives were undeniably important acts of global exploration, and these individuals were engaged in strategically placing Hawai'i in global geographies—the processes that occupy the center of this book. But they have received the lion's share of historical attention.[13] Thus, aside from the profile of Ka'iana (because he was the best documented of the early travelers and because his story illustrates the book's themes so well), this book does not dwell on the history of the ali'i.

This book follows the story of how Hawaiians explored and generated conceptions of global geography through a series of seven chapters. Chapter 1 introduces certain key concepts in the way Kānaka Maoli understood the world prior to Western incursion. It delineates the importance of perspective in Hawaiian geographical thought, Kanaka knowledge of and interest in faraway places, and the conviction that in far-off places there were powerful and good things that Hawaiians could use for their own purposes. Chapter 2 traces the explorations of two early Kanaka voyagers overseas to demonstrate that from the moment that they caught sight of Cook's ships off their shores, Kānaka actively sought to understand the contours of the globe and what the outside world might mean for them. Chapter 3 looks at the Kanaka encounter with Christianity from the 1810s forward and seeks to correct standard narratives of missionary endeavor. It centers Native Hawaiians' active use of Christianity and the search for knowledge from the outside for

Native Hawaiian purposes. It further contends that for Kānaka, exploring the outside world through Christianity meant reconnecting to other Pacific Islanders, not acquiescing to Western colonialism. Chapter 4 examines global geography as it was taught in the schools and presented in Hawaiian-language textbooks. These textbooks were riven with contradictions. As translations from American volumes, they echoed their source texts' racism and colonialism. Yet because Native Hawaiians' skills were needed to prepare the translations, they also reflected indigenous perspectives that rendered Hawai'i as central rather than peripheral. Chapter 5 looks at the circumstances of Kanaka laborers in New England and in California, emphasizing the ways that Kānaka Maoli entered into indigenous communities and other communities of color and engaged with the notion of race—a crucial category that was fundamental to Americans' understanding of world geography. Chapter 6 considers the ways that Hawaiians used religion to pursue a central place in the world. Some Hawaiians saw different Christian denominations as organizing principles for competing conceptions of global power, but others deployed Hawaiian story and chant to affirm that sacred power imbued Hawaiian places. Chapter 7 traces writing about American Indian people in Hawaiian-language newspapers from the 1830s to the end of the century. It argues that over the course of the nineteenth century, Kānaka came to see a likeness between themselves and American Indian people, and that act of recognizing likeness was part of the process of thinking about the global category of the indigenous—a conversation that is very current today. As a whole, this book demonstrates that Kānaka actively engaged in the process of global exploration, that in the process they deliberately shaped their place in the world, responding to the challenges of Western colonialism.

He Malihini Au: Birthplaces and Bone Places, 2015

He malihini au mai ka 'āina 'ē
Malihini maoli nō, malihini 'ōiwi.
Hānau 'ia e nā iwi, na nā iwi
Ma nā kulāiwi o ke kula ākea o ka loko iloko.
Mai ka liko a ka makua
Lo'i 'ole, ka huli iloko o ka 'āina momona o nā Maoli 'ē a'e.
Malihini maoli au.

Malihini lāua
Hānau ʻia e nā iwi, na nā iwi
Ma nā kulāiwi o ke kula ākea o ke kahawai nui
Kahawai o ke kaua
O nā kauā
O ka ua ʻē: ka ua, ka hoʻohekili i ke ikiiki.
ʻAuwai ʻole, nā huli iloko o ka ʻāina momona o nā Maoli ʻē aʻe.
Malihini ʻōiwi.

"He malihini au mai ka aina e." "I am a stranger from a foreign land." In 1865, Samuel Manaiākalani Kamakau took up a pen and wrote those words for his people, the Kānaka ʻŌiwi (Native people) of Hawaiʻi. Kamakau's phrase (which is the inspiration for my poem, above) was the opening words of a major work: a history and geography of Hawaiʻi presented as a narrated tour of the islands. It was a profoundly Native work: it followed Native conventions of telling history by recounting the stories of wahi pana, and it was published as a series of newspaper articles in Hawaiian, the Native language of the Hawaiian Islands. Yet Kamakau began by declaring himself a malihini: a stranger or foreigner. This was a narrative conceit. Kamakau was a Kanaka Maoli o kō Hawaiʻi pae ʻāina—a Native person of the Hawaiian Islands. But he invoked a narrator that came from the land from which the forebears of the Kānaka Maoli had hailed. He adopted the metaphor of the malihini so that he could show Kānaka Maoli their own land through foreign eyes, that they might see it anew and know it better.[14]

"He malihini au mai ka ʻāina ʻē." "I am a stranger from a foreign land." In 2015, I type these words into a keyboard, and in my case, they are not a narrative conceit. This is a book about the history of Hawaiʻi, and even though I am of Native Hawaiian descent and identify as a Native Hawaiian person, I am a malihini to Hawaiʻi. Milwaukee is what is called in Hawaiian my one hānau, the sands of my birth. I have lived in the Midwest for most of my life. I have more Chinese and European ancestors than Hawaiian ancestors, was raised with little contact with Kanaka Maoli communities, and only learned Hawaiian as an adult. These factors do not mean I am not Kanaka Maoli. Kānaka Maoli generally figure Native belonging through what anthropologists would call bilateral lineal descent: any person with lineal descent from any Kanaka Maoli ancestor is Kanaka Maoli. Under such a reckoning, I am Kanaka Maoli, and my father and other family members have encouraged me to understand myself as such. But the facts of my upbringing do mean I am a particular kind of Kanaka Maoli: he malihini maoli. This term of my own invention means "a Native stranger" and can even be translated as "a real foreigner." Knowing this, I turn to my keyboard to write, just as Kamakau took up his pen a century and a half ago. It is my hope that by naming and knowing myself as a malihini maoli, and by writing from the perspective of a Kanaka from outside Hawai'i into which I was born, I can write with utility to Kānaka about our history. Although I would never claim a fraction of his authority, I look to Kamakau as an intellectual kupuna: a Kanaka Maoli historian from whom I, as a Kanaka Maoli historian, draw inspiration and a lesson about my kuleana, or responsibility and role. I write this book, then, with deep humility and sincere hope that my position as a malihini maoli and my training in U.S. history and indigenous studies position me to make a useful contribution by discussing Kānaka and their exploration of the outside world.

This book emphasizes the bonds between Kānaka and other indigenous people. This relationship was my point of entry into the study of indigenous history. I grew up in the country of the Potawatomi nation and within a few hours' drive of the Menomonie, Ojibwe, Oneida, Ho Chunk, and Stockbridge Munsee Band of Mohicans, and as a teenager I was already reflecting on our situation as Hawaiians in relation to that of American Indian people. This is a question that it is imperative that Kānaka in North America ask, as Hokulani Aikau (a Kanaka scholar raised on Goshute land in Utah) argues.[15] It was through my dissertation at the University of Wisconsin (on allotment and landownership in the Muscogee Creek Nation of Oklahoma) that I entered into indigenous studies. It was only with the emergence of a new generation of Hawaiian studies scholarship, and especially the 2004 publication of Noenoe Silva's *Aloha Betrayed,* that I came to believe that I might contribute to Hawaiian studies. And so I took up the study of the Hawaiian language, determined to reclaim a linguistic heritage and take up the charge that Silva laid out: to move beyond colonialist histories of Hawai'i that depend only on English-language sources, and to replace them with histories of Hawai'i that draw on Hawaiian-language sources to center the experiences and perspectives of the indigenous people of Hawai'i.

The research for this book began with a question: how did Kānaka in the nineteenth century write about the past, and how did those writings resemble or differ from Western historical thought and methodology? This initial research question was shaped by my perception of my kuleana as a Native Hawaiian academic historian. I was well placed to research and tell this story. But as I delved into the sources—especially the nineteenth-century Hawaiian-language newspapers, which contain many serialized histories—I was struck by the relationship between historical and geographical writing. Historical writing constantly referred to specific places and often taught history by talking about those places. Kamakau's serialized story mentioned above exemplifies this model. Kamakau told the history of Hawai'i through a tour of places in the islands, in part because of the established model of teaching history through the story of wahi pana. Similarly, serialized histories of England, of France, of Russia, and of other countries contained within them constant references to far-off places. Most of these were undoubtedly translations from English-language works, and so this evocation of geography sprang from a different tradition than Kamakau's decision to tell history through geography. Still, it became clear to me that in all sorts of historical writing, geography was at issue.

As I considered this fact, I realized that for Hawaiians in the long nineteenth century, there was a particular urgency to certain geographical questions that matter to all nations. What is the nature of the world they inhabit? What is their place in the world? And how can they work to shape their position in the world to defend their own interests? As I reflected on those questions, the project shifted, and I arrived at what I think is a better understanding of my position and kuleana in regard to Hawaiian studies as a Native outsider, as a malihini maoli. From that position, as I read the words and traced the actions of Kānaka Maoli in the nineteenth century, I found that the exploration of the world is the inheritance of Native Hawaiians.

A note on spelling: contemporary Hawaiian orthography makes use of diacriticals to express the glottal stop and vowel length, but nineteenth-century texts only occasionally used apostrophes or dashes to express the glottal stop and only very rarely attempted to indicate vowel length. This poses some challenges, as glottal stops and vowel length are very important for discerning what word the author intended. Therefore, in order to avoid ascribing mistaken meanings to Hawaiian texts, I have tried to maintain their original spelling in quotations, although I have used what is now standard Hawaiian orthography in passages of my own writing. Names of people, akua (gods), and places have posed a particular challenge. Except in the case of the most well-known names where I can be relatively sure of correct spelling using diacriticals (such as Kalākaua, Kū, or Lānaʻi), I have made an effort to retain the spelling used in source texts rather than using updated orthography. I have made this choice out of concern that I might introduce incorrect renditions of names.

For an explanation of the term Kanaka (plural, Kānaka) and its usages, please see note 1 for this chapter.

All translations from the Hawaiian are my own, unless otherwise identified.

1
LOOKING OUT FROM HAWAI'I'S SHORE

THE EXPLORATION OF THE WORLD IS THE INHERITANCE OF NATIVE HAWAIIANS

Mai Kahiki mai ka wahine 'o Pele,
Mai ka 'āina o Polapola,
Mai ka pūnohu a Kāne,
Mai ke ao lalapa i ka lani, mai ke ao 'ōpua.

Lapakū i Hawai'i ka wahine 'o Pele.
Kālai i kona wa'a, Honua-i-ākea,
Kō wa'a, e Ka-moho-ali'i,
Holoa mai ka moku.

From Kahiki came the woman Pele,
From the land of Polapola,
from the rising reddish mist of Kāne,
from clouds blazing in the sky, horizon clouds.

Restless desire for Hawai'i seized the woman Pele.
Ready-carved was the canoe, Honua-i-ākea,
your own canoe, O Ka-moho-ali'i,
for sailing to distant lands.[1]

These are the opening words of an ancient song known as "Ka Huaka'i a Pele" (Pele's Voyage). Today, Pele is the most celebrated of the Hawaiian gods. Many Kānaka Maoli (Native Hawaiians) honor this beloved and yet fearsome akua (god). Her fiery lava has built up the islands, is building them still, and has the power to lay waste to the land. Tourists catch a glimpse of this power in images of the "volcano goddess" on souvenirs and from stories that rangers at Volcanoes National Park recount. In a sense, to many people, both Kānaka and foreigners, Pele is Hawai'i.

And yet this song reminds us that at her origin, Pele was not Hawaiian. She came from Kahiki, and more specifically from "the land of Polapola." The song signals that Polapola is sacred: it is wreathed in a mist of the major god Kāne, a mist that is red, the color of many sacred things. But the song also signals that Polapola is a physically real place: it was Pele's homeland, linked by the ocean to the land of her destination. When we render the place names in the song into their more familiar forms, we learn that Pele hailed from the very real island of Bora Bora (Polapola) in Tahiti (Kahiki). From there, she traveled 2,500 miles north to her new home in Hawai'i. She made the voyage aboard a canoe that was "ready-carved" for her and her shark-god brother, Kamohoali'i.

The phrase that names Pele's canoe, Honua i ākea, tells us something about the "restless desire" that drove Pele to Hawai'i. In Hawaiian, "honua" means "earth" and "ākea" means "broad, wide, spacious, open, unobstructed" and "public" in the sense that it is something that all can see and know. In Tahitian, the cognates of these words, "fenua" and "ātea," have the same meanings.[2] Joined together in this way, the phrase-name of Pele's canoe suggests (in the languages of her homeland and of her new home) how the Hawaiians who sang her song saw the Earth. It was a broad and expansive place, a world to behold, open for all to see and know. And so with a restless desire for Hawai'i, Pele boarded a vessel named for the breadth of the Earth, a vessel that was carved for "sailing to distant lands."

Spurred by this same belief that the world was a wide-open place to seek and understand, Native Hawaiians engaged in an intense process of global exploration in the century after foreigners appeared in Hawai'i in 1778. Some sailed to distant lands, as Pele had done. Others explored the world without leaving home, by engaging with foreign people; poring over, translating, and writing books about far-off countries; and in many other ways embracing world exploration. But why did Kānaka turn to exploring the globe so quickly and enthusiastically? In part, they did this because the Hawaiian exploration of the world did not truly begin with Cook's arrival. Kānaka already had a long history of exploring the earth and reflecting upon it.

Exploration is the motive power that brought the ancestors of the Kānaka Maoli to the archipelago from islands far to the south. Exploration is embedded in the sacred stories that explain the creation and working of the world to Kānaka Maoli. Exploration is retold in the songs and stories that explain the ocean around them and the many lands beyond the horizon. Those same stories and songs, and the Hawaiian language, reveal that prior to the arrival of Haole (Westerners) in 1778, Kānaka already had a deep and broad knowledge of the world, particularly of the ocean that was their home. At the very least three hundred thousand people, but more likely seven hundred thousand and perhaps as many as a million, inhabited eight islands stretching from Hawaiʻi to Niʻihau, practicing intensive agriculture and aquaculture, fishing the waters around them, and supporting an elaborate political and religious structure. Although they lived far from other lands and had been out of direct contact with places beyond the Hawaiian Islands for three centuries or more, the outside world was not unknown to them. Rather, Kānaka understood the world in ways that centered their own perspective and encouraged them to look out from that perspective with confidence to seek still greater knowledge. A heritage of exploration favored a drive toward exploration.

This chapter introduces certain key concepts in Kanaka Maoli world geography—that is, in the way Kānaka Maoli understood the world. It explores only those concepts that most directly affected the ways Kānaka engaged the renewed access to faraway places that came with the arrival of outsiders in 1778: the importance of perspective in Hawaiian geographical thought, the knowledge that there were faraway places much like the Hawaiian Islands, an interest in those places, and the conviction that in distant lands there were powerful and good things that Hawaiians could use for their own purposes. These ideas would fundamentally shape how Kānaka would embrace the outside world and how they would explore it in the decades after European incursion. This chapter offers only a glimpse of the depths of Kanaka knowledge of the world as recent Kanaka scholars have explored it—the grounded knowledge of the islands, the mental cartographies of the land, the conceptualization of land, sea, and sky.[3] All of these are essential to understanding Hawaiian geography, but they are somewhat apart from the purpose of this book, which is to understand the history of Kanaka Maoli explorations of global geography. For that story, the crucial thing to know is that Kanaka exploration of the world after European incursion was the continuation of a process of the Hawaiian exploration of the globe that began long before Captain Cook ever set foot on a ship.

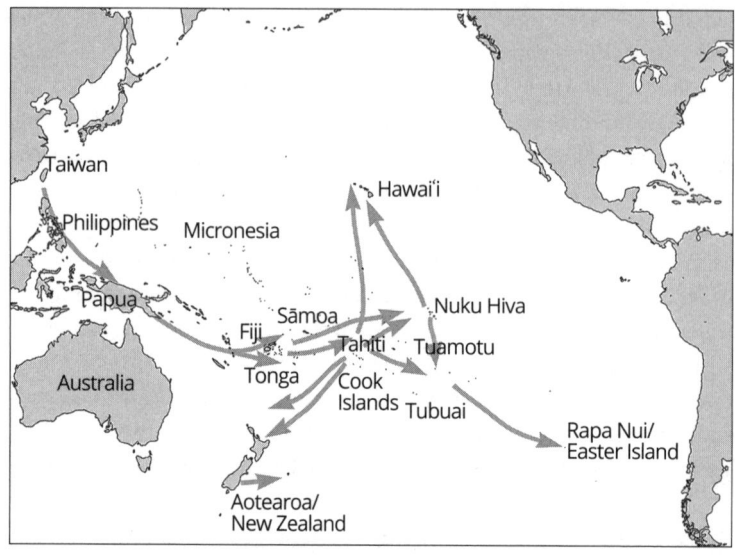

Routes of the exploration and settlement of Polynesia. Map by Matt Lindholm.

The exploration of the world is the inheritance of Kānaka Maoli because their ancestors settled the pae ʻāina Hawaiʻi (the Hawaiian archipelago) in the culmination of an extraordinary history of seeking. Beginning about four thousand years ago, men and women fanned out from Asia—probably the place we now call Taiwan—and across the rolling swells of the Pacific. Pushing forth across the sea in ocean-going vessels called waʻa in Hawaiian, *vaʻa* in Tahitian, *waka* in Māori, *vaka* in Marquesan, Tongan, and Rarotongan, and similar cognates in other languages, these men and women came to settle Hawaiʻi, Sāmoa, Tonga, Fiji, Tokelau, Hiva Oa, Tahiti, Vanuatu, Niue, Rapa Nui (Easter Island), Moʻorea, Huahine, Raʻiātea, Bora Bora, Aotearoa (New Zealand), Tuvalu, Fatu Hiva, Nuku Hiva, Rarotonga, Mangaia, Mangareva, Kiribati, and dozens of other islands. Dozens of archipelagos became the homes of people who shared common origins, similarities of language, and similarities in their beliefs about the world and the proper way to live in it. In the late eighteenth century, Europeans came to call these many islands "Polynesia"—a colonial term invented for colonial purposes.[4] The peopling of Hawaiʻi was part of a heroic late stage in the Polynesian exploration of the ocean world, beginning perhaps around

300 CE, perhaps several centuries later. For hundreds of years, astonishing long-distance canoe voyages maintained direct contact between Hawai'i and the far-off archipelagos that included Tahiti, Sāmoa, Nuku Hiva, and Hiva Oa, two thousand miles to the south. Then, perhaps around the year 1300, direct contact with those distant lands came to an end. Scholars speculate that a changing climate may have altered currents and winds enough to disturb established patterns of navigation, or that contact ceased when Hawai'i had achieved full political, religious, and economic autonomy. Remarkably, Hawai'i seems to have been out of direct contact with outsiders until the arrival of the Englishman Captain James Cook and his crew four or five centuries later, in 1778.[5]

At that point, direct exploration began anew. Judging from songs and stories that were recorded by Kānaka in the nineteenth century (after American missionaries first introduced written language to the islands in the 1820s) questions about their ancestors' migrations to Hawai'i were of interest to Kānaka, but they were just one part of a very large body of knowledge. A close reading of these written works permits us to reconstruct some fundamental concepts of Kanaka Maoli geographical thought as it existed prior to European incursion. These concepts help us understand why Kānaka acted as they did in the decades after Cook's first landing.

In Cook's own time and up to the present day, accounts of that encounter have frequently suggested that Kānaka were utterly ignorant of the outside world. Cook himself implies that before he arrived in Hawai'i, Kānaka believed they lived alone in a watery world. Regarding his first encounter with Kānaka Maoli (whom he called "Indians") off Kaua'i in 1778, he wrote, "I never saw Indians so much astonished at the entering [of] a ship before." Cook reported that "their eyes were continually flying from object to object, the wildness of their looks and actions fully expressed their surprise and astonishment."[6] Cook implies that he had revealed a new world to them—not only the ship, but the world at large that it represented. Cook's narrative and the notion that Kānaka were ignorant because they were isolated attributes extraordinary power to the arrival of Westerners and their knowledge. The implication of utter ignorance of the outside world has been leveled at other Pacific Islanders, as well. The anthropologist Greg Dening writes that, for the people of the Marquesas, "their islands were their whole world. There was no outside world."[7] A Hawaiian history textbook echoes this idea when it declares that because "Hawaiians lived for hundreds of years isolated from the rest of the world," they "knew nothing of events in Europe or the Americas or Asia, or even on the other Polynesian islands in the Pacific from which they had come."[8] While it is likely true that Kānaka were unaware of events elsewhere in the Pacific after the end of the interarchipelagic canoe voyages, it is essential to note the passage's emphasis that separation from the outside world resulted in an *ignorance* of the outside world. When Cook arrived, godlike he revealed to benighted Hawaiian Indians a world they never suspected lay beyond the horizon.

The problem with this account is that it is not true, and its untruth serves colonialist purposes. Rather than tearing away the curtains of ignorance that supposedly blinded Kānaka, the arrival of Cook and other Haole confirmed many ideas about the outside world that were preserved in the classical oral literatures of Hawai'i. Undeniably, Cook's landing at Kaua'i transformed the history of Hawai'i. It did so, however, not by revealing a world previously unsuspected, but by reinitiating direct engagement with a wide and broad world that Kānaka already thought about, talked about, sang about, and told stories about.

To understand that, and to take into account previous Kanaka Maoli knowledge of the world, we must first consider what it means in this context to *know* about the world. Academic discourse commonly distinguishes between rational knowledge that is justified by truth and irrational belief that is not. Like all social and epistemic binaries, this knowledge/belief binary creates and legitimates hierarchy: it constructs knowledge as superior to belief. This binary also correlates to a whole range of other hierarchical binaries: while knowledge is associated with the Western, the male, the schooled, the modern, the scientific, the adult, and the white, belief is associated with the non-Western, the female, the unschooled, the archaic, the mythological, the childish, and the nonwhite.[9] To talk about Kanaka Maoli belief, then, would carry two liabilities: first, it would reinforce the many hierarchies that subjugate Kānaka Maoli (and all other people who end up on the losing end of the binaries mentioned above), and second, by constantly implying that Kanaka geography is untrue, it would interfere with our undertanding of Kanaka geography on its own terms.

For this book, knowledge instead relates to certainty. The remaining pages of this chapter are dedicated to describing crucial things that

Kānaka Maoli *knew* about the world far from their shores before Cook sailed into the waters of Hawai'i. They were certain of these things about distant and unseen places just as they were certain of things about close and visible places. Here, we would do best to return to the Hawaiian language, in which the word 'ike expresses a meaning that encompasses the English words "see," "know," and "knowledge." To 'ike something that is unseen is to be as certain of it as if it were visible. Thus one cannot translate 'ike as "belief." A belief is a mana'o—an intention or a belief that is recognized as possibly true or untrue. Our purpose is to consider Hawaiian 'ike, not mana'o; knowledge, not belief.[10]

Far from believing that they were the sole inhabitants of the only islands in the ocean, Kānaka Maoli knew they were one part of a large and populated world. The rich and extensive oral literature of Hawai'i teaches much about the nature and contours of the broader world. Nineteenth-century Kānaka could draw on a rich body of story and song that recounted travel between Hawai'i and other lands in Oceania.[11] This body of knowledge begins with accounts of the first Kānaka who came to Hawai'i. Some mele and mo'olelo recount how Hawai'iloa, identified variously as a fisherman, a chief, and a mariner from the lands near Kahiki-honua-kele, discovered Hawai'i after having traveled there by wa'a. He settled his family and others on the islands. Later, other ancient voyagers in wa'a came from other islands and settled in Hawai'i.[12] Kānaka, then, descend from ancestors from other places—which implies the existence of other places inhabited by humans and reachable by canoe.[13]

In order to recover a sense of the geographical knowledge and explorations of Kānaka prior to the incursion of foreigners, in this chapter I draw on mele and mo'olelo that were written down in the nineteenth century. Hawaiian had always been an exclusively oral language, but when Haole, working with Hawaiians and Tahitian missionaries, created a way to write the language with Roman characters in the 1820s, Kānaka vigorously embraced literacy. With that tool and with remarkable determination, Kānaka preserved a far larger body of their oral literature in writing in their own language than almost any other indigenous people who suffered the onslaught of colonialism.[14] These texts by Kānaka offer both tremendous opportunities and challenges for learning about older ideas. One can fairly ask if the versions of these mele and mo'olelo that were written in the 1800s accurately reflect the ideas of people in years prior to the Western incursion. Given that Kanaka intellectual life was dynamic and changing in response to changing circumstances, the historical context shaped writing by Kānaka. For example, a Christian condemnation of previous ideas and practices is apparent in writings about ka poʻe kahiko (the people of old) by Davida Malo, Kepelino, and less famous writers. But in terms of understanding Kanaka geographical knowledge, the mele and moʻolelo upon which this chapter draws do provide usable evidence. The great concern of nineteenth-century Kanaka writers was to record them accurately. This fact, along with the many points of agreement between different texts by different authors, suggests that the general contours of geographical thought that the texts contain conform to the shape of older ideas prior to the arrival of outsiders in 1778.

While untold numbers of songs and stories were not recorded in writing and have been lost, the hundreds of mele and moʻolelo that we do have amply demonstrate that Kānaka Maoli knew themselves to be inhabitants of a broad world full of lands and waters and beings. They knew, moreover, that the islands and their people were connected to global geography by genealogy, by the gods and other powerful beings, and by the movement of people, birds, and other living things.

GEOGRAPHY AND GENEALOGY

To understand Kanaka understandings of the world, it is best to start with their island home and the ocean in which it lies. The pae ʻāina (archipelago, literally meaning "group of islands") was not limited to the eight largest islands where Kānaka made their homes: Hawaiʻi, Maui, Kahoʻolawe, Lānaʻi, Molokaʻi, Oʻahu, Kauaʻi, and Niʻihau. It also included the many smaller islands not far from the main islands, where people did not live full time but that were familiar nonetheless: Molokini, Lehua, Mokuʻumeʻume, Kaʻula, and others. That pae ʻāina was surrounded by ka moana, the ocean. Ka moana was a fundamental fact of Hawaiian existence. It bounded the shore of each island, and it linked each shore to all the shores of the world. Neither wall nor bridge, ka moana was a field of moving and rolling space that one might navigate to arrive at the world's many lands.

Kānaka looked out across the ocean at a world that they knew extended far beyond the horizon. From Oʻahu, one can see neither Molokaʻi i ka hikina (Molokaʻi in the east) nor Kauaʻi i ke komohana (Kauaʻi in the west), yet Kānaka knew they are there. Similarly, Kānaka knew that farther beyond the horizon to the east was a land called Kahikikū and farther beyond the horizon to the west could be found Kahikimoe.[15] They knew that far away lay many lands from which ancestors, gods, and useful plants had come and to which Kānaka had once traveled. Kānaka understood they had a genealogical as well as spatial relationship to these and other places in the world—that is, they had ancestors and relations elsewhere in the world. After all, some of the most important moʻolelo narrate how akua wahine (female gods) gave birth to the islands. (One hesitates to call them "goddesses," as the English term risks a diminutive connotation in comparison to the word "god." In Hawaiian, deities are akua, whether they are male or female, and marking them as male or female is optional.) Perhaps no song or story portrays the genealogical bonds between the Kānaka and foreign lands more vividly than the chant known as "Ka Mele a Pakui" (The Song of Pākuʻi). In the time of Kamehameha, the first mōʻī (king) to rule over the archipelago as a whole, Pākuʻi was a great kahuna, a highly trained expert knowledgeable of a vast array of story, song, prayer, and especially moʻokūʻauhau, or genealogy. One gains a sense of his position, and of the respect Kānaka accorded history and genealogy, by the way a nineteenth-century Hawaiian historian described him: Pākuʻi was "he kahuna . . . , he kakaolelo no loko mai o ka papa mookuauhau o na kahuna nui o ka oihanakahuna" ("a priest . . . and a historian belonging to the board of historians and genealogist of the order of the priesthood").[16] Pākuʻi dedicated much service to Kamehameha in the field of battle (in 1795 he was taken captive in the Battle of Nuʻuanu) and in the court (in 1797 he composed a famous mele inoa, or name song, for Kamehameha's son and heir, Liholiho).[17]

The archipelago's ties to foreign lands were familial and genealogical. "Ka Mele a Pakui" recounts the origins of the islands with a narrative of romantic and sexual intrigue that merits close attention for what it reveals about Hawaiian geographical thought. In ancient times, an akua kāne (male god) named Wākea and an akua wahine named Papa mated and produced a series of island-children:

> O Wakea Kahiko Luamea,
> O Papa, o Papahanaumoku ka wahine,
> Hanau Tahiti-ku, Tahiti-moe,
> Hanau Keapapanui,

Hanau Keapapalani,
Hanau Hawaii;
Ka moku makahiapo,
Keiki makahiapo a laua.

Wākea, son of Kahiko, the abyss.
Papa-who-births-islands was the female,
Kahikikū and Kahikimoe were born,
Keʻāpapanui was born,
Keʻāpapalani was born,
Hawaiʻi was born;
[Hawaiʻi was] the first-born island,
The firstborn child of those two [Wākea and Papa].

The song thus begins by listing the births of Keʻāpapanui (also sometimes transcribed as Keʻāpapanuʻu), Keʻāpapalani, Kahikikū, and Kahikimoe, all of which are presumably islands. It emphasizes, however, that Hawaiʻi was the firstborn island-child. It goes on to list younger siblings to whom Papa then gave birth, notably the islands of Maui and Kanaloa (now known as Kahoʻolawe). Following these births, Papa, the islands' mother, traveled away to the distant land of Kahiki. In her absence, Wākea turned his attentions to other female gods:

Haalele Papa hoi i Tahiti,
Hoi a Tahiti Kapakapakaua.
Moe o Wakea moe ia Kaulawahine
Hanau o Lanai Kaula.
He makahiapo na ia wahine.
Hoi ae o Wakea loaa Hina,
Loaa Hina he wahine moe na Wakea,
Hapai Hina ia Molokai, he moku,
O Molokai a Hina he keiki moku.

Then Papa left and returned to Kahiki,
Went back to Kahiki, to Kapakapakaua.
Wākea slept with Kaʻula wahine [woman]
And Lānaʻi [the child of] Kaʻula was born.
He was that wahine's [woman's] firstborn.

Then Wākea returned and found Hina,
Hina was gotten as a wahine for Wākea to sleep with,
Hina became pregnant with Moloka'i, an island;
Hina's Moloka'i is an island-child.

Wākea thus fathered two new islands (Lāna'i and Moloka'i) by two different female gods (Ka'ulawahine and Hina) while Papa, the mother of his other island-children, was away. At this point, Pāku'i reports, a kōlea (a bird known in English as the Pacific golden plover) by the name of Laukaula flew to Kahiki and told Papa about Wākea's infidelity, throwing her into a fury:

> Haina e ke kolea o Laukaula
> Ua moe o Wakea i ka wahine.
> O ena kalani kukahaulili o Papa.
> Hoi mai Papa mai loko o Tahiti;
> Inaina lili i ka punalua;
> Hae, manawaino i ke kane, o Wakea.

> The kōlea Laukaula declared
> That Wākea had slept with the woman.
> Papa's anger was fierce and fiery.
> Papa returned from within Kahiki;
> She was angry and jealous of the rival;
> Was wild and evil-minded toward her kāne [male, man], Wākea.

Thus enraged, Papa returned to the Hawaiian Islands where she mated with the male god Lua (a young god in his "leaf opening days"). The couple conceived the island-child O'ahu-a-Lua (Lua's O'ahu), generally referred to simply as O'ahu. At this point, Papa reconciled with Wākea:

> Hoi hou aku no moe me Wakea.
> Naku Papa i ka iloli,
> Hoohapuu Papa i ka moku o Kauai
> Hanau Kamawaelualanimoku,
> He eweewe Niihau,
> He palena o Lehua,
> Ha panina Kaula.

> [Papa] went back and lived with Wākea.
> Papa was restless with child-sickness,
> Papa conceived the island of Kaua'i
> And gave birth to Kamāwaelualanimoku.
> Ni'ihau is of the same lineage
> Lehua was a border
> And Ka'ula the very last.[18]

The song thus narrates that Papa gave birth to four more children by Wākea, the large islands of Kaua'i and Ni'ihau, and the smaller islands of Lehua and Ka'ula. Pāku'i thus makes plain that the pae 'āina of Hawai'i is made up of siblings and half siblings, born of the female gods Papa, Hina, and Ka'ulawahine, and fathered by the male gods Wākea and Lua. The islands were a family, and kinship also bound the Hawaiian Islands to far-off lands. As the start of the song recounted, Papa and Wākea had other children-islands, including Ke'āpapalani, Ke'āpapanui, Kahikikū, and Kahikimoe.[19]

The content and message of the song are very much connected to the rise of Kamehameha to supremacy over the entire archipelago. Like all histories and geographies, Pāku'i's song teaches us about the author's stated subject but also the author's time and the sources upon which the author drew. The song's political content is particularly apparent in its emphasis on the kinship of the islands to one another. Remember that Pāku'i was a kahuna in the court of Kamehameha, the chief who was by conquest and diplomacy uniting all the islands under one mō'ī for the first time. Naming the islands as kin, and naming Kamehameha's native island of Hawai'i as the eldest sibling, serves to legitimate Kamehameha's efforts to unify the islands and rule them.

Lest one imagine that Pāku'i was merely inventing a narrative to serve his mō'ī, however, it must be emphasized that while Pāku'i was the composer of the song, he drew upon sources that were known to other kāhuna: the enormous body of Hawaiian geographical, historical, and genealogical knowledge as preserved in mele and mo'olelo. Pāku'i's account agrees with many other stories of the birth of the Hawaiian Islands, notably a creation chant by Kaleikuahulu, who was born on Moloka'i in 1725, probably about two or three generations before Pāku'i. Like Pāku'i, Kaleikuahulu was a renowned historian, genealogist, composer, and priest, and like Pāku'i he served in the court of Kamehameha, perhaps as a senior partner to the younger kahuna.[20] And like Pākuʻi, Kaleikuahulu had to draw upon historical and genealogical sources that would have enjoyed legitimacy among their kahuna peers, the aliʻi (chiefs), and the Kānaka more broadly. Because Hawaiian stories and songs were preserved and transmitted only in oral form until the 1820s, there exist no accounts of the islands' genealogy that were written down prior to Kamehameha's reign. Nonetheless, the broad acceptance of these songs strongly suggests that they were in accord with older songs and stories, and the relations between islands that they describe were not merely inventions in the service of Kamehameha's ascendency.

A GEOGRAPHY OF SPECIFIC, NAMED, PHYSICALLY REAL PLACES

The far-off places mentioned in songs and stories were, in the ʻike of Kānaka, specific and real places. This becomes clear when we look at the places mentioned in the "Mele a Pakui" that are not in the Hawaiian archipelago. Despite their close kinship to the Hawaiian Islands, Keʻāpapanui, Keʻāpapalani, Kahikimoe, and Kahikikū islands lie far from their siblings. In a well-known translation of the song, Keʻāpapanui and Keʻāpapalani are translated as "the foundation stones" and "the foundation of heaven" and refer to structures that hold up the earth and the heavens. The names of the remaining siblings, Kahikikū and Kahikimoe, raise thorny issues of translation. Davida Malo (one of the earliest, most productive, and most widely read of the nineteenth-century Hawaiian scholars) states that Kahikimoe is the circle of the horizon and that Kahikikū is the area or ring of the sky above the horizon.[21] And yet they are more than celestial terms. Other sources describe them as huge and distant lands, with Kahikikū to the east and Kahikimoe to the west. Their very names mark them as distant: kahiki (as a general term or adjective) is a term somewhat like the English word "overseas" or "foreign" that can refer to any distant, foreign place. The anthropologist Marshall Sahlins asserts that Kahiki, the unseen, and the sky are the same thing. However, this mistakenly implies that Kahiki (as a proper noun) is simply the mysterious "beyond." Kahiki is also a particular place where both gods and people can live, and sometimes is the place we call Tahiti today. Kahiki appears as the proper name of a specific foreign place in many Hawaiian moʻolelo and mele: remember that Pākuʻi's song informs us that Papa was in Kahiki when she learned of Wākea's infidelities. Other moʻolelo recount that Hawaiʻiloa, the first person to come to the Hawaiian Islands, sailed from Kahiki. Stories concerning other ancient travelers such as Moʻikeha do the same. Such usages do not refer to just any generic foreign place: after all, other mele and moʻolelo describe lands, like Nuʻumealani, as lying in the area of the Kūkulu o Kahiki (the Pillars of Kahiki). One place can only be near another if both are specific locales, so the name Kahiki clearly can refer to a specific place.[22] The same can be said of Kahikikū and Kahikimoe: the names refer to specific places, not abstract realms.

Some of the distant lands described in mele and moʻolelo are not found on maps today (at least under names we can identify), but others are easily identifiable. The "Song of Pakui," for example, mentions Nuʻuhiwa and Polapola, which are cognates for the Marquesan and Tahitian names for Nuku Hiva and Bora Bora.[23] These places and many others appear in stories about the movement of people and gods back and forth between the Hawaiian Islands and distant places, especially Kahiki. Hawaiʻiloa, the first Kanaka to come to the islands and who gave his name to the largest island, traveled from Kahiki. Moʻikeha, hero of an epic of ocean navigation, sailed from the Hawaiian Islands to Kahiki. Pāʻao, the priest who brought to Hawaiʻi the foundations of what became its temple-centered and hierarchical system of worship, had traveled from Kahiki. Some accounts maintain that the island of Oʻahu is named for Ahukinialaʻa, a chief who came from Kahiki.[24] Kānaka knew that these many voyages to and from distant places had shaped their society: people, plants, and a religion from far-off lands left their stamp on the islands. One story explains that when the aliʻi Olopana came to the islands from Kahiki, he settled his people on Maui, Molokaʻi, Oʻahu, and Kauaʻi, and this is why people on those islands spoke differently than people from Hawaiʻi Island.[25] Among the many distant lands that Kānaka knew about, Kahiki loomed particularly large.

Still, Kahiki was just the beginning of the places they knew about. If one were to assemble a gazetteer of distant lands and places by pouring through published Hawaiian songs and stories, one would have just begun if one included Kahiki, Kahikikū, Kahikimoe, Nuʻuhiwa, Polapola, Hāmoa, Nuʻumealani (or Nuʻumehalani), Wawaʻu, Upolu, Pukalia-iki, Pukalia-nui, Alala, Pelua, Palana, Holani, Kuina Ulunui, Unuili,

Melemele, Hiʻi-kua, Hiʻi-alo, Hakalauai-po, Moanawaikaloo, Kapakapakaua, Ulupaupau, Keolewa, Haenakulaina, Kauaniani, Kalakenuiakane, Kuaihelani, Kapaahu, Moaulanuiakea, Keʻāpapanuʻu and Keʻāpapalani, Kahiki-nui-kaialeale, Kānehunamoku.[26] Some of these refer to place names we can identify in the Pacific; others do not. Such a list could go on and on, and it would have been compiled only from the fraction of the classical Hawaiian historical, genealogical, and geographical heritage that eventually made it into print.

KE KŌLEA AND KE ALOHA: POWERFUL AND GOOD THINGS FROM MYSTERIOUS PLACES

Given their distance from foreign lands, it is perhaps not surprising that a migratory bird would be both proof of the existence of far-off lands and a means to reflect on those lands' meaning. As keen observers of the environment, Kānaka knew that a number of bird species spent part of their lives in Hawaiʻi and part of their lives elsewhere. The most celebrated of these was the kōlea—the very bird that traveled from Hawaiʻi to Kahiki to tell Papa about Wākea's infidelities. A number of ʻōlelo noʻeau (proverbs and wise sayings) refer to the kōlea arriving bedraggled and thin in Hawaiʻi in August, only to fly off again in the spring, fattened up. (Kānaka use these proverbs today to suggest how foreigners come to Hawaiʻi to enrich themselves.) Kānaka knew the kōlea was flying to some destination and, since no one in Hawaiʻi had ever seen a kōlea hatchling, deduced that it must reproduce there. Though there is no evidence that Kānaka had direct knowledge of Alaska where the kōlea hatches, the nesting place had to be a physically real place from which the bird could migrate.

Knowing that the outside world existed, Kānaka yearned to know its mysteries. The kōlea symbolized much of the wonder of Kānaka about the faraway world they did not see but by deduction and historical memory knew to exist. In the words of one ʻōlelo noʻeau, "ʻO ka hua o ke kōlea aia i Kahiki" ("the egg of the plover is laid in a foreign land," or the egg of the plover is laid in Kahiki).[27] The kōlea's hatching place represents something that must logically exist but is beyond the ken of humanity. The same theme appears in another story, in which the mōʻī named ʻUmi was seeking the bones of Pae, a deceased priest. Pae's sons told ʻUmi that in accordance with aliʻi practice, they had hidden them to prevent their desecration. To make clear to ʻUmi that he would never find the bones, the sons made use of the figure of the kōlea: "O nā hōkū o ka lani ka i ʻike iā Pae. Aia a loaʻa ka pūnana o ke kōlea, loaʻa ʻo ia iā ʻoe." Mary Kawena Pukui translates this as "Only the stars of heaven know where Pae is. When you find a plover's nest, then you will find him."[28] This was the mysterious reality, but also the power of the distant world: to be known to exist, and yet to remain unseen. Indeed, the bird linked the Hawaiian archipelago and faraway places in some moʻolelo and ʻōlelo noʻeau—as in the story of the kōlea flying off to Kahiki to inform Papa of Wākea's affairs with other female gods.

This was unwelcome news for Papa, of course, but for Kānaka, the kōlea also represented how distant places could be the sources of good and powerful things. This is vividly demonstrated by the words of the hula kōlea, one of a number of hula in which kneeling dancers imitate the motions of animals to the accompaniment of a related song:

> Kolea kai piha!
> I aha mai nei?
> Ku-nou mai nei.
> E aha kakou?
> E ai kakou.
> Nohea ka ai?
> No Kahiki mai.
> Hiki mai ka Lani,
> Olina Hawaii.
> Malaʻelaʻe ke ala,
> Nou, e ka Lani.
> Puili pu ke aloha.
> Pili me kaʼu manu.
> Ka puana a ka moe?
> Moe oe a hoolana
> Ka haliʻa i hiki mai;
> Ooe pu me aʼu
> Noho pu i ka wai aliali.
> Haiʼna [sic] ia ka pauna [sic].
> O ka hua o ke kolea, aia i Kahiki.
> Hiki mai kou aloha, maeʻele au.

A kōlea at high tide!
What is it saying?
It beckons with its nodding head.
What shall we do?
We shall eat.
Where does this food come from?
From Kahiki.
When the Chief comes,
Hawai'i makes merry.
The path is clear and smooth
For you, o Chief.
Embrace the beloved.
I cling to my bird.
What is the meaning of this dream?
Lie down and listen
To the fond recollection that comes forth:
You are here with me,
We remain together in the crystal clear whiteness.
Repeat the chorus:
The egg of the kōlea is laid in Kahiki/a foreign land.
When my love approaches, I am benumbed.[29]

The song operates at a number of levels, deploying kaona, or multiple meanings, which are a device important in Hawaiian literature. The most important kaona can be found in the kōlea itself, which in this context represents the migratory bird, but also love and eroticism. The shorebird at high tide nods its head in a way reminiscent of a lover's beckoning. Prompted by this implied invitation, the song asks, "E aha kakou?" (What shall we do?). The answer, "E ai kakou," raises multiple interpretive possibilities. The most direct is "we shall eat." This leaves open the possibility that we will eat the body of the kōlea, but also that we (including the kōlea) will eat together. But because the word for eat ('ai) sounds nearly identical to the word for sexual intercourse (ai), the song can also be heard as a call to lovemaking. This interpretation is supported by the lines that follow, with their references to clear and smooth paths, embracing the beloved and clinging to the kōlea, and dreams and fond recollections of remaining together in clarity and light. In either case, "e ai kakou" is neither an invitation to the isolated lovemaking of two sweethearts nor for them to dine privately together.[30] Hawaiian grammar makes clear that the poem invokes a *collective* celebration, not a private tête-à-tête. The Hawaiian language features four pronouns that translate into English as "we." The pronoun that the song uses here is "kākou." Kākou refers only to groups of three or more people and is only used when the person addressed is included in the group. Thus it is a collective ʻolina (joyous merrymaking) that the song calls forth.

At first glance, none of this might seem to relate closely to Kanaka geographical thought, but the song emphatically connects to geography through the references to the migratory kōlea that frame the poem. It asks "nohea ka ai?" (from whence comes the food?), a reference to the kōlea, or possibly to the food that sustains the kōlea. The answer is provided: it comes "from Kahiki." The poem returns to the point near the end, when it repeats the proverb "O ka hua o ke kolea aia i Kahiki" (The egg of the kōlea is laid in Kahiki). This is not merely a lesson in a bird's natural history. It is also a call to a collective celebration of love and a meditation on where love comes from. Like the kōlea from Kahiki, love is of mysterious origins. The mysteries of the bird and of love and of distant lands are summed up by the line "Hiki mai kou aloha, maeʻele au" (When my love approaches, I am benumbed)—a powerful statement of affection, as māʻeʻele (numbness) is commonly used to refer to "a strong internal glow of love."[31] Good and powerful and beautiful things—like the kōlea, like love—can come from distant and unseen lands.

This was true also of cultivated plants. From the plants they used, Kānaka knew that good and powerful things could be brought from distant lands and Kānaka could make them useful for their own purposes. Kānaka knew that a number of the plants that they most valued, including some of daily importance and of great ritual and spiritual power, were grown from stock taken from elsewhere long ago. Stories narrated how the coconut was first brought to Hawaiʻi from Kahiki by Apua and Aukelenuiaiku. In addition to its spiritual significance as a kinolau (embodiment or form) of the major god Kū, the coconut tree was invaluable to Kānaka for fruit that provided food and oil, husks that gave coir for twine, and trunks for building and making drums.[32] Aukelenuiaiku also introduced the ʻōhelo ʻai from Kahiki. This plant is important for its edible berry, but also for the medicinal properties of its leaf buds, leaves, and fruit and the fact that it is sacred to Pele, the god of the volcano.[33]

Several stories recounted how people and akua brought different varieties of ʻulu (breadfruit) to Hawaiʻi. Moʻolelo refer to a man from the Hawaiian Islands named Kahai who traveled to Upolu, far to the south in Sāmoa, and brought back a specific cultivar of ʻulu. This tree, useful for its food, the medicinal properties of its bark and sap, and its wood, is also a kinolau of the gods Kū and Haumea.[34] Another variety of ʻulu with exceptional qualities was introduced to Puʻuloa (Pearl Harbor) by fishermen from that place after a great wind blew them off course, all the way to Kunehunamoku in Kahiki. This was a land where gods dwelled, a land that sank below water if humans came to see it—but apparently remained at the surface for the lost fishermen. There they found a wonderful ʻulu, which they brought back to Puʻuloa. The female god Haumea later propagated ʻulu across the islands, so that others might benefit from it.[35] ʻAwa (kava), the roots of which are used to make a calming drink that is sometimes consumed in ritual manner, was brought from Kahiki by Oilikukahena.[36] Similarly, the "Kaʻao no Kana a me Niheu" (Legend of Kana and Niheu) says that a hedge of "ʻūlei of Nuʻuhiwa" and kī (ti) surrounded the house of Kapepeekauila. Tellers and hearers of the story would have known that the plants were carefully selected for the hedge, as they had protective powers. The name of the plant (ʻūlei of Nuʻuhiwa) itself preserves the memory of its origin in a distant land, Nuʻuhiwa.[37]

"MYTHICAL LANDS"?

Nuʻuhiwa is readily identifiable as Nuku Hiva, a major island 2,300 miles to the southeast of Hawaiʻi. Upolu (the place of origin of Hawaiian breadfruit) is also easily found: it is the most populous island of Sāmoa, 2,500 miles to the southwest of Hawaiʻi. In contrast, Kunehunamoku and Kahikikū are not found on contemporary maps. The fact that they are not on maps, that gods dwelled there, and that wondrous things happened there does not mean, however, that they were "mythical," "mystical," or "mythological" lands, as Americans treated them throughout the twentieth century.[38] In Hawaiian global geography, the presence of gods and other-than-human beings in far-off lands does not make a space different from home. It makes it like home. The Hawaiian Islands and their waters were also spaces where a diverse array of other-than-human beings (akua and kupua and moʻo and others) dwelled and through which they moved, and this hardly meant the islands were mythical. Most famously one can visit the crater of Kīlauea, dwelling place of Pele, god of the volcano. But many other real, visible places that stories associate with the deeds of gods and sites where stories say gods are still present were (and are) still all around Kānaka Maoli. Kānaka could still dance hula at the site at Hāʻena, Kauaʻi, where the volcano god, Pele, saw Lohiau dance the hula and fell in love.[39] They could visit the beach at Oneawa in Kailua, Oʻahu, where the female god Hiʻiaka landed with her beloved companion, Wahineʻōmaʻo.[40] They could walk though Kaliuwaʻa Gorge on Oʻahu, site of many of the exploits of Kamapuaʻa, the pig god.[41] Mele and moʻolelo were filled with references to these and many other akua and kupua and moʻo in places that were visible, physical, and nearby, not just unseen, metaphysical, and distant.

Moreover, the *movement* of gods and people between Hawaiʻi and far-off lands similarly established connections between places near and far. Witness the story of Nāmakaokahaʻi, a female ocean god who was Pele's elder sister and rival. In a fight on the island of Maui, Nāmakaokahaʻi killed Pele, whose mighty bones forever mark the landscape of the Hāna district at the hill called Kaiwiopele (Pele's bone). (Other stories make it clear that Pele did not die in a permanent or human sense, as she continues to be active in the world.) Furthermore, recalling the song of Pele's travels that opens this chapter, we can see how the movements of gods between far-off places and Hawaiʻi link the islands to a broader world geography.[42]

A GEOGRAPHY OF PERSPECTIVE

Kānaka knew about far-off places, reflected upon them, and valued the connections of Hawaiʻi to them. This distinctive knowledge of the world was matched by a distinctive way of understanding space: a consistently perspectival way of looking at the world. Kānaka understood geography from points of view. Kānaka did not describe the world as an abstract truth, but as a world seen from a perspective—a point that this chapter emphasizes because it would shape the way Kānaka engaged with the very different ideas about geography that Westerners would bring.

The perspective of the sun in motion defined Hawaiian terms for the directions in which the ocean extends in its vastness. The sun rises in ka hikina, "the arriving." It sets in ke komohana, "the entering." As it proceeds across the heavens, to its right lies ka ʻākau (the right), which in English is styled as "north." To its left lies ka hema (the left), which English calls "the south." These four perspectival directions describe space beyond the shore, out in ka moana. On land, these terms can be used as well, often to reference districts such as Kona ʻĀkau and Kona Hema (North and South Kona).[43]

Here we must stop ourselves to emphasize a crucial point: ka hikina does not *mean* "the east"; ke komohana does not *mean* "the west"; ka ʻākau does not *mean* "the north"; and ka hema does not *mean* "the south." Seamlessly translating Hawaiian into English in that way would conjure what Lawrence Venuti calls an "illusion of transparency" that hides profound differences in ways of understanding the world.[44] In English, north, south, east, and west are the *cardinal* directions and on maps they are generally the reference points for geographical description. For example, the familiar system of mapping locations on Cartesian coordinates (think of a city map with a numbered and lettered grid, or a world map showing latitude and longitude) describes spaces relative to absolute notions of north, south, east, and west.[45] In contrast, Hawaiian terminology reveals a *perspectival* orientation. It differs at a fundamental conceptual level from such categorical descriptions as cardinal direction. One of the very first Kānaka to use written language, a man named ʻŌpūkahaʻia, emphasized the sun's perspective in understanding east and west. In an 1818 description of the Hawaiian language, he rendered "eastwardly" phonetically as "ka hikina ka lā" (the coming [of] the sun) and "westwardly" as "komohana lā" ([the] entering [of the] sun). When ʻŌpūkahaʻia translated "northwardly" and "southwardly" he switched perspective, describing them relative to himself. He glossed northwardly as "o kai" (toward [the] sea) and southwardly as "o uka" (toward [the] uplands).[46] Fascinatingly, these two did not describe north and south relative to ʻŌpūkahaʻia's position when he wrote the word list. He produced it in Cornwall, Connecticut (where he was living for reasons that will become clear in chapter 3). There, the sea is to the south and the uplands lie to the north. Where would ʻŌpūkahaʻia's terms make sense? In much of Kohala on Hawaiʻi Island, where ʻŌpūkahaʻia had lived as a boy. This is the reference point he used. ʻŌpūkahaʻia therefore defined directions from individual perspectives (even remembered perspectives): that of the sun and that of the individual person standing on the land.

This is of the utmost importance for this book because it demonstrates the nature of Hawaiian ʻike about global geography. When we recall that ʻike means both "to see" and "to know," we are brought back to the point that just as one literally sees from one's own perspective (where one's eyes are located), so too does one know the world from one's own perspective. This underscores a crucial truth: all geographical knowledge is perspectival and relative, including one's own. Since there can be no objective stance from outside of space, this permits the centering of the speaker or agent. Kānaka Maoli looked out at the world from their own islands, because it was natural for them to do so. In 1869, the Hawaiian writer Samuel Kamakau would playfully explain this as follows: "Ina e ku ke alo o ke kanaka i kahi a ka la e puka mai ai, alaila, ua maopopo aia kona lima hema ma ke kukulu akau, a o kona lima akau ma ke kukulu hema, a o kona alo ka hikina, a ma kua ke komohana" (If a man stands facing the place where the sun emerges, then, his hema [south, left] hand is on the akau [north, right] direction, and his akau [north, right] hand is on the hema [south, left]).[47] The joke, of course, is that hema and ʻākau are easily thought of as antonyms, but perspectivalism upends that expectation: one's ʻākau (north, right) can be to the hema (south, left). With his deft phrasing, Kamakau expresses the truth of Hawaiian perspectivalism and permits the contrast to Western cardinal directions to come into sight. In looking out from Hawaiʻi, Kānaka Hawaiʻi centered their home in their own worldview.

Before the foreigners came, what did it mean to look out from the shores of Hawaiʻi, Niʻihau, or the other affluent and densely populated islands that stretched in a chain between them? We cannot know with certainty what or how people in the past thought, but a close examination of Hawaiian language, story, and song does reveal that Kānaka had every reason to have the desire and self-assurance to explore the world. Because their language expressed a fundamentally perspectival way of understanding space, Kānaka could look out at the world from a standpoint that was confidently centered in their own islands. Because stories and songs recounted the deeds and genealogies of ancestors who had come from afar and because they told about distant places and about the people and gods that lived there, Kānaka could look out at a world that was unseen but not truly unknown. Because stories and songs suggested that distant places were the source of many good things—nourishing coconut, soothing 'awa, protective 'ūlei of Nu'uhiwa, and even the benumbing love whose origins were as mysterious as the hatching grounds of the kōlea—Kānaka could look out from their islands and expect that still more precious things were to be found far away. After Cook's arrival, this sureness of perspective, heritage of exploration, and knowledge that distant places could be the source of precious things meant that Kānaka could look at the world with the confidence and interest to explore it.

2

PADDLING OUT TO SEE

DIRECT EXPLORATION BY KĀNAKA IN THE LATE EIGHTEENTH CENTURY

In the fourth year of Kahahana's reign as aliʻi ʻaimoku (ruling chief) of Oʻahu and Molokaʻi, Kānaka looked out from the Waialua and Waiʻanae coasts of Oʻahu and saw vessels so enormous and bristling with tree-like masts that Kānaka would come to call them moku, meaning "islands." Kānaka knew that their ancestors and the people of Tahiti, Sāmoa, Nuku Hiva, and other islands had traveled back and forth across the Pacific. Some would have heard stories that reported on the arrivals of strangers from foreign places that had come to Hawaiʻi years and decades before.[1] Still, this was an extraordinary event. Before anyone could canoe out to examine them, the moku moved on, sailing northwestward. The next day a group of Kānaka was at sea off Kauaʻi, fishing with heavy lines near Waimea. Mopua and the other fishermen caught sight of the moku. Knowing that this was a matter of the utmost importance, they hurried back to shore and rushed to the court of their aliʻi (chief), Kāʻeokūlani, who was the son of the great Kekaulike, mōʻī (loosely translated as king) of the islands of Maui, Lānaʻi, and Kahoʻolawe and brother to the current mōʻī of those islands. Informed of the dramatic development by the fishermen, Kāʻeokūlani, his court, the kāhuna (priests and other experts), and many others gathered at the shore and scrutinized the vessels. Kāʻeokūlani dispatched a party on canoes—two aliʻi named Kiikiki and Kaneahoowaa, the kahuna Kuohu, and some men to paddle. They were to contact the moku and return to report to him about the strangers.[2]

The exploratory party paddled out and boarded the moku to learn what they could. They observed protocol. Having discerned which man was the aliʻi of the moku, Kuohu and Kiikiki stepped forward before him, Kuohu wearing his palaoa, a whale-tooth pendant that marked his station, and both in regalia: "ua kaei ia i ke oloa ma ka hope, a ua kaei ia i ka mea ulaula ma ka ai a pupu ma la lima hema" (they were girded by fine white kapa [bark cloth] at the back, and a fine red kapa was gathered at their necks and drawn tightly in their left hands). They murmured a prayer, grasped the arms of the aliʻi of the strangers, and bowed. They then declared the situation noa—meaning, in this case, that they lifted ritual restrictions on the encounter. After receiving a gift from the aliʻi of the Haole, they returned to shore. They reported to their mōʻī about the people and things on the moku and assured him that it was safe to go aboard. Accordingly, Kāʻeokūlani went to investigate for himself.³

These actions by Kāʻeokūlani, the other aliʻi, and the kahuna upon their first encounter with James Cook's ships bear close attention because they typify Kanaka Maoli engagement with the outside world: Kānaka went out into the world and actively engaged it. Kānaka did not wait for the world to come to them any more than Kāʻeokūlani waited for the Haole to paddle to shore. As soon as Mopua and the other fishermen brought him the news, Kāʻeokūlani went into action. Likewise, in the first years after Cook's arrival in Hawaiʻi, Kānaka did not passively wait for the strangers to describe the nature of the outside world to them. For their own reasons, and in keeping with their own heritage of exploration and curiosity and confidence in engaging the outside world, Kānaka Maoli went out into the ocean to seek knowledge of the strangers and the world far from their shores. Moreover, they explored the world from their own premises and for their own purposes. The aliʻi Kiikiki and the kahuna Kuohu observed their own protocol in their own waters. It was they who declared the situation noa, making it possible for conversation to begin. It was they who informed their mōʻī, Kāʻeokūlani, that his direct communication with the people aboard the moku could commence.

This account, drawn from a historical account written by Samuel Manaiākalani Kamakau and published in a Hawaiian-language newspaper in 1865, refutes accounts of the encounter between Cook and Kānaka as written by Westerners. In doing so, it recasts the way we think about Kānaka and Europeans. To use the wording of the historian Pekka Hämäläinen, it "turns the telescope around" in the study of indigenous–European relations: rather than looking at Kānaka from the point of view of Cook and his men, it looks at the outside world through indigenous perspectives. Hämäläinen writes that reversing the gaze through the telescope will "allow us to look at Native policies toward colonial powers as more than defensive strategies of resistance and containment."[4] Native people pursued agendas far beyond defense. This is particularly true in Hawai'i in the first decades of Hawaiian relations with Haole: Kānaka Maoli were not the passive objects of exploration, but the active and deliberate agents of exploration. Kānaka ambitiously and confidently pursued the knowledge of the outside world that would serve their purposes, as individuals and, over time, as a people. And here, the image of the gaze through the telescope is particularly apt because of its nautical connotations and especially because of its emphasis on sight. In Hawaiian, the word 'ike means both knowledge and sight. Kānaka went to the ships, and they saw (ua 'ike) new things. In the process, they gained knowledge ('ike). And that knowledge meant power, for both individuals and the society.

Unlike the account that opens this chapter, standard Western recountings of Cook's "discovery" of Hawai'i draw only on English-language accounts by Englishmen—especially a journal written by Cook and edited after his death. These standard narratives tell a story that begins with Captain James Cook. They situate Cook in the history of English and European exploration and trade (only sometimes mentioning empire) and portray him as the active explorer of Native people. While the English emerge from history and have distinct identities, Kānaka Maoli at Kaua'i seemingly exist outside of history until his arrival and have no distinct group or individual identities—even though Cook claims (implausibly) that his officers could easily converse with the "Indians," whose language was clearly akin to Tahitian, of which they knew a bit.[5] As a group, Cook writes about the people of Hawai'i as yet another group of "Indians," one of many such groups he encountered as he intruded upon lands that were unknown to him. Cook refers to no Kanaka at Kaua'i by name. In Cook's account, there is no Mopua fishing, just some "natives" on canoes that gaze fearfully up at the ship. There are no Kā'eokūlani and Kiikiki and Kuohu. There is little sense of protocol, just the following: "Some of them repeated a long prayer" and others "sung and made motions with their hands."[6] To return to the issue of 'ike, of sight and knowledge, Cook set the pattern that other Haole followed. He could not see the Kānaka at Kaua'i clearly, as Western fantasies of discovery and superiority obscured his vision.

Histories of Cook in Hawai'i generally skim over the events off of Kaua'i to hurry to Hawai'i Island. The rush to Hawai'i Island serves to center Cook and other Englishmen in the narrative even more. On Hawai'i, Cook's sailing master (the famous William Bligh) became the first Haole to set foot on the Hawaiian Islands—a story that is recorded with care, while the name of the first Kanaka setting foot on a Western ship goes unmentioned. On Hawai'i, Cook would interact with key figures in transformations that would bring one ali'i, Kamehameha, to supremacy over all of the islands. On Hawai'i, Cook would be killed in a way Westerners found barbarous. All of these events on Hawai'i place Cook and the Haole at the center of attention in Hawaiian history. In what are still the most widely cited narratives of this encounter, by Ralph Kuykendall and Gavan Daws, Hawaiians are the passive object of exploration, Mopua, Kiikiki, and Kuoho are never mentioned, and Cook and other Westerners stand at the center of the story.[7] In contrast, the account that begins this chapter depends on a Hawaiian-language account by a Kanaka, Kamakau. It gestures toward the Hawaiian historical and political context of these events. And most of all, the account emphasizes Kanaka initiative in exploration.

This chapter will depend fairly heavily on Kamakau, because this historian provides us with the richest available accounts of events in these years, accounts based on extensive research. Granted, Kamakau's work is not infallible, and his emphasis on political history differs from this book's focus on social, cultural, and intellectual history. Yet by reading Kamakau's histories of this time alongside what we know about Kanaka society from other historians' accounts of the late eighteenth century, we can begin to explore what is crucial to us: the actions of Kanaka people understood in Kanaka terms. Those terms permit us to gain insight into Kanaka initiatives in global exploration. The other primary sources, predominantly accounts of English captains and ships' officers who observed or interacted with Kānaka, can then help us to trace actions and events. Their descriptions of Kānaka's actions and guesses at Kānaka's motivations can then be read through what Kamakau and other sources tell us about Kānaka at this time.

This chapter demonstrates that from Cook's arrival in 1778, Kānaka followed the model established by Kā'eokūlani by going out and actively seeking to understand the contours of the globe and what the outside world might mean for them. It emphasizes the Hawaiian context for this story, especially the contention between the ali'i and mō'ī of different islands for supremacy. It then traces the way that several Kānaka set forth on Western ships to explore the outside world, paying particular attention to the stories of a woman the English called "Winee" and a man named Kaʻianaʻahuʻula. The stories of "Winee" and Kaʻianaʻahuʻula (generally shortened to Kaʻiana) have been told in English, but in ways that obscure Kanaka perspectives and center the fantasies that Englishmen elaborated around them and around Hawaiian history.

One crucial initial area of Kanaka exploration was to discern what was the nature of the strangers, and especially if Cook was what Englishmen called a "god." This question has garnered vastly disproportionate attention in historical accounts, but properly understood, it demonstrates a fundamental pattern in the Kanaka exploration of the world: Kānaka used moʻolelo (stories) and their own society along with direct observation in evaluating the world they went out to see.[8] Knowing how these strangers were like or different from themselves was an essential first issue for Kānaka to explore. To Kānaka, it would have been evident from the deference his officers and men showed him that they considered Cook to be a highly ranked aliʻi. Kamakau reports that some Kānaka initially wondered if Cook might be an akua—a word that includes a very broad range of beings in Kanaka cosmology. At this time in Hawaiʻi the highest ranked of the aliʻi, such as Keōpūolani, were considered to be akua, as were the children she would bear by Kamehameha.[9] Thus Cook would quite plausibly have been an akua of some sort, and descriptions from the time and shortly after suggest that Kānaka in his presence performed the sort of ritualized demonstrations of deference—the moe kapu (prostration), for example—that were appropriate for some aliʻi of extremely exalted status. Because his arrival at Hawaiʻi coincided with and mirrored expectations of the Makahiki, a time particular to the akua Lono, some Kānaka apparently questioned if he might be Lono. But according to Kamakau, it took only one day for the Kānaka on Kauaʻi to decide that the foreigners were men rather than akua. On Hawaiʻi Island, the decision seems not to have been long in coming, either: on Cook's return to the island in 1779, Kānaka executed him for violently usurping the right to impose rules and decide punishment. Thus Kānaka explored the nature of the Haole from Kanaka premises and via their own observation: they judged what they saw (an exalted aliʻi) in the context of their understanding of the world (such an aliʻi might be an akua), then by continued exploration adjusted their judgment (deciding he was a human interloper and usurper) and acted accordingly (by killing him).[10]

Kamakau writes that in these early days of encounter, Kānaka similarly explored the material culture of the Haole according to their knowledge of the outside world that was embedded in moʻolelo. Kānaka saw that the Haole had many coconuts on their ship. They knew that moʻolelo described Niuolahiki, the far-traveling coconut tree akua that could move about and stretch out its limbs. They wondered if the Haole had gotten the many coconuts on their ship from that akua. No, they had simply brought the coconuts from other islands. Kānaka saw enormous hides on the moku, much too big to come from dogs or pigs, the only large nonhuman land animals in Hawaiʻi. Moʻolelo spoke of the akua Kūʻīlioloa who took the form of a gigantic dog. Kānaka wondered if that akua was the source of the large hides on their ship. No, the hides came from foreign animals called bulls.[11] The hypothesis that Cook was an akua has received inordinate attention from Westerners because it has flattered Western sensibilities. In fact, it was one avenue of exploration among many.

That process of exploration would pause for nearly a decade after Cook's two visits to the islands in 1778 and 1779, because no Haole ships came to Hawaiʻi in those years. Beginning in 1786, Western ships again began to call at Hawaiʻi to obtain water and fresh food and as part of the process of imperial exploration. Two British and two French vessels arrived in 1786. In the following decade, between three and seven Western ships came to the islands each year.[12]

EXPLORATION AND KANAKA POLITICS: ʻIKE AND POWER

As soon as those ships appeared, Kānaka again took the initiative in pursuing knowledge about these strangers and far-off places. A second great era of Hawaiian exploration of global geography was in motion. A central feature of the Hawaiian context for that exploration was a complex political landscape in which different aliʻi were contending for authority on individual islands and in groups of neighboring islands. These contentions were in process prior to the incursion of the Haole, but their arrival brought new opportunities in the struggles for leadership. Those struggles can be understood as tensions between the formal structure of Hawaiian society and the ambitions of individuals in it—a tension elucidated by the scholarship of Lilikalā Kameʻeleihiwa. In its formal structure, Hawaiian society was a hierarchy with the mōʻī or aliʻi ʻaimoku at the apex serving as intermediaries between the people and the akua. As we have seen, some of the aliʻi were themselves considered to be akua. Below the mōʻī were numerous levels of aliʻi and kāhuna. The higher ranks of aliʻi and kāhuna advised the mōʻī, while lower ranks served as intermediaries between their superior and the vast bulk of the people. That large population was the makaʻāinana, who derived their name from their role in society: they were ma ka ʻāina—on the land. It was they who cultivated the extensive crops of taro, sweet potato, banana, sugarcane, and breadfruit; it was they who built and tended the highly refined fishponds; it was they who fished the sea; it was they who made the kapa. In short, it was they who fed the society and clothed it and rendered the various forms of tribute (ʻauwaeʻāina, hoʻokupu, uku, and ʻauhau) that flowed up the hierarchy. For their part, the highest echelons of society offered tribute to the akua to ensure that all would be pono (in proper balance, correct, right) and the lāhui (the society as a whole) would be safe and prosper.[13] At the time of Cook's arrival in 1778, four ruling chiefs held sway over different parts of the archipelago. Kāʻeokūlani (whose name is often rendered by the shorter form, Kāʻeo) ruled the westernmost of the major islands, Kauaʻi and Niʻihau. Moving east, Kahahana was the ruling chief of Oʻahu and Molokaʻi. Kahekilinuiʻahumanu (Kahekili in its short form) ruled Maui, Lānaʻi, and Kahoʻolawe, and Kalaniʻōpuʻu was mōʻī of Hawaiʻi Island.

But in none of these places was power static and for none of these men was their position granted, because individual initiative churned the formal structure of society. Individual aliʻi sought to improve their position in the hierarchy or even reach its summit. The four aliʻi ʻaimoku in 1778, like other male and female aliʻi nui (high chiefs) in the islands, owed part of their ruling power to formal structure as it was manifested in the degree of chiefly rank and mana (sacred power) they inherited from their ancestors. But inherited rank was only one claim to power; so, too, were the accomplishments of an individual and his or her success at consolidating power by cultivating loyalties and allegiances and through intrigue and warfare. As Kameʻeleihiwa has shown, the path to increased mana often lay through sex and war, by coupling with a mate with high rank or defeating an aliʻi nui.[14]

Yet we must also note that knowledge was as crucial as sex or war for gaining, and especially retaining, power. Aliʻi depended on knowledge: their own and that of kāhuna and advisors who could guide them in gaining and retaining their position. Kāhuna had the genealogical knowledge to help (or undermine) an aliʻi's claim to rank. Kāhuna and advisors had the strategic expertise to assist an aliʻi in governance and war. Kāhuna were generally aliʻi themselves but had received particularly extensive training that had given them a command of the moʻolelo, mele (songs), pule (prayers), rituals, and skills relevant to their own areas of special expertise—whether they be genealogy, the worship of an akua, the making of canoes, the construction of buildings, or any number of other fields.[15] Describing the chiefly society of Hawaiʻi as one of hereditary rank should not blind us to this central fact: knowledge was essential to gaining and exercising power before and after the Western incursion into Hawaiʻi.

Moreover, Kānaka recognized that knowledge was dynamic and changing—an important point that contradicts anthropological, historical, and popular notions that confine them to the category of "traditional." The arrival of the Haole meant that there were new things to know about the wide world far from Hawaiʻi, and Kānaka wanted that knowledge. As chapter 1 explains, Kānaka celebrated their heritage of exploration in story and song, and moʻolelo and mele gave Kānaka every reason to believe that there were good and useful things that could come from far away. Add to this that the exploration of distant places offered individuals access to knowledge, and that Kanaka society placed a premium on knowledge, and the spur to Kānaka to explore the world was a powerful one and keenly felt. In 1778, this new phase in the Kanaka Maoli exploration of the world began by paddling out to the Haole's ships and boarding them. Soon, however, it also meant traveling on those ships to explore places no Kanaka had seen.

That process can be understood by tracing the travels of two Kānaka who were separated by rank and by gender. One was a makaʻāinana woman. The other was a man and an aliʻi of the very highest rank. And yet they met on the coast of China, they sailed on the same foreign ship far from Hawaiʻi in the South China Sea, and they formed a very close bond. The story of that woman and that man is the clearest way to gain a sense of why and how Kānaka explored the outside world, and how those explorations changed the way they looked at Hawai'i and its relations with foreigners.

KA WAHINE ON THE *IMPERIAL EAGLE:* A MAKA'ĀINANA LADY'S MAID TO NOOTKA AND MACAO

The woman remained on the ship because she wanted to. The English called her "Winee," but that was not her name. The English did not take the time to record her name. They did record the names of powerful chiefs both male and female, but she was not a chief. The label they placed on her was probably just a mispronounced approximation of the Hawaiian word "wahine." The word means "woman," but Haole gave it a derogatory tone when they used it to name Hawaiian women. Similarly, "kanaka" (meaning "man" or "person" or sometimes "Hawaiian") almost became a slur when used to name Hawaiian men—a slur like calling American Indian women "squaw" and African American men "buck." The source documents that mention her, all written by Englishmen and Englishwomen, never grant her the dignity of her name. To recapture some of her human dignity, we can at least call her Ka Wahine, The Woman. The Haole demeaned Ka Wahine, but still she stayed on the ship, and she sailed on it to see and know extraordinary places.

On May 24, 1787, Ka Wahine and others pulled up in canoes alongside a vessel named the *Imperial Eagle* in Kealakekua Bay off Hawai'i Island. They came aboard and traded fish to the sailors, mostly Englishmen, who were under the command of Charles T. Barkley. The *Imperial Eagle* had been sailing off shore for five days, trading nails and other items for the product of Hawaiian labor: sweet potatoes, hogs, salt, plantains, coconuts, rope, and more.[16] In the nine years since Kā'eo and his retinue first explored Cook and his moku off of Kaua'i, five Western ships had come and gone from the islands.[17] Although the English feared sharing Cook's fate, executed as he was by Kānaka for arrogating to himself the authority to punish and even kill Kānaka, the Western sailors needed Hawai'i. It was the only island group in the North Pacific. The Westerners depended on it as a waystation and a provisioning point for food and water as they developed a lucrative trade in furs and other goods between northwestern North America and China. The Haole vessels brought sudden and unpredictable bouts of Western violence to

"Wynee, a Native of Owyhee, One of the Sandwich Islands." Lithograph by J. Walter in Meares, *Voyages Made in the Years 1788 and 1789*. Just as the name "Wynee" almost certainly did not resemble her real name, this portrait probably does not reflect Ka Wahine's actual appearance. It appears to be based on "A Young Woman of the Sandwich Islands" (1785) by John Webber of the Cook expedition. Courtesy of the James Ford Bell Library, University of Minnesota.

Hawai'i—men were killed for taking an axe or a small boat as Westerners aimed to enforce their sense of property on the Kānaka. The Haole vessels brought terrible diseases. Tuberculosis, typhoid, smallpox, syphilis, gonorrhea, and other diseases devastated this population with no immunities to the foreign microbes Westerners passed them with their goods and their physical contact.[18] Still, the vessels were useful for the aliʻi and the makaʻāinana. Kānaka provisioned the Haole with hogs, vegetables, fish, salt, and fresh water, obtaining in return iron goods, woven fabrics, and other useful and decorative items of foreign make. This trade had begun with Cook's very first stop at Kauaʻi, and so did the exchange of female sexual labor to Western sailors for these same goods.[19] Given the outcome of her visit to the ship (employment by the captain's wife), it seems unlikely that sexual exchange was the reason that Ka Wahine was present on the *Imperial Eagle* in 1787. She came to trade goods and, like the many other Kānaka who hurried out on canoes at the sight of Western ships, probably also to explore the Haole's ship for herself.

In any case, when the other Kānaka who had come to the *Imperial Eagle* climbed off the ship and into their canoes, Ka Wahine stayed. The captain recorded in his log, "One of the natives remained on board, signifying an intention to go in the ship."[20] She was kept onboard as a laborer: Ka Wahine became the personal servant (the "lady's maid") of Frances Trevor Barkley. The ship's captain had met and wed this seventeen-year-old daughter of an English minister while in port in Ostend, Belgium, and she traveled aboard the ship.[21] The question of what Barkley called Ka Wahine's "intention"—that is, her agency—is important. Worldwide in the seventeenth, eighteenth, and nineteenth centuries, Westerners showed no compunction about kidnapping and enslaving the kind of people with which they classed Kānaka Maoli—dark-skinned people from Africa, North and South America and the Caribbean, and the islands of the Pacific. In fact, in coming decades, Europeans and Americans would appropriate the word "kanaka" as a term for enslaved Pacific Islanders who were captured in the "blackbirding" trade in the South Pacific.[22] Europeans seized the word "kanaka" just as they seized the bodies of South Pacific Islanders. In his terse wording about Ka Wahine's "intention," Barkley himself seems to be signifying that he had not enslaved her. Rather, Ka Wahine chose of her free will to "go in the ship"—not just to enter it but also to go *elsewhere* in it. Once the *Imperial Eagle* had concluded its trading in Hawaiʻi's waters, she sailed away from Hawaiʻi.

This would be a voyage of firsts. Frances Barkley was the first European woman to visit the Hawaiian Islands. Ka Wahine was, according to historian David A. Chappell, the first Kanaka woman to travel abroad on a European vessel.[23] On it, she served Frances Barkley as they traveled eastward across the Pacific, where she would become the first Hawaiian woman to visit the country of the Nuu-chah-nulth, on what the English called Vancouver Island. There, Charles Barkley traded with Nuu-chahnulth for otter furs they had hunted and processed. Once Charles Barkley finished his trading, Ka Wahine crossed the Pacific again, serving Frances Barkley on the long voyage across the ocean to China, where Charles would sell the pelts. Charles moved about in port, but Frances stayed in the home of a Mr. Cummins, an English businessman who lived in the Portuguese-controlled port city of Macao.[24] As Frances Barkley's personal servant, Ka Wahine probably lodged in that house with her while Charles Barkley sold his furs. When he had sold the furs, he sailed the *Imperial Eagle* off toward India and Britain. Ka Wahine did not join them. Frances Barkley later wrote in her memoirs that "Winee" chose to return to Hawai'i rather than continue on with the Barkleys. Another English captain offered a different explanation for Ka Wahine's remaining in Macao: the Barkleys left their servant in a foreign land because she had grown seriously ill.[25] Perhaps they judged that the cost of her keep now outweighed the value of her labor.

Ka Wahine was not the only Kanaka in Macao at the time. At least three others were there, as well, and they would all board the same ship to return to Hawai'i. Two of the Kānaka were referred to cursorily by another English captain as "a stout man and boy from Mowee [Maui]."[26] Like Ka Wahine, they were probably people of humble origin who had arrived in Macao as laborers. These were some of the very first of thousands of Kānaka who would work the Westerners' ships for decades to come. As a Kanaka laborer on a Western ship, Ka Wahine was unusual only for her gender and for the early date of her departure. Captain John Meares recalled that "numbers" of men "pressed forward" onto his ship "with inexpressible eagerness" to travel with him, and other captains found the same. As the eighteenth century gave way to the nineteenth, Kanaka laborers aboard Western vessels would become increasingly commonplace.[27] A few women traveled, as servants, as the companions, concubines, or wives of foreign officers, and a few as chiefly travelers to explore the world.[28] Most Kānaka who traveled on these ships, however, were men traveling as workers.

The names that the Haole called them tells us much about their entry into capitalist labor markets: to their employers, they were laboring bodies, their laboring value determined by age and gender and what

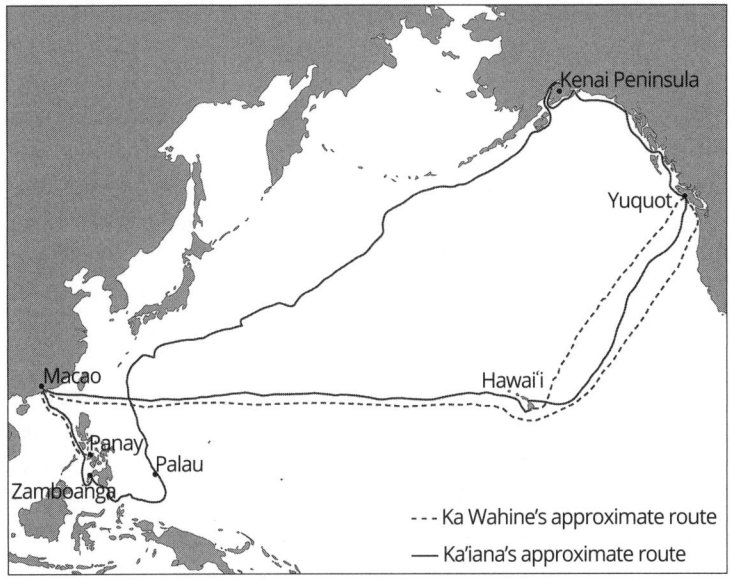

Routes of Ka Wahine and Kaʻiana. Map by Matt Lindholm.

Westerners of this time were coming to call "race." In his account of his voyage, Meares's partner and employee William Douglas called the boy from Maui by no name but "the boy" and called the man Tawnee—very likely an attempt at Tāne, but also a reference to his tawny skin.[29] Tāne (spelled kāne today) means "man." Like Ka Wahine, Tawnee's entire identity to the English was, therefore, gender and age. Aboard the ships (and in the emerging capitalist system of which they were essential vehicles), the labor expected of workers depended in large part on the sex and age and race of their bodies: boys, "stout" men, and women performed different tasks. Describing and even naming them by sex and age—woman (Winee) or man (Tawnee) or "boy"—signals what they meant to these early modern labor markets: to the English, they were laboring bodies. And in the labels assigned them, one senses also the label of racial difference and inferiority Westerners placed on those bodies.

The bodily labeling of these Kānaka might lead one to surmise that, because they had no other means of survival, they had been reduced to laboring abroad. This would almost certainly be a mistake, though an understandable one: later in the nineteenth century, when many

Kānaka lost access to the land that had sustained them, they had no other option but to labor for wages. In the twentieth century, dire circumstances in occupied Hawai'i forced many Kānaka into diaspora. Those later generations of Kānaka were proletarians, people forced to sell the labor of their bodies because they had no access to land or capital or other means of support. But Kānaka at the turn of the nineteenth century could still choose to remain part of a noncapitalist economy. The Hawaiian economy of noncapitalist production and distribution under different levels of land divisions (most notably the ahupua'a) was still operational at this time.[30] It would not have forced them out to sea as proletarians in search of a market for their bodily labor.

Why, then, did Ka Wahine and the "stout man and boy from Mowee" go to sea? Western sources do not tell us. They show us what served Westerners' purposes (the bodily labor of subjects categorized by gender, age, and race), but those sources are mostly blind to what did not serve Westerners' purposes (Kānaka's own motivations and what travel abroad meant to them and their society). This signals us a key question to ask as we think about the lives of Kānaka that we encounter through Western sources. What were their motivations? This will be difficult, almost impossible, to discover for three of the four Kānaka that were in Macao late in 1787: "Winee" and the "man and boy from Mowee."

The fourth, however, was famous among and well documented by both Kānaka and Englishmen: Ka'iana'ahu'ula, an ali'i from Kaua'i Island. Ka'iana was a man of exalted rank by dint of mo'okū'auhau (genealogy): his grandfather was Keawe, the mō'ī of Hawai'i and the product of illustrious ancestors; his mother was Kaupekamoku, who descended from ruling families of O'ahu, Maui, and Hawai'i; his close relations included some of the most powerful men and women in the archipelago.[31] As such, he was probably among the twenty or thirty most highly ranked of the hundreds of thousands of people in the islands. Yet status was not static in Hawai'i, and Ka'iana's efforts to reach an even higher station had already made him a key figure in the archipelago's politics when he boarded an English ship in late August 1787. His story intersects significantly with that of Ka Wahine.

It is not just the events of their lives that overlapped. Both Ka Wahine and Ka'iana provide excellent examples of two ways that different Kānaka would explore the world over the coming century and more—as humble laborers and as ali'i travelers. Moreover, both demonstrate how Haole failed to understand Kānaka and Kanaka acts of exploration, despite intense physical proximity with Ka Wahine and intimate relations with Kaʻiana. By telling their stories together, these themes emerge, and we can get some sense of the activities and motivations of Ka Wahine, who is less well documented than the powerful Kaʻiana.

KAʻIANA: A LEARNED ALIʻI ABOARD THE *IPHEGENIA*

Kaʻiana's explorations of the Westerners began almost immediately upon their arrival in Hawaiʻi. In Westerners' accounts, "Tyanna" (one of several ways they transliterated Kaʻiana) was a chief from Kauaʻi, the island where Kānaka first explored Cook on his voyage in 1778. In 1779, when the *Resolution* returned to Kauaʻi, a man named "Taiana" asked for and received passage on Cook's vessel to the island of Niʻihau. Normally Kānaka made voyages between the islands by oceangoing canoe, but presumably the passage gave him the chance to examine the Haole more closely. The exploration that "Taiana" had in mind was just beginning, however: according to the journal of Cook's medical officer, he asked to stay aboard and travel to Britain, but Cook denied him passage.[32] It is not certain that this "Taiana" was Kaʻianaʻahuʻula, the one that later would famously travel abroad. David G. Miller points out that it may have been his cousin Kaʻiana Ulupe.[33] If this "Taiana" was not the famous traveler, it is yet another reminder that many Kānaka were eager to explore the world.

Eight years later, in 1787, Kaʻianaʻahuʻula was on Hawaiʻi Island when Captain John Meares arrived in August to trade for provisions for a trip to China. The British Empire was the fabric of Meares's life. He was born in British-occupied Ireland in 1756, the son of a functionary of the imperial court in Dublin, and from the age of fifteen on, he had made his life on the military and merchant marine ships that allowed Britannia to claim to rule the waves. As a proud representative of the commercial arm of that empire, Meares expressed no surprise when he wrote that many Kānaka "wanted to accompany us to *Britanee,*" that is, Britain. But Meares selected only one passenger. He wrote, "Tianna, a chief of Atooi [Kauaʻi], and a brother of the sovereign of that island was alone received to embark with us, amid the envy of all his countrymen."[34]

Why did Meares select Kaʻiana alone to travel on his ship with him? Kaʻiana fascinated Meares, and that fascination may explain Meares's choice. Meares's account of the voyage returns several times to Kaʻiana, whom he describes in rapturous terms that place Kaʻiana squarely in the realm of the noble savage. "He was near six feet five inches in stature, and the muscular form of his limbs was of an Herculean appearance." Nor is it merely his physique that Meares admired. Kaʻiana carried himself as befit a lord, his "carriage" was "replete with dignity, and having lived in the habits of receiving the respect due to superior rank in his own country, he possessed an air of distinction." Meares admires the nobility of what we now know as the noble savage. In most noble savage writing, the untrammeled dignity of the indigene sprang from the supposedly egalitarian freedom of "savage" society. For Meares, however, an implied similarity between hierarchy in Britain and in Hawaiʻi stands at the origin of Kaʻiana's much-admired nobility.[35]

But condescension infuses these admiring tones in a way that encapsulates how Western hierarchies functioned in regard to Kānaka for many years to come. On the one hand, Meares specifically refuses to term Kaʻiana a savage. He writes that this aliʻi possessed "capacities" that would "forbid any enlightened reason from applying the name of savage to any human being, of any colour or country, who possesses them." But on the other hand, this statement itself reserves to Westerners (the implied owners of "enlightened reason") the right to judge who is and is not a savage.[36] Moreover, Kaʻiana's "colour and country"—that is, the fact that he was a brown man from the kind of country that Europeans like Meares were busy colonizing at the time—determined the specific ways Meares and other Haole misunderstood Kaʻiana.

What most ennobled Kaʻiana to Meares were "those capacities which education might have nurtured into intellectual superiority"—that is, an untutored mind.[37] In Meares's narrative, Kaʻiana appears as a brilliant and inquisitive child: when arriving at Macao in October 1787 and seeing all the vessels there, he frolics with joy at learning they had arrived in Britain. Meares describes himself as gently explaining where they actually are, whereupon Kaʻiana quickly recovers from the disappointment and explores the Western ships, "their internal arrangements, with all of the various apparatus they contained." Meares describes Kaʻiana as exhibiting a childlike dejection at the sight: the vast display of technological superiority dwarfs Kanaka capacity. He again recovers, though, in what Meares depicts as Kaʻiana's humble spirit of service, "the same spirit which urged him to quit his native country, in order to return with knowledge which might instruct, and arts that might improve it." To Meares, prime evidence of his learning was his learning about Europeans and their ways. Kaʻiana learned to wear "the dress of Europe with the habitual ease of its inhabitants" and even "manifested no common degree of intellectual exertion, by discriminating, as occasion offered, between the people of the several European nations." Meares's account of Kaʻiana paints an admiring picture of the gifts that curiosity and European tutelage could bring to the innate but untrained intelligence of a noble savage.[38]

But Kaʻiana was anything but untutored or childlike, and recognizing Meares's failure to understand that lays the foundation for a truer understanding of Kaʻiana and of the Kanaka Maoli project of global exploration. That understanding can only proceed from Kanaka Maoli perspectives, which can best be found in the nineteenth century in Hawaiian-language sources. Samuel Manaiākalani Kamakau, the historian whose account of the first encounter of Kāʻeo and his court with Cook off Kauaʻi begins this chapter, again allows us to get beyond European misperception. He describes Kaʻiana, Meares's childlike and untutored mind, as follows:

> O Kaiana, he alii naauao,
> a ua kuhohonu kona akamai ma ke kaakaua,
> ma ke kuhikuhi puuone,
> ma ka papa huli honua,
> ma ka oihana kahuna,
> ke kakaolelo,
> he kilo aohe e nalowale na olelo huna,
> he kalaiaina no pohokano,
> he mailo i ko na'lii ano,
> he akamai i ke kuauhau kupuna.[39]

> Kaʻiana was a learned chief,
> and he had profound knowledge of directing warfare,
> of divination,
> of the determination of land boundaries
> and in the priesthood,
> was trained in counsel,
> a reader of omens who had not forgotten the secret words,
> frugal in kālaiʻāina [appointing lands and thus their tribute to chiefs],

gaunt in the manner of the aliʻi,
expert in ancestral genealogy.

Simply put, Kaʻiana was one of the most highly trained intellects of his generation of aliʻi, and his training was in the skills that were pertinent to a man who was raised to rule. War, chiefly genealogy, the secret words and rites of the priesthood and divination, counsel, land divisions and the attendant issues of who received tribute from and who had control over land districts—these were ways that power was gained, kept, and exercised in Hawaiʻi. Kaʻiana mastered these skills. It is significant, moreover, that the historian Kamakau describes Kaʻiana this way. In his prolific writings in the mid-nineteenth century, he offered brief descriptions of scores of the most important aliʻi for generations past. Few dwell on the naʻauao (learning) and akamai (intelligence) of a chief as does this sketch of Kaʻiana. Noelani Arista identifies this enumeration of Kaʻiana's qualities as a helu. This genre of composition, John Charlot explains, was a form of listing that was important in narration, could be chanted, and was "very close to poetry." In the rhythmic, reiterative power of his helu, Kamakau invokes Kaʻiana as a paragon of chiefly excellence. In his intelligence and his learning, Kaʻiana was exceptionally gifted, but in things that mattered to aliʻi more generally.[40]

Similarly, Kaʻiana's drive to explore the globe should be understood as emblematic of the Kanaka exploration of the world. Kānaka Maoli did not begin their intellectual lives or their exploration of the world with the arrival of the Haole any more than Kaʻiana boarded Meares's ship as a naïve and untutored child. This was a man of learning, and we have every reason to believe that he wanted to learn more: that his quest to learn about the world via exploration was a quest that was important to him in its own right, driven by the will to know. Furthermore, we must consider the political meaning of Kaʻiana's exploration. He was a ruling chief, and boarding that ship would give him the opportunity to further his extraordinary knowledge. New knowledge—about the places these ships sailed to and from, about the people aboard these ships, about the people in foreign lands—had become available and was now clearly pertinent to his future. Most Kānaka Maoli, and especially makaʻāinana like Ka Wahine, had far less power and far less knowledge of the techniques of power than Kaʻiana. Though we know next to nothing about Ka Wahine, we know that even the makaʻāinana had the knowledge (both technical and ritual, although the distinction may be specious in Hawaiian society at this time) that was most pertinent to what Kānaka called their 'oihana—their trade or profession. Fishing and farming and other maka'āinana activities had their prayers and rites just as surely as the priesthood did. Indeed, the word 'oihana pertains as much to the maka'āinana crafts of fishing and farming ('oihana lawai'a and 'oihana mahi'ai) as to the priesthood ('oihana kahuna).[41] Kanaka individuals as diverse as Ka'iana and Ka Wahine had knowledge and sought more knowledge because they needed it and wanted it. The exploration of the world was part of that seeking. Meares could not see this in Ka'iana, and the Barkleys could not see this in Ka Wahine, but we must see it in both to understand the Hawaiian exploration of the world. Texts authored by Kānaka Maoli are essential to understanding why Kanaka traveled abroad and explored the world.

LOVE AND POWER: MEARES AS KA'IANA'S AIKĀNE

Hawaiian-language texts are similarly essential to understanding the nature of Ka'iana's relationship with Meares and how Ka'iana came to be the only Kanaka Meares allowed to travel on the ship among the many who asked. Meares's account hints at homoeroticism and romantic love but leaves the reader wondering. Meares writes about Ka'iana not only with rapturous admiration for his features, height, strength, and musculature, but also with a striking affection and tenderness. He describes Ka'iana as both "Herculean" (literally godlike) and lovely. In his published account of his voyages, Meares includes an engraving of a portrait of Ka'iana that was made in China (most likely at Meares's expense, as Ka'iana apparently had no money to commission it). The artist, he says, captured Ka'iana's face well, "but found the graceful figure of the chief beyond the powers of his genius." Ka'iana, we are told, alternately dressed beautifully in Western clothing and handsomely in the malo (loincloth), 'ahu'ula (feathered cloak), and mahiole (feathered helmet) of an ali'i. Despite his imposing physique, according to Meares he possessed "the delicacy of constitution that discriminates the chiefs from the common people." When it was time for the two to part in September 1788, Meares reported that Ka'iana wept bitterly. Meares takes care to describe himself as having preserved the cool demeanor that British class and gender expectations required of him but alludes to the depth of his feeling by allowing that it was only with great difficulty that he "mastered his emotions" and held back tears. Thus Meares uses tropes that in the Western tradition hint at sexual attraction and romantic love, but just as important, he never names that relationship as sexual or romantic. In a Western account, one is therefore presented with a question: Were Kaʻiana and Meares lovers?[42]

Kamakau presents the reader with an unproblematic answer: yes, they were. Meares is described as "kana aikane a hoa aloha hoi." Meares was "his aikāne and friend, too." An aikāne is a romantic same-sex friend, generally and unproblematically assumed to be a sexual partner. Decades after Meares and Kaʻianaʻs aikāne relationship, the term came to mean "friend," under the influence of American Christian missionaries. Since then, scholars have occasionally debated if sexual relations were always part of aikāne relationships, pointing out that such romantic and emotionally intimate friendships need not have involved sexual relations. This is true, but it misses an essential point: the sexual nature of the relationship was assumed. Some aikāne relationships may not have been sexual, just as some marriages in Europe and the United States were not sexual, but the assumption was that they were. To Kānaka, there was nothing illicit or even surprising about sexual relations between aikāne. The word is formed by the elements ai, meaning "coitus," and kāne, meaning "man." Given that Kamakau says Meares was Kaʻianaʻs "aikāne *and* friend," it is even more certain that this was a romantic relationship and even more likely that it was sexual. Hawaiians had no notion that aikāne were gay or homosexual in the sense that those words are commonly used today; they were not members of a distinct minority population set apart by an enduring same-sex orientation. Rather they were engaging in one kind of sexual relationship, and in their lifetimes they would probably take part in relations with both men and women. What seems a problematic conjecture of allusion in Mearesʻs writing (were they lovers?) in Hawaiian is a simple fact.[43]

It is telling that Kamakau writes that Meares was "his" (that is, Kaʻianaʻs) aikāne and friend ("kana aikane a hoa aloha hoi"). Although evidence is not complete or conclusive, aikāne relationships appear generally have been between people of close but differing ranks. The possessive pronoun "his" suggests that Kamakau understood Kaʻiana to occupy the role of the senior or more highly ranked member of the relationship, and Meares the junior or lower-ranked role. (Both men were about thirty-two or thirty-three at the time, so status more than age would have been the differential in this relationship.[44]) This does not mean that Meares understood himself in the inferior role; after all, his account posits Kaʻiana as a physically impressive but intellectually childlike noble savage that he introduces to the wonders of the wider world. (Like the European assumption that Kānaka took Cook for a god, Meares overestimated the awe in which Kānaka held him. He believed that "natives" from Vancouver Island to Hawaiʻi called him "Noota" in honor of his vessel the *Nootka*, whereas Kamakau actually refers to him simply as Kane, which in this context is probably a Hawaiianized version of his given name, John.[45]) Meares may have believed that he was the captain of the relationship as he was of his ship, but in Kanaka Maoli terms as written by Kamakau, Kaʻiana was the leading partner—and Kanaka Maoli terms were the perspective from which Kaʻiana proceeded as he engaged with foreigners like Meares. Even if Kamakau is in error, and Kaʻiana was Meares's aikāne (i.e., if the couple understood Meares to be more highly ranked in the relationship), the most important point would remain: Kaʻiana's mode of entry into foreign travel and exploration was a relationship that was Kanaka Maoli in its structure and that Kānaka Maoli understood as such. Far from abandoning Kanaka premises to gain entry into Haole society and structures, Kaʻiana explored the world from his own indigenous premises.

He also explored the world for his own purposes, and those were to gain knowledge and to build his power. Kaʻiana's background and his later career must complicate Meares's statement that Kaʻiana traveled "in order to return with knowledge which might instruct, and arts that might improve" his society.[46] Kaʻiana was a powerful and learned man. Much of his power came from his knowledge, and this trip offered a chance to better cultivate that knowledge. That knowledge and power could indeed serve a broader social interest, but given the rivalries for rulership in these years, we would do well to construe social interest through the ambitions of Kaʻiana and the aliʻi he served upon his return, Kamehameha.

EXPLORATION, GENDER, AND CLOTHING

The first thing that both Kaʻiana and Ka Wahine explored was not far-off places, but the Haole themselves. The great bulk of their time was undoubtedly spent with the British. Kaʻiana spent fifteen months away,

Ka'iana dressed in his mahiole and 'ahu'ula, with spear. Lithograph by J. Walter in Meares, *Voyages Made in the Years 1788 and 1789*. This image is probably based on the portrait of Ka'iana made by Guan Zuolin (Spoilum) in Macao and the portrait "A Man of the Sandwich Islands, with His Helmet," made by John Webber in 1779 in Hawai'i. Courtesy of the James Ford Bell Library, University of Minnesota.

departing September 2, 1787, and returning December 6, 1788. Only about five months, a third of the total time away, was spent on land at Macao or anchored off of Zamboanga (on the island of Mindanao in the Philippines) and Yuquot (in the Nuu-chah-nulth country in present-day British Columbia).[47] Ka Wahine similarly spent much of her time

John Meares, Esq. Lithograph by J. Walter in Meares, *Voyages Made in the Years 1788 and 1789*. Courtesy of the James Ford Bell Library, University of Minnesota.

away from Hawai'i aboard ship. Thus they had ample opportunities to examine the material and social realities of the ships. The only accounts of these voyages are those that British officers penned, and they emphasize events on land. Little is said of Ka'iana's time aboard ships, and Ka Wahine barely enters the narratives. Still, reasonable inferences suggest that both would have examined the ships quite closely. Accounts by Western sailors describe Kānaka who came aboard searching the alien environment with their eyes and usually their hands as they explored the Haole's ships and the Haole themselves.

Ka Wahine was the personal servant of Frances Barkley and would have spent much of her time with her and with her things. Apparently, Ka Wahine learned about the gendered material culture of Western women, given what she chose to bring back to Hawaiʻi for her own use and as gifts: a mirror, a basin, a porcelain bottle, a gown, a hoop, a petticoat, a cap, and "a great variety of articles."[48] How she had obtained these things—perhaps as gifts or pay from Frances Barkley, perhaps as barter—is not clear. What is obvious is that she had quickly gained a sense of the gendered material culture of the West and took hold of it for herself.

Are we to gather from this that Ka Wahine was inculcated with, and even perhaps embraced, the restrictive gender norms of the West? That conclusion would be too hasty. Instead, what we can see is Ka Wahine exploring and mastering the gendered material culture of the Haole, acquiring cosmopolitan status goods, and bringing them home for her use and for gifts. Ka Wahine died during the return voyage to Hawaiʻi, and on the day she died, she entrusted her goods to Kaʻiana: the mirror, porcelain basin, and bottle as gifts to him; the hoop, petticoat, gown, and cap as presents for "his wife"; and the rest of her goods for him to convey to her own father and mother as gifts for the family. Foreign goods like these were still extremely rare in Hawaiʻi. Moreover, they were generally in the hands of aliʻi who had traded for or received them from foreigners.[49] They would therefore have been particularly remarkable gifts to Ka Wahine's parents, who would have been from the same (presumably) makaʻāinana background as she. The gendered nature of these goods is clear, of course. Then again, it would have been strange if Ka Wahine had not chosen items with care to their gender. Gender roles in Hawaiʻi differed from the British gender roles of Frances and Charles Barkley, and there was more room for flexibility and ambiguity in the Hawaiian gender system, but gender was still a central organizing principle of Kanaka society. In exploring the foreign world through a gendered lens, Ka Wahine and other Kānaka were exploring it on their own terms.

This was true also of Kaʻiana, whose wardrobe choices reveal the deliberateness and confidence with which he explored and engaged with the foreign world. Meares reports that he learned to name and wear British-style garments with fluency. Like Ka Wahine, the use of clothing was one of the ways he demonstrated a command of the gendered expectations of the Westerners among whom he moved. And yet at times, Kaʻiana chose to dress differently. Multiple accounts describe Kaʻiana exploring Macao by foot, startling all those around him with the sight of a Kanaka Maoli over six feet tall striding through the polyglot trading town attired in a feathered cloak and helmet and carrying a spear. (This is also how he dressed when he sat for a portrait in Macao.) The ʻahuʻula and mahiole are not just Hawaiian attire; they are specifically chiefly regalia and are worn for effect on special occasions only. These precious garments featured motifs made of tens of thousands of tiny red, yellow, green, and black feathers completely covering a netting of cord made from the olonā plant. It took generations to harvest the feathers for each cape and helmet, as expert trappers snared specific species of upland birds (ʻōʻō, mamo, ʻapapane, and ʻiʻiwi), plucked a few feathers of the right color, and then released the birds. For Kaʻiana to display himself in such striking and precious garments was a deliberate act of stepping outside norms that he fully understood—especially when he wore nothing else. (Nathaniel Portlock noted that when he wore no "other dress" than loincloth, cloak, and helmet it was "scarcely modest.") Paired with a spear, this was a masculine bodily and sartorial display that was calculated to impress with its drama. Kaʻiana made clear that while he traveled among the British, he was a Hawaiian aliʻi. This type of deliberate choice of how to dress was perfectly in keeping with the importance of physical performance and attire in chiefly society and showed that Kaʻiana explored the streets of Macao on his own terms.[50]

MACAO: A WORLD OF NATIONS

While Portlock describes Kaʻiana as "generally walking about wherever his inclination led him," no accounts mention anything of Ka Wahine's time there.[51] We could speculate on the many novel things they must have seen, heard, smelled, tasted, and touched that were surprising and different from what they knew in Hawaiʻi: a crowded city, different forms of architecture, strange domestic spaces, animals such as horses and cattle that did not exist in Hawaiʻi, new foods, unfamiliar technologies, and much more. Speculation is necessary in history, but if we guess too much at their experiences, we run the risk of falling into the trap that Cook and Meares and the Barkleys and Portlock and the other Westerners made for themselves: we would not see Kaʻiana and Ka Wahine, but only what we expected to see. It is better to start with what, for all their limitations, Englishmen's accounts tell us about Kaʻiana's explorations of Macao.

From their writings, we learn Kaʻiana drew a crucial lesson in global geography from Macao: the world was full of foreigners from different places, who understood themselves to be different from one another, had different ways, spoke different languages, pursued their own interests, and were often rivals for power. This was a lesson Kaʻiana could not have fully appreciated before his visit there. In the language of Westerners of his time, he was seeing a world of "nations." Here we must understand that term more broadly than we might today. "Nations" included what could be called the vested members of particular nation-states and imperial powers (the British, the Portuguese, the Chinese, and so forth) and also the diverse world of their empires and the non-nation-state people who moved through them. Kaʻiana witnessed such people and the vulnerability that nonnational status meant for them when "a company of poor Tatars" paddled up on sampans to the vessel of an English captain who was hosting "a sumptious entertainment" to which Kaʻiana had been invited. Kaʻiana asked what they were doing there. When he learned "that they were beggers who came to supplicate the refuse of the table," Kaʻiana was appalled. He declared that no one was reduced to begging for scraps of food on Kauaʻi. He insisted that food be brought immediately, and according to an Englishman who was present, took great "pleasure and satisfaction" in distributing the victuals in equal shares among the "Tatars." Englishmen of the time used the term "Tatars" to refer to a number of peoples, especially Central Asians, so it is unclear who these people were. What is clear is that these "Tatars" were not among the nations that enjoyed power and wealth in Macao. Their poverty thus bore witness to the vulnerability of some of the many different peoples that populated the city.[52]

Kaʻiana, Ka Wahine, and other Kānaka had very little experience with what Westerners would call national differences (whether we think of nation in terms of long-term and established differences of culture, of language, of ancestry, or of political belonging) although mele and moʻolelo did contain within them the sense that different places might be populated with different kinds of people. The Hawaiian Islands were divided into units that have sometimes been called kingdoms—the lands ruled by various aliʻi ʻaimoku or mōʻī. Some differences of accent,

This view of the beach at Macao circa 1787 portrays the heterogeneity of the city: a Chinese man converses with a Portuguese friar while two men, whose bared torsos and toga-like garments suggest they are neither Chinese nor European, sit in the background. Detail from an engraving by Louis-Joseph Masquelier from a painting by Gaspard Duché de Vancy in La Pérouse, *Atlas du Voyage de La Pérouse*. Courtesy of the James Ford Bell Library, University of Minnesota.

word usage, story, song, and practice did (and still do) exist across the islands, and at least one moʻolelo attributed differences of accent to the linguistic influence of the settlers from Kahiki who arrived with the aliʻi Olopana after the islands had long been populated. Still, there was not a sense of national differences dividing Kānaka linguistically or culturally. Thus Kānaka did not have a sense of separate nations among themselves and had very limited experiences of it among outsiders. In the ten years since Cook's arrival in 1778, only five foreign ships had come to the islands, all of them under British command.[53] Crews included Englishmen, Scots, and non-British sailors, but the limited interaction that Kānaka had with the foreigners is unlikely to have made many (or perhaps any) Kānaka aware of how Haole divided themselves up into nations. In contrast, Kaʻiana was in long-term close contact with the Haole. He may have begun to learn about the world of nations aboard ship (the crew was mostly but not entirely English), but at Macao he studied it intensely and with deliberation.[54]

In Macao, Meares writes, Kaʻiana brought "no common degree of intellectual exertion" to understanding the nature of the differences between the people around him. The tiny and densely packed city in which Kaʻiana and Ka Wahine arrived was one of the most astonishingly cosmopolitan places on the globe in 1787. As they walked through Macao, they would have seen Chinese, most of all, but also people from all around the world. The trade port was one of the first global cities in the making of a global world. The Chinese who lived there and the Portuguese who claimed dominion over it shared the winding, cobbled streets of this pseudo-Portuguese port with the city's English, French, Dutch, Spanish, and other free European inhabitants, people from Central Asia and what we now call the Middle East, and slaves from Africa, China, Japan, and the islands of Southeast Asia.[55]

Meares reported that by exertion Kaʻiana taught himself the skill of "discriminating, as occasion offered, between the people of the several European nations, whom he daily saw, and those of England, whom he always called the men of *Brittanee*." Whether he looked to national costume or physiognomy or some other clue to determine national difference is not clear. The phrasing ("the several European nations") is also revealingly ambiguous, because Britons of the time often distinguished between themselves and people from continental Europe. Could Kaʻiana tell the English from the continental Europeans, or could he discern the differences between different European nationalities, as well? Kaʻiana was based in Macao's English element. He had arrived on an English ship. Meares placed him under the supervision of his English chief mate, Mr. Ross. He spent a lot of his time, and perhaps was lodged, in the home of one of the most prominent English merchants in Macao, John Henry Cox, who dealt in British manufactured goods, North American furs, Asian opium, and other goods drawn from the era's increasingly dense web of global trade.[56] According to Portlock, Kaʻiana was popular with the Englishmen "from whom he received invitations and every mark of civility."[57] Given Kaʻiana's English social base in Macao, the divide between continental Europeans and the English may have been one of his first lessons about the world of nations.

It wasn't the only one, though, because Kaʻiana was already emulating English beliefs about the inferiority of people from non-European nations, specifically the Chinese. While there is no reference to Kaʻiana's encounters with the Indians, Indonesians, Malaysians, or Africans in Macao, English accounts report repeatedly that Kaʻiana "entertained the greatest contempt" for the Chinese, indeed, that he considered them "with a degree of disgust which bordered on extreme aversion." Kaʻiana's views were a particularly vehement mirror of the racism of Europeans toward the Chinese. The expressions were both physical and cultural. He had "contracted a prejudice," one assumes from Englishmen, "against the form and shape and manner of their persons" as much as "against their practices and customs." His objection to "their bald heads, distended nostrils, and unmeaning features" was more than intellectual. According to English sources, the Chinese "raised in his mind the strongest sensations of contempt."[58] To the English, Kaʻiana's expressions of racism were noteworthy for their passion.

Then again, the very restrained English describe him as extreme in his expressiveness generally, so Kaʻiana may just have been bringing his usual expressiveness to the way he emulated English racism. The clearest evidence of this was the objections that Kaʻiana raised to the Chinese "shutting up and excluding the women from the sight of all strangers." In this, he was echoing English claims to civilizational superiority. European men of the time (and after) claimed that the supposedly freer position of women in European society than in non-European societies was a proof of European superiority.[59] In his objection, Kaʻiana made clear that he was an astute student of English racism. He merely expressed himself more freely. Racism was part of the geography of nations that Kaʻiana studied at Macao.

RITUAL AND POWER

While race and national difference were entirely new to Kaʻiana, precise religious ritual was essential to the practices of chiefly Hawaiʻi. In Macao, Kaʻiana studied it assiduously like the learned aliʻi that he was, and there he would get a sense of religious difference around the world. Although tiny, Macao boasted a Catholic cathedral and at least two other Catholic churches. The English captain Nathaniel Portlock reported that Kaʻiana frequently found his way to these churches and, by close observation, learned the ritual of worship at the mass. He "always observed the manner, motions, and attitudes of the congregation," learning when to stand, when to kneel, "appearing very studious to imitate them by exact conformity to all their actions, gestures and behaviour."[60] Portlock emphatically insisted, however, that Kaʻiana was *"no professed papist."* To an Anglican from anti-Catholic England, it might have been essential to note that Kaʻiana, whom he admired, had not made a "profession"— that is, a statement of Catholic faith.[61] But to Kaʻiana, a learned chief expert in the ʻoihana kahuna, faith may not have been a central issue. Ritual perfection was. At Hawaiian temples (heiau and luakini) and other sites of worship in Hawaiʻi, a perfect ritual could bring a fertile harvest or a victory in war. An error or an interruption could bring the execution of the offender. It was as an expert in Hawaiian ritual, not a convert to Christian belief, that Kaʻiana studied the Catholic mass.

Macao would also have given Kaʻiana a sense that the outside world was divided into different religions as it was into different nations. The East India Company, chartered by the English Crown to trade in Macao, paid for Anglican chaplains in Macao, where they said services in the chapel and buried the Anglican dead.[62] In 1787, the Macao mosque was only fourteen years old, and while Kaʻiana must have seen it, it does not appear in the descriptions of his voyage.[63] Given his movements about town, Kaʻiana would also have encountered the landscape of Chinese worship. In a town of under three square miles that was confined to a tiny peninsula and hemmed in by walls and water, he could scarcely have missed it. At least a half dozen grand temples such as the Kun Iam Tong and the A-Ma temples towered over neighboring buildings. Smaller temples like the Lin Fung Miu, the Lin Kai Miu, and the Hong Kung Miu served as major landmarks.[64] Even given Kaʻiana's reported contempt for the Chinese and thus his unlikeliness to emulate worship at these places, such surroundings would have reinforced the notion that this was a world of religions as well as nations.

Did Ka Wahine learn these lessons or different ones? The lack of sources on her experiences is frustrating, but some reasonable conjectures are possible. She was sick, female, and a servant, and these facts would have much more sharply limited her experiences in Macao in comparison to Kaʻiana—a healthy man and a chief who was popular among English gentlemen. Ka Wahine had come on board with the intention of accompanying the Barkleys to England but was too ill to continue the voyage. It is hard to imagine she had much strength to explore the city. Moreover, as a servant to Frances Barkley, she would have mostly remained close to her employer even when she was feeling well. As a captain's wife, Barkley occupied the status of a lady, at least in Macao. Englishmen may have enjoyed distinguishing the freedom of their women as compared to Chinese women, but in reality British convention would not have permitted Frances Barkley to explore the streets, markets, and churches of Macao as Kaʻiana did. Ka Wahine may have been sent on errands for the Barkleys. If so, her servant status may have given her some room to explore. Still, it is hard to imagine that this amounted to freedom to roam. Undoubtedly she saw many things, many places, and many kinds of people and gained knowledge from them. But the most intense site of her exploration was probably the household in which she worked. Because she was sick, that site was possibly a pallet or bed in the Cummins home where the Barkleys lodged. Despite these restrictions, given how profoundly it differed from a home in Hawaiʻi, it still may have been something to explore.

NATIONS AND RACE

Though he had more room to move, Kaʻiana also faced restrictions: only a narrow isthmus connected Macao to the mainland, and it was blocked by a wall with a gate. By Chinese law, no non-Chinese could go through the gate and enter the rest of China from Macao. That wall embodied some of the most important lessons in world geography that the city offered for a person in Kaʻiana's position. Macao was a Western outpost but existed under Chinese sovereignty. It existed only at the leave of the imperial government, and its very purpose was to concentrate and isolate the commercial, religious, and cultural presence of the foreigners.[65] A century before Kaʻiana's arrival, foreigners had had more freedom to move. Western Christian missionaries had made their way into China itself, and Western and Muslim Indian merchants had gained footholds on the coast. In response, the Chinese government pushed back, limiting foreigners to specified small areas of specified port cities, including Macao and nearby Guangzhou (Canton). Commonly referred to in Cantonese as *gweilo* (ghost fellows), the Westerners were confined. Within the boundaries of Macao, they could disembark, move about, and carry on their trade, even the illegal but tolerated trade in opium.[66] Yet always, as the gate on the isthmus testified, the Chinese maintained their claim to authority over the *gweilo*.

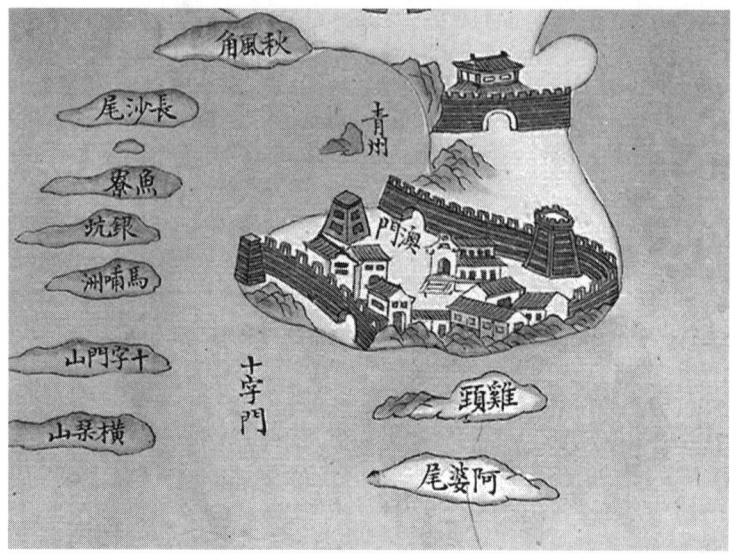

Chinese and Europeans both knew that the wall between them on the isthmus was foundational to what Macao meant. Although they depict Macao quite differently, the Chinese map and the French map both prominently show that wall in the upper right corner. Details from "Qi sheng yan hai tu" (Coastal map of seven provinces), circa 1820, above, and "Plan de la Ville et du Port de Macao," from Bellin, *Le Petit Atlas Maritime,* 1764, at right. Courtesy of the Library of Congress, Geography and Map Division.

The English chafed at the Chinese assertion of sovereignty in their own country. It went against the Westerners' sense of their right to power and to freedom, their sense of superiority over non-Europeans. Indeed, this may have been a subtext of the English disapproval that Kaʻiana learned and then voiced about Chinese women's lack of a right to move about freely. For the English the restrictions on their own movements likely seemed demeaning and feminizing: Did the Chinese men take them for Chinese women, locking them up behind a gate in this manner? Kaʻiana repeated the English outrage back to them with his own vehemence, the expressiveness that always surprised them. This suggests that (at least in the presence of the English, but perhaps in his own mind, as well) Kaʻiana may have believed his situation to be akin to that of his English friends and his English aikāne.

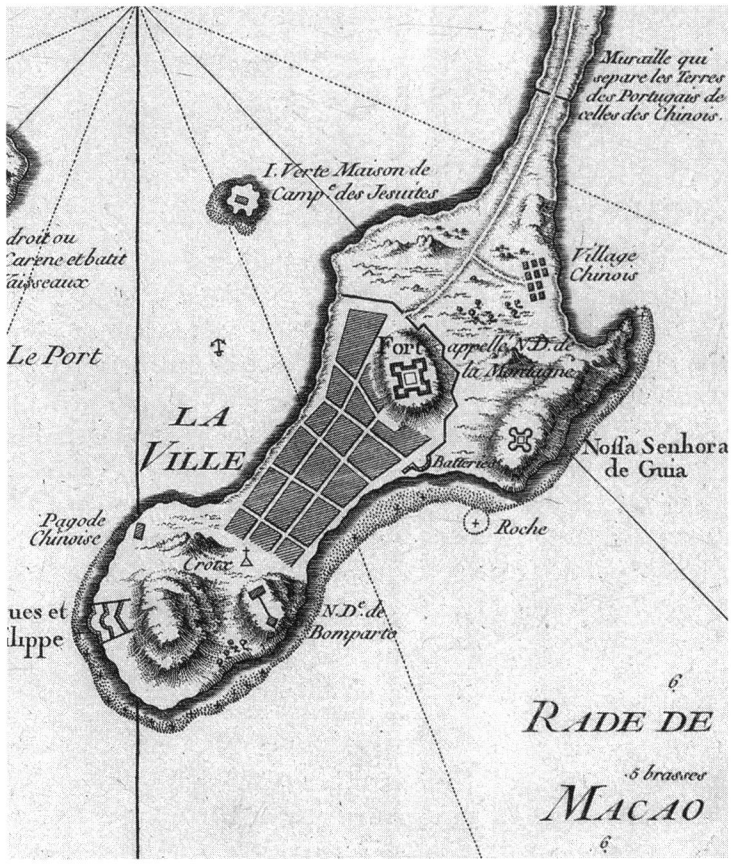

For a Kanaka Maoli, this was a mistake, a false lesson in the alignments of global geography. In this first decade of contact with the Westerners, the Kānaka did not yet know that in the next century they, their children, and their grandchildren would face Haole intent on dominating their islands culturally, economically, and politically. Seen in that context, for Ka'iana to identify with Westerners chafing at Chinese restrictions would be to look at the gate on the Macao isthmus from the wrong side: from the side of a *gweilo* trying to get into China, not the Chinese trying to contain the Westerners' drive to dominate. On the remainder of the voyage, Ka'iana would have every opportunity to unlearn this mistake. His actions in coming years suggest that he had gained a far better sense of this essential lesson in global geography: a difference of interests separated him from the Haole much as the gate separated China from the *gweilo*.

KA WAHINE AND KAʻIANA: AFFECTION AND BONDS TRANSCENDING STATUS

On Sunday, January 22, 1788, John Smith piloted the *Felice* and its companion, the *Iphegenia*, away from Macao's Taipa Harbor, making way to North America.[67] Aboard there was a more diverse complement of passengers and crew than Meares had had when he arrived: Ka Wahine (who was very sick), Kaʻiana, the unnamed man and boy from Maui, and at least one other Pacific Islander traveled in the ships. So too did Comekala (also referred to as Callicum in the sources), a Nuu-chah-nulth chief who had traveled with Meares from Vancouver Island to Macao. Finally, the ship counted among its crew a number of Chinese. The Chinese had impressed Meares for two reasons: the quality of their labor in shipbuilding and carpentry and the cheap wages he could pay them. When he had been in Nootka to trade for the furs he had just sold, he had noted the abundant timber available there and decided to try to build a ship in Nootka—or, more precisely, have workers make him a ship there. These Chinese would be those workers. The presence of this more diverse crew would be part of Kaʻiana's explorations.

Bad weather at sea soon after their departure provided Kaʻiana the chance to learn more about Westerners and the limits that non-Europeans could place on them. Not far out of Macao a storm hit with winds so severe that the mast of the *Iphegenia* needed to be replaced. While the sprung mast was an inconvenience, geopolitics posed a more serious problem. In a crippled ship, they were out of reach of Macao or Canton, and "there was no friendly port nearer to us than Batavia," the Dutch outpost in what is now Indonesia.[68] In the end, the situation was so dire that they decided to make for the Spanish settlement of Zamboanga (Samboingan), at the southern dip of the island of Mindanao, in the Philippines.[69] As they limped along the southern edge of the Philippines, Kaʻiana would have had ample opportunity to reflect on the limits on the power of his English friends.[70] With a crippled ship and a crew beset with scurvy (surprisingly soon after leaving port), the ships needed to anchor and send men ashore for a tree trunk for a mast and fresh food. They moved slowly past Luconia in what is now Malaysia, then Lubang, Mindoro, and the Calamian Islands in the Philippines. In plain sight lay forested coasts, with trees aplenty from which a mast could be cut. Villages with "well cultivated" fields that made for "a picturesque and fruitful scene" could easily have supplied the ship with the fresh fruits needed to cure the scurvy.[71] Yet Meares did not dare to send men ashore because he was concerned that the people of these islands would not welcome such a trespass. He was not even sure that things would go well at the Spanish fort and settlement on Mindanao, given the rivalries between the British and the Spanish. In such circumstances, the limits on European power would have been apparent.

They were probably also frustrating to Kaʻiana, because Ka Wahine's illness was increasingly dire, and he had grown close to her. No chance to seek harbor meant even more difficult conditions for Ka Wahine. Kaʻiana gave her his "constant attendance" as she withered to what Meares called "a living spectre." The aliʻi exerted himself so much in tending to her that he also caught a fever and needed to be confined to bed. Ka Wahine died on February 5, 1788, off the central Philippine island of Panay. She was buried at sea with Christian prayers, yet Kaʻiana and the man and boy from Maui may have been the only ones present who knew her true name.[72] Kaʻiana's grief overwhelmed him; the crew were "under very painful apprehensions" about his health. Meares believed Kaʻiana's grief was the expression of "that delicacy of constitution" that he thought was "peculiar to the great men of his country."[73]

Meares was mistaken. In his mistake, he again failed to see Kaʻiana and Ka Wahine. Profound mourning and expressions of grief were (and still are) a mark of Kanaka life and death. In Kaʻiana's time, the passing of a loved one brought on long and loud wailing that lasted for days in scenes of shared grief. Kānaka were known to smash out teeth in mourning, to cut the hair from the sides of their heads, to cut off an ear, or to have the tip of the tongue tattooed. Today, the physical marking of mourning has changed, but the profundity of Kanaka grieving remains. Kaʻiana's mourning was not an aberration. It was appropriate. But at the same time, for an aliʻi of such an exalted rank to cast himself into deep mourning over a presumably makaʻāinana woman suggests that their experiences far from home changed something: it may have given Kaʻiana a stronger sense of the Kānaka as one people. Men and women of Kaʻiana's high rank were rare, and historical accounts suggest they were distant from the maka'āinana. Many ali'i of Ka'iana's rank were surrounded with strict kapu (sacralizing restriction). A person like Ka Wahine probably would never have dared draw near them, let alone speak to them. And yet Ka'iana stayed by Ka Wahine's side, tending to her in her last days, and mourning her dramatically. This suggests the ways that being away from Hawai'i, and the lessons in world geography that he was drawing from his experiences, may have changed the way Ka'iana thought about what being Kanaka meant. His mourning suggests that his experiences led Ka'iana to de-emphasize the differences of rank that separated him from Ka Wahine, and to emphasize his connection to her as a Kanaka. Those bonds among Kānaka stood in distinction to the affinities that had grown between Ka'iana and his English aikāne and friends.

MAGUINDANAO: EUROPEAN BONDS AND THE LIMITS PLACED ON EUROPEANS

The bonds that brought Europeans together, and their conflicts with non-Europeans, were a lesson in global geography that was underscored when Meares finally found a place to repair the mast of the *Iphegenia*. On February 7, 1788,[74] the two ships passed through the strait that separates the islands of Basilan and Mindanao and approached the Spanish fort at Zamboanga with caution. Ka'iana's aikāne thought that the fortifications looked "indifferent" but raised the ship's colors to signal its identity. Spanish priests came out in a boat rowed by "Malayans," flying a white flag, demonstrating no intent to attack. After determining the peaceful intentions of the Englishmen, the Spanish allowed them to come ashore but warned them against cutting wood in the forest. The Spanish fort (encircled by high stone walls and a moat) and the small colonial town (defended by a palisade and a moat) were entirely surrounded by the Sultanate of Maguindanao. The Sultanate of Sulu had nearly taken the fort thirty years before. The Spanish warned that without their protection, the men of the *Iphegenia* would have no defense from attacks by the people of Maguindanao. In fact, when a party was sent to cut timber, one of the Chinese carpenters was captured, never to be recovered.[75] The limits on European power were palpable.

In contrast, relations between the Spanish and English demonstrated the ties between Europeans. The next day, another small boat approached the *Iphegenia*. A Spanish officer delivered a "polite invitation" from the Spanish governor to an "entertainment" (a festive dinner to be held in honor of the visiting vessels). The officer reiterated the warnings about "the perfidious character of the natives of the island" and invited them to cut wood closer to the village, within the effective boundaries of Spanish influence. The good will between the ship and the fort was signaled with a ritual exchange of nine-gun salutes by the cannon. The next day, the English went ashore where they "were regaled, after the Spanish fashion, with sweetmeats and cordials" followed by "a very handsome repast." The British returned the favor the following day, hosting the Spanish for as lavish a dinner as they could manage. It was all very cozy—Meares found there was nothing like food and wine "to annihilate the force of religious distinctions" between British Protestants and Spanish Catholics.[76]

The easy annihilation of these distinctions actually reinforced another powerful distinction that Europeans were constructing in this period: the line between European nations who wished to colonize and the Native people of the places they were trying to colonize. That line was complicated in Zamboanga (and elsewhere in the Spanish Empire) because Spaniards built their empire on a graduated hierarchy of races and of people of mixed ancestry—not just a binary opposition between Europeans and non-Europeans. Taxonomies of *castas* (castes) placed people with more European ancestry over those with less and also created hierarchies among different non-Europeans—indigenous Filipinos (termed *indios,* meaning "Indians"), Chinese, Africans, etc. This created a nuanced system of domination, inclusion, and exclusion: Meares noted that among the officers in the colonial militia who attended the dinner party on shore there were "natives of Manilla, and others of Maguindanao, whose complexions were so dark as to approach very nearly the blackness of the African."[77] They were racialized, but they were also what Meares calls "gentlemen"—a signal that they were culturally incorporated into colonial society. What caught Meares's attention was the presence of these dark-skinned men and the positions of relative authority that they held, but it is crucial to recognize that race structured their roles. Racial ranking animated the system.

Precisely how much of this Kaʻiana could have absorbed in his stay at Zamboanga is an open question. He would have been present for the dinner on board, may have been present for the dinner on shore, and definitely came ashore to see the Spanish fort. He knew about Haole national

Detail of an inset on a Spanish map representing racial and cultural "castas" of the colonial Philippines. Left to right: a "tall rustic," a Spaniard, a "black Negro" born in the Philippines, "Indians cockfighting," and "Ateas, or wild men of the countryside." Pedro Murillo Velarde, "Carta hydrográfica y chorográfica de las Yslas Filipinas," 1734. Courtesy of the Library of Congress, Geography and Map Division.

differences: he could distinguish some European national differences by observation. He knew about European religious differences: he attended Catholic mass when his Protestant English comrades would not. He knew something about the divisions that Europeans drew among themselves. He had mastered English anti-Chinese racist discourse in Macao. Given all of these lessons learned, Ka'iana was in a good position to witness these friendly Spanish–English relations and see that Europeans made the differences between themselves and non-Europeans a crucial criterion in elaborating hierarchies of power, whether they were binary hierarchies (as in much of the British Empire) or graduated ones (as in Spanish-occupied Zamboanga).[78] The category "non-European" is, of course, a crude way to group the great bulk of the world's population. But the European hierarchies that Ka'iana witnessed treated that category as a reality because it was a foundation of their own power.

Seen in that light, and because Ka'iana was a powerful man intent on power, the ability of the Sultanate of Maguindanao to place limits on Europeans (as the Chinese did in Macao) was probably obvious to him. Kaʻianaʻs life shows that he was a man acutely aware of power, and as Meares put it, the Spanish claim to the "entire dominion" of the Philippines was a "mere assertion." The real seat of authority would have been as obvious to Kaʻiana as it was to Meares, who wrote, "The sovereign of Magindanao is a powerful prince." With many chiefs recognizing his superiority over them, the sultan had the power to constrain the Spanish, despite all their pretentions to power, behind moats, palisades, and walls.[79] Like the sultan, Kaʻiana came from a family of what Europeans would call "powerful princes" who had many chiefs below them that recognized their superior rank and authority. The limits that non-Europeans could place on Western incursions would be an added lesson in understanding the global geography of nations that Kaʻiana was studying.

OTHER ISLANDERS: HATOHOBEI AND PALAU

This informative visit to Mindanao had repaired the mast but delayed the voyage, and the *Iphegenia* had proved to be a cumbersome and slow vessel. Meares declared that the ships would split up. He would hurry on the *Felice* to reach Nootka in time to buy the best furs before other traders arrived, while William Douglas would pilot the *Iphegenia*, with Kaʻiana aboard. Douglas was instructed first to take Kaʻiana and the man and boy from Maui back to Hawaiʻi, then sail to Nootka to join Meares. Kaʻiana was eager to return home. Douglas faced even more delays, however: for all the comraderie of the British and Spanish, the Spanish governor demanded payment from Douglas. After three days negotiating that price, and faced with the slowness of the *Iphegenia*, Douglas decided that he would ignore Mearesʻs instructions to sail first to Hawaiʻi to drop off the Kānaka. Instead, he would sail directly to the Aleutians to trade and then to Nootka to meet Meares. Because of Douglasʻs choice, Kaʻianaʻs eagerly awaited return home would be delayed more than nine months.

This unilateral change of plans, along with the conduct of Douglas and his men on the voyage across the Pacific and on the coast of North America, reinforced an essential lesson about the shape of the world: Kānaka and other non-Europeans could not place their trust in agreements or trade with Europeans. The first place that the *Iphegenia* weighed anchor was the island of Hatohobei (which Douglas called Johnstoneʻs

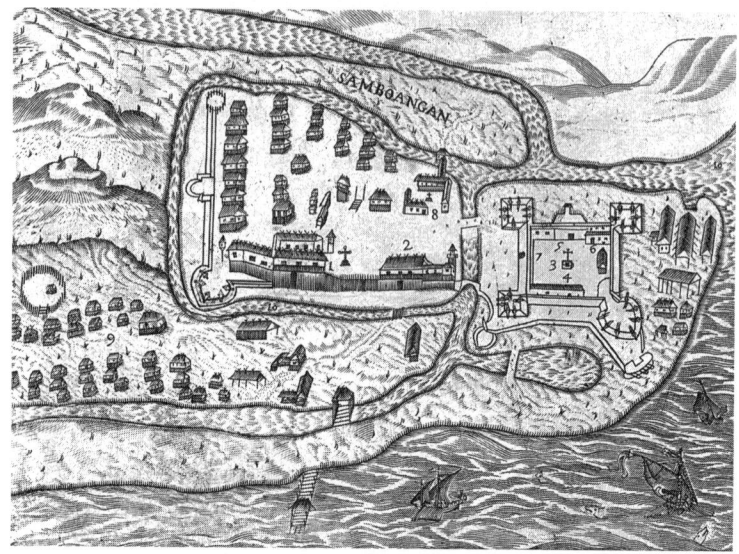

Detail of an inset on a Spanish map showing the fort and colonial settlement at Zamboanga, with palisades, moats, and stone walls. At left is a "village of Lutaos," the Native people of that part of Mindanao. Pedro Murillo Velarde, "Carta hydrográfica y chorográfica de las Yslas Filipinas," 1734. Courtesy of the Library of Congress, Geography and Map Division.

Island and is now often referred to as Tobi, the southernmost state of Palau). Douglas immediately recalled Hawaiʻi. His ship was met by people paddling out on canoes "exactly the same as those of the Sandwich Islands [Hawaiian Islands]; and the people not only displayed the same activity in the water as the Sandwich Islanders, but made use of several expressions which Tianna readily understood." Given that Hawaiian and Tobian are only very distantly related languages, this may have been a stretch or a passing resemblance. A more substantial resemblance that might have struck Kaʻiana was the nature of trade: the people of Hatohobei traded coconuts and taro for iron goods from the ship. Hawaiʻi, then, was only one place where people traded with Europeans in this way. That lesson would be reinforced as the *Iphegenia* continued northward. But before they could trade again, much of the crew fell sick, as did "Tawnee," the man from Maui. Although Kaʻiana cared for him as he had Ka Wahine, the pathogens that were claiming the lives of so many Kānaka at home and at sea took his life as well. Two of the four Kānaka that had left Macao together were now dead. Like Ka Wahine, "Tawnee" was sent "to a watery grave." With a sickly crew, Douglas piloted the *Iphegenia* toward the main islands of Palau, to the north, in search of "roots"— probably food crops to feed his men, but perhaps medicines to heal them. When the ship arrived at Palau, again the ship was met by canoes. Following what was apparently the protocol of trade, Douglas's men lowered a bucket tied to a rope down from the deck to the canoes containing "a few nails." In trade, the Palauans sent up what he deemed to be only "an inconsiderable number of cocoa-nuts." Douglas, who expected "natives" to be docile and dominable trading partners, decided that their frugal trading meant they were determined to "play the rogue." He resorted to intimidation, firing a musket over their heads, and sailed on.[80]

What could these events have meant to Ka'iana? He had now witnessed the nature of interaction between Europeans and non-Europeans in a number of places—Hawai'i, Macao, Maguindanao, Hatohobei, and Palau. Among these places, the ones where trade most resembled relations in Hawai'i were Hatohobei and Palau. As Douglas's intimidation of the Palauans demonstrated, this was a system in which Hawaiians would find themselves subject to Europeans' demands and Europeans' violence. The people with whom Europeans traded had some real power: they inhabited the lands Europeans traveled to, and they had the things (foods, furs, medicines) Europeans wanted. Given that fact, Europeans used intimidation to weigh the scales of power in their own favor. Intimidation was a way of expressing that what Europeans wanted in a situation was simple supremacy, not the complexities that Chinese and Maguindanao power created in Macao and Zamboanga.

NOOTKA, NATIVE AUTHORITY, AND THE COMPLEXITIES OF POWER

In Nootka, Ka'iana would witness another place where power was more complex than the Europeans wished. Nootka, like Macao and Zamboanga, was a rich site for studying world geography. It was also a site that Ka Wahine had witnessed because she had stayed there first before the Barkleys sailed to Macao. In thinking about the lessons of Nootka, then, we can think about a place where both Ka Wahine and Ka'iana learned more about the world.

The British understood the Nuu-chah-nulth people no better than

Cook understood his encounter with Kāʻeo and his men off Waimea, Kauaʻi. Indeed, the first encounter between the Englishman and the Nuu-chah-nulth had been quite similar. Cook arrived at Yuquot four years after 1774, the first time Europeans saw it. Like the Kānaka at Waimea, the Nuu-chah-nulth did not wait for the strangers to come to shore but instead paddled out to explore them. As Cook sailed into the bay, at least thirty-two canoes came out to the ship, surrounding it. Like the aliʻi and the kahuna who had come out to Cook's ship at Waimea, a chief named Moachat drew up to the *Discovery* in a canoe. Like them, he formally addressed the strangers. And as at Waimea, the British failed to grasp the importance of his discourse. As in Hawaiʻi, there was violence as the British tried to impose their own ideas of property in places where they had no dominion: Nuu-chah-nulth took iron articles, an anchor, and a watch, and Cook applied "force" to get them back.[81] Neither Ka Wahine nor Kaʻiana witnessed those events, which happened a decade before they arrived there. But the relations that characterized this early encounter—Nuu-chah-nulth wariness and care to pursue their own interests, British sense of their own superiority and entitlement—still obtained in the years when Ka Wahine and Kaʻiana arrived at the island the British called Nootka.

The name of the place was actually Yuquot, and Ka Wahine and Kaʻiana witnessed it at the height of the wealth and power of the people who lived there. On this island off of Vancouver Island lay the summer village of the Mowachaht Nuu-chah-nulth (now part of the Mowachaht/Muchalaht First Nation). About 1,500 people inhabited its twenty longhouses in the summer. The place was the summer site of a confederacy of which Maquinna (also known as Taxawasip) was the ranking chief. The large size of the their settlement, their substantial buildings (longhouses measuring thirty yards long on a side), and the plentiful food they had to trade testify to the power and affluence of the Mowachaht.[82]

At Yuquot and other sites along Vancouver Island, Ka Wahine and Kaʻiana saw a place where Native people had shaped the terms of trade to their own benefit, demanding far more in trade and tribute-like gifts than was the norm in Hawaiʻi. Europeans' journals and reports make clear that Maquinna and his people were tough in their negotiations with the traders. Furthermore, Maquinna had made himself the intermediary for some trade with other Natives of the region. Consequently, the fur trade had increased the wealth of both the chief and the Mowa-

Maquinna ("Maquilla"), the chief of Yuquot, and Comekala ("Callicum"), the Yuquot man who traveled with Ka'iana and Meares. Lithograph by J. Walter in Meares, *Voyages Made in the Years 1788 and 1789*. Courtesy of the James Ford Bell Library, University of Minnesota.

chaht people at Yuquot. Furs were exchanged not only for beads or small quantities of nails, but also for goods that cost the Europeans more money and effort. Meares kept his on-site ironworkers and smiths busy making finished goods specifically to accommodate the changing demands of the Nuu-chah-nulth. On top of this, in hopes of keeping access to furs, the British made gifts for which they received no payment. Gifts were substantial: in August 1788, the Nuu-chah-nulth chief Waccanish and his people at Clioquot received firearms, ammunition, a dozen swords with brass hilts, a copper kettle, coats heavily adorned with buttons, items made specifically "to suit the fancy of the women," and a copper still-head whose valuable metal the Nuu-chah-nulth could reuse. Meares mocked the Nuu-chah-nulth for what he called their "infantine" tastes, but his actions make clear that they were hardly children to be taken advantage of. Meares knew he needed to please them. In other words, Maquinna and some others on the Northwest Coast had arranged favorable terms of trade with the Europeans. Among the Nuu-chah-nulth, Maquinna is still remembered for skillfully negotiating the terrain of rival trading partners (English and Spanish) and thinking through what Native leadership would mean in the changed circumstances that European incursion had brought.[83] On the heels of what Ka'iana had seen in Macao and Zamboanga, this served as yet another demonstration that non-Europeans could place limits on Westerners' power and make use of Westerners' presence.

At Nuu-chah-nulth, Ka'iana was present for events demonstrating that Meares was unable even to control his own crew. Intimations of mutiny had swirled through the *Felice* since shortly after it left Zamboanga, and active rebellion broke out in Nootka, when Meares came gun-to-gun with the mutineers on the *Felice*. At issue was the ship's itinerary. The crew was "absolutely haunted" by "the idea of being eaten" by the Nuu-chah-nulth and wanted to return to Hawai'i, which they described chiefly in terms of bodily comforts and sexual opportunities: they wanted to return to the "voluptuous abodes" and the "gratifying pleasures" of the islands. The confrontation took place shortly before the *Iphegenia* with Ka'iana aboard finally rejoined Meares and the *Felice* in Nootka. But because he was so keen to return home, Ka'iana would have probably learned that Meares only regained his ship by promising a voyage directly to Hawai'i after Nootka. Even more dramatically, Meares expelled eight mutiny leaders from the shelter of the ship and consented to Nuu-chah-nulth stripping them of their Western clothing and enslaving them.[84] Meares, like Cook and like many other Europeans with which Kānaka dealt, tended to overestimate his power over others and the awe in which others held him, but it is unlikely that Ka'iana made the same mistake. The limitations on the Haole were simply too evident.

KAʻIANA AND THE NUU-CHAH-NULTH

Just as evident was the Haole disdain for the Nuu-chah-nulth, and either Kaʻiana echoed British contempt for them as emphatically as he had taken up British disdain for the Chinese, or Meares simply read his own beliefs onto the man he admired and desired. Meares reported that Kaʻiana "held the customs of Nootka in detestation; and could not bear the idea of their cannibal appetites; without expressing the most violent sensations of disgust and abhorrence." The Nuu-chah-nulth were not cannibal—in fact, Coll Thrush has shown that they themselves feared that the British ate human flesh. The idea that Nuu-chah-nulth consumed human bodies came to Kaʻiana from the British, and the racist ideas Meares attributes to Kaʻiana mirror his own closely. To Meares, whereas Kānaka were noble savages, "North Western Americans" (that is, Indians of the Northwest Coast) were merely savages. "There was no comparison," he wrote, between the Kānaka, who were "not only clean to the extreme in their food, but also in their persons and houses," and the Native people of the Northwest, whom he calls "most filthy brutes," devouring food of "a most disgusting nature," and "cannibals." Meares declared Kānaka innocent of cannibalism, saying they made burnt sacrifices of humans to the gods, but did not eat them. In Meares's colonial vision, the people of the Northwest were beyond hope, but the qualities of Kānaka held out hope for them of what Meares considered an honor: they "may one day be ranked among the civilized subjects of the British empire."[85] There is little likelihood that becoming the subject of a British emperor was a hope that Kaʻiana shared. After all, he was a man raised to rule who sought to increase his own power. Still, we can justifiably imagine that Kaʻiana was fully conscious of his aikāne's racist and imperialist beliefs and was able to think about their ramifications for himself and Kānaka as a whole. This was an important geography lesson, because Meares's beliefs were ones that would become powerful among Haole: that Kānaka were uncommonly noble savages who would benefit from Western imperialism.

Kaʻiana had already had the opportunity to observe two organizing principles of Western racist and colonialist thought and to draw his own conclusions in global geography from them. At the one hand, there was the principle of a graduated hierarchy of peoples and races. This is what Meares evoked with his hope for British imperial authority at the top, civilized imperial subjects like Kānaka in an intermediate position, and allegedly irredeemable savages at the bottom. This position valorized Kānaka over other people, and thus Kaʻiana had interest in echoing it—as he did when showing disdain for Chinese and Nuu-chah-nulth. But Kaʻiana had also observed that in the minds of Europeans, a binary underlay this hierarchy: European versus non-European. He had been shown this at Macao, at Zamboanga, and at Yuquot, and his fervent expression of British racist ideas in China and North America make it hard to believe he did not note this binary. The fact that this binary underlay the hierarchy led to one central fact about European ideas about global geography that Kaʻiana and other Kānaka would need to think about as they assembled their own ideas about global geography: Europeans believed themselves superior to non-Europeans and believed they had the right to treat them as they wanted. In Macao this meant selling opium despite Chinese law outlawing it; in Mindanao it meant seizing land for a fort despite the resistance of the Sultanate of Maguindanao; and in Yuquot this meant simply seizing the food and other supplies they wanted when Nuu-chah-nulth would not supply them, as Meares's employee William Douglas had done. Given that binary pattern with European in power over non-European, the intermediate position as noble savages or civilized subjects was a fragile promise indeed.

In the binary, which side would Kānaka be on? The question was crucial, but the wording is tricky. If being on a side meant identification and alliance, it does not appear that Kaʻiana thought he was on the side of the Nuu-chah-nulth any more than he was on the side of the Chinese. But if being on a side meant finding oneself placed by others on one half of a European/non-European divide, Kaʻiana would know that in the minds of the English, he did indeed fall on the same side as them. The word "Indian" is crucial here. To Meares and other Englishmen of the end of the eighteenth century, Kaʻiana and the Kānaka Hawaiʻi were Indians, as were the people of Mindanao in the Philippines, as were Nuu-chah-nulth and all the rest of the indigenous inhabitants of North America. Meares calls Kaʻiana an "amiable Indian," writes about dangerous Indians in Maguindanao, and discusses his trade with the Indians of Nootka. Kaʻiana's English was probably rough, but this was the word that he heard the English use for his people and others. The word "Indian" served as an indicator that could reveal to Kaʻiana where Haole believed Hawaiʻi and the Kānaka fit into the global geography he studied as he traveled in and around the Pacific. This does not mean that Kaʻiana came to actually believe that Kānaka were "Indians" or savages, noble or otherwise. But it was a term that was a clue to where Kānaka would find themselves positioned by Haole. This was crucial for thinking through how to interact with the Haole. As Kaʻiana had seen, the Haole treated the people they called "Indians" with disregard and, like Douglas, took from them the things they wanted. If they saw Kānaka as Indians, then Kānaka would have to be very wary in their dealing with the Haole.

This crucial final discovery added to the others Kaʻiana made as he explored the outside world. This was a world of many nations and empires, and other entities that were sovereign but outside of the nation and empire system—places like the Sultanate of Maguindanao and the lands of the Mowachaht Nuu-chah-nulth. He had discovered and explored a world of religions, and in the churches of Macao, he studied new rituals like the kahuna naʻauao (learned expert) that he was. Kaʻiana discovered that, while the Europeans believed themselves to be both superior to others and very powerful, walls and gates and palisades in Macao, Zamboanga, and Yuquot bore concrete testimony that non-Europeans had the ability to constrain their ambitions and limit their powers. Terms of trade varied, and the Nuu-chah-nulth received more for their furs than the Kānaka received for their hogs and vegetables because Nuu-chah-nulth like Maquinna had shaped the market to serve their interests. In a final lesson, Kaʻiana learned that Europeans classed him and other Kānaka with Maquinna and the Nuu-chah-nulth as Indians. The Haole believed themselves entitled to seize land and goods from them and treat Indians unfairly in trade. From this, Kaʻiana could draw a final conclusion: Europeans needed to be dealt with firmly and with wariness, as the useful things they had were sometimes only provided at a very high price.

BACK TO HAWAIʻI: PUTTING LESSONS IN WORLD GEOGRAPHY INTO ACTION

Kaʻiana spent over a year away from the islands, departing September 2, 1787, and returning in October 1788. Only about five months was spent on land at Macao or anchored off of Zamboanga (on the island of Mindanao in the Philippines) and Yuquot (in the Nuu-chah-nulth country in pressent-day British Columbia).[86] What Kaʻiana learned about world geography served him well after he returned home. It enhanced his power in Hawai'i's rapidly shifting political enviroment. It prepared him to deal with Europeans in a sophisticated manner, as we can see from his interactions with another English captain, George Vancouver. By the time Vancouver sailed into the waters of Hawai'i in 1793, much had changed for Hawai'i and for Ka'iana. After his tearful farewell to Meares, Ka'iana had risen to new heights of power as a close associate of Kamehameha. The two ali'i saw that they could be helpful to one another.[87] Kamehameha's ambitions to power gave Ka'iana incentive to become allied to him. As for Ka'iana, his study of the world via travel enhanced his own already extraordinary learning in Hawaiian matters and made him valuable to Kamehameha. As Kamakau narrates it (using the spelling "Tahiti" to refer to "Kahiki," referring to foreign lands), Kamehameha saw that Ka'iana was a potentially important advisor who had gained powerful knowledge abroad. During Ka'iana's "holo ana i Tahiti i na makahiki ekolu me ke ano alii, me kana aikane a hoa aloha, oia o Kapena Kane" (trip to Tahiti for three years in the manner of a chief, with his aikāne and friend, Captain Kane [John Meares]), he had witnessed "[h]e mau makahiki kaua nui ia o na aupuni o Europa a me ko Amerika" (years of great war for the nations of Europe and America). These experiences combined with his other skills to make him "makaukau i ke kaua, a me ka makaukau i na loina o ke ki pu ana" (prepared for war, and prepared to oversee the shooting). Kamehameha came to a decision: "koho aku la o Kamehameha ia Kaiana i Pukaua a i Alihikaua Nui nona, a e hele e kaua me Keoua Kuahuula" (Kamehameha chose Ka'iana as his general and Great War Strategist, and went and made war with Keōua Kū'ahu'ula).[88]

Kamakau rightly emphasizes the value of Ka'iana's knowledge of the world, although that voyage was less than three years and the knowledge was less narrowly military than Kamakau suggests: Ka'iana had seen no military engagements, although he had seen the arms and fortifications at Macao, Zamboanga, and Yuquot. Similarly, Kamakau (and a number of British and American writers of the nineteenth century) might be mistaken when he writes that Ka'iana gave Kamehameha muskets and cannon he had purchased for him overseas. As David G. Miller has pointed out, no account from Meares or Portlock (or, for that matter, Douglas) suggests that their ali'i passenger was acquiring arms or transporting them to Hawai'i.[89] But Kamakau's central point remains valid:

Kamehameha saw Kaʻiana's knowledge gained abroad and believed it would be useful to him. Thus the knowledge that Kaʻiana had sought through his explorations was already proving useful to him by tying him to a rising figure in the life of the islands. Kamehameha made Kaʻiana one of his punahele, his "favorites" and close advisors. Kaʻiana's knowledge had enhanced his power in the court of Kamehameha to such a degree that the mōʻī's kuhina (advisory council) were alarmed and jealous and conspired against him.[90]

Thus we can see the political power of knowledge about the world. Meares writes that Kaʻiana returned to Hawaiʻi "with a mind enlarged by the new scenes and pictures of life which he had beheld, and in the possession of various articles of application, or comparative magnificence, which would render him the richest inhabitant of his native islands."[91] The statement is misleading when we consider it in the light of Kamakau's explanation of what drew Kamehameha to Kaʻiana. Meares exaggerates the importance of the items Kaʻiana brought back and underestimates the power and specificity of the knowledge Kaʻiana had gained about global geography—specifically, about relations between Europeans and non-Europeans in Asia, the Pacific, and North America. The words "scenes and pictures" only hint at what he had gained: he had learned by observation and analysis lessons in global geography. Exploration of global geography mattered to Kaʻiana, demonstrating the influence of the Kanaka explorations of global geography on the shape that Hawaiian history took. Kaʻiana's travels familiarized him with the strategic and economic alignments of power in the Pacific, with the rising power of a divided West, with the potentials for different ways of relating to and benefiting from the incursion of Westerners.

Those lessons were quite apparent in Kaʻiana's actions in the years after his return from abroad, and particularly the independent course he struck in his relations with Haole. Kaʻiana's knowledge about Haole motivations and the shallowness of their commitments to Native people were eminently apparent in the way he acted differently than other aliʻi toward the Englishman George Vancouver and Kamehameha's advisors, the Englishman John Young and the Welshman Isaac Davis. Young and Davis were sailors who had left their ships, been taken captive, and become close counselors to Kamehameha.[92] As Vancouver visited Hawaiʻi Island in 1793, Kamehameha, Keʻeaumoku, and other aliʻi showered him with much service and extremely valuable goods—hogs, yams, kalo (taro), and other food. Kaʻiana himself was forthcoming with a generous gift on the occasion of his first visit with Vancouver: a mahiole, the same kind of feathered helmet that Kaʻiana had sometimes donned in Macao. (That mahiole is likely one of those that are now in the Vancouver Collection of the British Museum in London.) He also wanted to give Vancouver fifteen hogs, among the finest that Vancouver had seen in the islands, but the captain regretfully declined them, as the other chiefs had been so generous that there was no room left on the ship. The exchange over the hogs raised the ire of Keʻeaumoku and Kamehameha, who were present—it raised the issue of who would gain the loyalties of Vancouver and perhaps benefit from that relationship. The other aliʻi were willing to compete for that relationship.[93]

Kaʻiana was not: he ceased at that point making gifts and refused to perform deference to Vancouver. His giving and withholding of gifts was perfectly calibrated to signify his own position. As chiefly regalia, a mahiole was a fitting gift from a chief to a chief. The British valued hogs only for the salt pork they turned them into, but hogs—and especially very fine hogs such as the ones Kaʻiana gave—carried ritual significance for Kānaka. Hogs were not for the makaʻāinana and not for daily consumption. They were for chiefly feasting (particularly male chiefly feasting) and sacrifice. Perfect hogs that were free of blemish (especially puaʻa hiwa, black hogs) were valued offerings at luakini.[94] Thus in this first encounter, Kaʻiana made gifts that spoke of his own exalted position as an aliʻi and kahuna and of the wealth of the lands placed under his command by Kamehameha, from which he could draw pigs of that high quality. He was speaking a language of gifts that was only partly intelligible to Vancouver, who did not fully understand the meanings of mahiole and puaʻa hiwa. His gifting carried kaona, to employ the terminology of Kanaka oral and written literature as elucidated by Brandy Nālani McDougall and Georganne Nordstrom. Kaona, sometimes termed "double meaning," is a literary device with a powerful social function. By including a reference that conveys a particular message to part of the audience but is invisible to others, it constitutes two groups: insiders who understand and outsiders who don't.[95] This is a clue to a crucial fact. Kaʻiana's gift giving was a performance aimed at least as much at the other aliʻi as it was at the Haole, who could not fully grasp the meanings of these gifts.

After he had made these generous and impressive first gifts, Kaʻiana stopped. Again, his actions demonstrated his expression of his position in regards to the Haole and the other chiefs. The other aliʻi continued their gifts to Vancouver, but Kaʻiana basically cut off the Englishmen until Vancouver's departure. Having made a point about his status and the wealth of his lands with his original gifts, by not gifting further he demonstrated that he was not courting favor with the Haole and did not need to prove his status to his fellow aliʻi. He continued to visit Vancouver but came empty handed when he did. He only resumed giving gifts upon Vancouver's departure on Sunday, March 9, 1793, when he brought Vancouver what the Englishman derided as "about half a dozen small ill fed hogs." Vancouver had no room or need for them due to the generosity of the other aliʻi who were courting this relationship but felt obliged to give Kaʻiana gifts in return—a "scarlet cloak[,] axes and a variety of other useful articles." Again, Kaʻiana asserted his own position of dignity and authority, declaring the gifts insufficient. When Vancouver refused to give more, Kaʻiana did keep them, but he had made his point. These gifts did not recognize him properly and certainly not in the same manner as Vancouver honored Kamehameha. Vancouver rightly saw that this was part of Kaʻiana's growing rivalry with Kamehameha, which would end in Kaʻiana joining the other side in the struggle for supremacy over the islands. And yet there was more to it than that: Vancouver was struck that Kaʻiana's "conversation was in so haughty a stile and so unlike the general conduct of all the other chiefs."[96]

Kaʻiana acted differently than "all of the other chiefs," and used the language of gifts differently, because he understood the nature of the Haole's gifts and alliances better than the other chiefs did. He knew that to be Hawaiian was to be Indian in the Haole's eyes. The Haole liked to maintain good relations with Nuu-chah-nulth with gifts and trade, but when Douglas found it more convenient to simply seize things, he did. Why, then, should Kaʻiana or any Kanaka invest too highly in good relations with the Haole? Kaʻiana knew that Maquinna and other chiefs on Vancouver Island had succeeded in obtaining good terms of trade and generous gifts of tribute from the Haole, and if Kānaka were to get the same, courting favor through repeated lavish gift-giving was probably not the way to do it. Furthermore, the lessons in the geography of power that Kaʻiana had learned in his travels certainly would have undermined any sense of awe at the powers of the Haole. Kaʻiana had seen the British and other Europeans penned behind walls in Macao and had witnessed them hide behind walls in Zamboanga and Yuquot. He had seen Meares threatened by a mutiny, and his men stripped and enslaved by Nuuchah-nulth. He had seen the way that other people obtained the good things they wanted from the Haole and simultaneously contained them. In sum, Kaʻiana could see (ʻike) the Haole better than they could see him, and his knowledge helped to give him power. Vancouver could see only an Indian that was beset by "envious pride," and pointed to his "turbulent and aspiring disposition." And doubtlessly, Kaʻiana was prideful and aspiring, and was on increasingly bad terms with Kamehameha. But he was also naʻauao (learned), and part of his learning was due to the travels he had made. In fact, Kaʻiana may have sought to protect the value of his knowledge of the Haole and the outside world. By this time, Kamehameha had two British advisors, Young and Davis. Because they brought him valuable inside knowledge of the Haole, Kamehameha had rewarded them with lands, as he had Kaʻiana. Kaʻiana wanted to get rid of Young and Davis, perhaps considering that they made the knowledge he had gained through exploration less rare and thus less precious.[97]

Kamehameha extended his authority westward across the island chain through a series of battles and alliances. In the last battle for supremacy in the islands, at Nuʻuanu on Oʻahu in May 1795, Kaʻiana switched sides. He abandoned Kamehameha, aligning himself with Kalanikūpule, who was fighting Kamehameha in an effort to retain control of Oʻahu. In the words of Ka Mōʻī Davida Kalākaua, the seventh monarch of the unified Hawaiʻi that Kamehameha would forge, "After giving to the conqueror his best energies for years, and faithfully assisting in cementing the foundations of his greatness, he turned against him on the very eve of final triumph." Kalākaua notes that some people attributed this change of sides to sheer treachery, while others believed that Kaʻiana thought that "by blood Kamehameha was not entitled to the sovereignty of the [island] group" and sought "to open to himself a way to supreme power."[98] In the battle, Kaʻiana was killed and his body laid in offering to Kamehameha's war akua, Kūkāʻilimoku.

Kaʻiana, Ka Wahine, and other Kānaka had not waited for the Haole to come to shore. They paddled out to explore them. Like the man and boy from Maui, and the many other Kānaka to travel abroad on Haole ships, Kaʻiana and Ka Wahine went one step further. They traveled far from home to explore the outside world, as was part of their heritage. Their explorations cannot be considered typical of other Kānaka's experiences, but they were emblematic of the explorations of a people who eagerly sought to understand the nature of the world beyond their shores, and how best to relate to that world. It was a world of power. To the exploration of it, Kānaka brought their confidence and their already considerable knowledge, and like Kaʻiana, they brought their interest in making the new knowledge they gained serve them. In the coming decades, Kānaka would continue to bring their knowledge and their own interests to the active exploration of the world beyond their shores. As part of their process of exploration, Kānaka would even make a place for Haole and their religion in Hawaiʻi.

3
A NEW RELIGION FROM KAHIKI

CHRISTIANITY, TEXTUALITY, AND EXPLORATION, 1820–1832

To earn geography books, the Kānaka paddled canoes laden with timber for twenty miles and built a missionary a house. To explore the world, the Kānaka would labor and study and voyage, because they wanted to know new things about the world for their own purposes and in their own way.

It was July 1832, twelve years after the first American missionaries had come to the islands. They told Kānaka they would die and be confined to a pit of fire if they followed their own religion, but their souls would live forever in peace and plenitude if they abandoned those ways and embraced the religion and ways of the Haole. Ephraim Spaulding, a native of Vermont and a graduate of Middlebury College and Andover Seminary, wanted a house at the town of Lāhainā on Maui so that he could preach this message there. On the afternoon of July 20, Spaulding gazed on his new home, built for him by Kānaka Maoli. The timber to frame it had come by canoe from twenty miles away, paddled by twenty Kanaka men. For his labors, each had earned a geography book. The labor to erect and thatch the house was paid in the same currency: a newly published geography textbook in the Hawaiian language that bore the title *He Hoikehonua, He Mea Ia e Hoakaka'i i ke Ano o ka Honua Nei, a me Na Mea Maluna Iho*—meaning "A Geography: A Description of the Nature of the World and of the Things upon It."[1]

Spaulding's arrival provides a useful lesson about the relationship of Kanaka initiative, Christianity, and textuality to the project of Hawaiian exploration of the globe. Spaulding could dwell in Lāhainā only because Kānaka had literally made a place for him, and the men who made that place did so to earn a book that described the world that many Hawaiians of their generation were exploring as they labored on ships around the Pacific and the Atlantic. When we remember that Spaulding represented in this context the coming of Christianity to Hawai'i, the building of his home centers Hawaiian initiative in that missionary process, but it also centers a desire for learning that went beyond the wish to learn about Christianity. By their own initiative, Kānaka made a place for Christianity in Hawai'i, and in large part their motivation was to learn and explore and gain knowledge about matters that we today would not consider to be specifically religious. These men did not labor to earn Christian scriptures, but a Hawaiian-language book that described the world, its lands, seas, nations, and peoples. Kānaka made a home for Christianity in their society because they wanted to learn things from missionaries. Some of those lessons were religious, and there is no reason to doubt the sincerity and depth of the dedication of many Kānaka to Christianity. But some of those lessons were not strictly religious, and they gave Kānaka the access they wanted to knowledge about the world.

But before examining the schools and the textbooks used in them (the subject of the next chapter), we need to understand the process that brought the missionaries to Hawai'i and motivated Kānaka to learn to read and write and then later to form and attend schools. Two stories will correct standard narratives of missionary endeavor in Hawai'i by centering Native Hawaiians' active seeking of knowledge from the outside for Native Hawaiian purposes—an act of Kanaka Maoli exploration. First, the life of a man named 'Ōpūkaha'ia who traveled abroad to explore demonstrates how Kānaka spurred missionary action. Second, the way that Kānaka Maoli were attracted to Christianity by their contacts with Tahitians demonstrates that exploring the outside world through Christianity meant reconnecting to other Pacific Islanders, not simply acquiescing to Western colonialism.

Narrating the early years of the Hawaiian relation with Christianity in this way reverses the geography and reverses the directionality of standard accounts of the advent of Christianity to Hawai'i. Assumed by historians as narrated by American Christians, Americans brought Christianity from the land of light to the land of darkness. But that is folly. It was a trained Kanaka spiritual adept whose explorations brought him to study Christianity and to believe that it held power for his people. It was Kanaka travelers—some with extensive religious training—who spurred and assisted the movement of the first Christian missionaries to Hawai'i. It was the creation of ties between Kānaka and Tahitians that brought Christianity's first successes in Hawai'i. And bonds with other Pacific Islanders created a push to evangelize in the Pacific. In other words, the history of the introduction of Christianity to Hawai'i and its early history there must emphasize Native initiative and connections to other Pacific Islanders. This does not mean we celebrate missionaries as artifacts of Hawaiian agency. After all, Americans became set on missioning Hawai'i. Nor does it mean celebrating Christianity's success in Hawai'i as an indigenous success story: Christian missionaries, foreign Christians, and missionaries' children especially were crucial to the loss of effective Hawaiian national sovereignty and the overthrow of the Hawaiian Kingdom. But it does mean recognizing that Christianity was a major object and vehicle for the Hawaiian exploration of the globe—of foreign people, places, and ideas. Centering Kanaka Maoli initiative in this way transforms the way we understand the early history of Christianity in Hawai'i, because we understand it from Hawaiian perspectives rather than Western ones.

Standard accounts of Christianity in Hawai'i start with the story of a handful of Kānaka Maoli, but quickly move on to place Haole initiative at the center of attention. In the 1810s, the vagaries of seagoing labor landed several "Sandwich Island youths" in New England. These accounts recall how one youth, whom Americans called Obookiah, underwent a profound conversion to Christianity, then was taken ill and died. The story of his piety and his death encouraged new Christian fervor in New England, and with the encouragement of other young people from Hawai'i, white Americans launched a missionary enterprise toward Hawai'i, the beginning of a major worldwide American missionary push. The New England Calvinists arrived in Hawai'i in 1820. Standard accounts emphasize that from the point of view of the missionaries, the timing was fortuitous, because shortly before their arrival in Hawai'i some ali'i (chiefs) had declared the complex and extensive religious and political system of kapu (sacralizing restriction) to be overturned. The missionaries enjoyed tremendous success as Kānaka converted in droves. The missionaries learned Hawaiian and, with the assistance of a number of Kānaka Maoli including Davida Malo, created the first written version of a language that had previously been exclusively oral, began translating the Bible into the language, and taught Kānaka to read. Kānaka threw themselves into the study of reading and writing. Soon, the Hawaiian nation was both largely Christian and literate, one of Christianity's great success stories in the nineteenth century.[2] This standard account dominates the writing on the history so completely that even explicitly anticolonial accounts of Hawai'i's past echo it. With good reason, these latter histories condemn

Christian ascendancy in the islands, rather than cheering missionary success, and point to the role of American missionaries and their descendants in the colonial domination of Hawai'i by the United States. Like anticolonial accounts of Western Christian missionary work elsewhere around the world, these histories depict it as a European and American initiative that did incalculable cultural, political, and economic violence to indigenous people globally. Note, however, that condemning the missionaries' influence in Hawai'i has sometimes elided into portraying Kānaka as powerless before them—or as Francine Du Plessix Gray writes, "this island people weakened by centuries of isolation lacked the psychological antibodies, as well as the physical ones," to resist the missionary onslaught.[3] Gray's formulation is an anticolonial history of missions, but not a particularly liberatory one.

We also need what some scholars would call a *decolonial* history of missions: one that does not assume that autonomous indigenous perspectives are restricted to opposition to missions.[4] The notion that Kānaka strategically engaged with missions and missionaries for their own reasons is present in essential scholarship by Lilikalā Kame'eleihiwa and Noenoe Silva.[5] It is deserving of greater elaboration. At first glance, the idea appears troubling, because it would appear to name Kānaka as encouraging the entry of a destructive force into Hawai'i. But there is a difference between pointing to the important role of indigenous people in encouraging Christian missions and blaming them for the destructiveness of the missions: Kānaka had no way of knowing how dangerous the missionaries and other Haole would become. It is essential to consider how Kānaka encouraged missionaries and, most important, what motivated them to make a place for missionaries, much as they built a home for Ephraim Spaulding in Lāhainā in 1832. The encouragement of missionaries was part of a Kanaka exploration of the world that sprang from deeply indigenous sources.

'ŌPŪKAHA'IA AND THE HAWAIIAN INITIATIVE IN MISSIONARY EDUCATION

There is no better place to start than with a Kanaka Maoli convert that non-Hawaiians have widely celebrated and profoundly misunderstood: 'Ōpūkaha'ia. Americans have misappropriated 'Ōpūkaha'ia's story from the time he was living in the 1810s to the present. In the nineteenth century, American Christians took the self-penned narrative of a trained kahuna (priest) and adult man who made a determined decision to travel overseas to learn and framed it as the story of a lost man-child far from home with inarticulate urges toward Christian enlightenment. In writing their accounts of ʻŌpūkahaʻia's life, Americans did violence to Hawaiian history, perverting it to serve the purposes of Americans who sought to create a missionary enterprise to Hawaiʻi. Recent American historians have done little better. The introduction to the sole edition of his memoirs that is still in print, published by a missionary society in 1967 and reprinted in 2012, opens by describing "a fifteen-year-old boy" who "looked out across Kealakekua Bay . . . and wondered about the handsome square-rigger he saw anchored there." Written in 1967, the introduction likens his yearning to rebel and run off to that of "his mid-Twentieth Century counterparts"—an apparent effort at demonstrating the relevance of his story for disaffected youth in the 1960s. "Unhappy with his lot, brooding almost to the point of suicide over the cruel events that had robbed him of his family, dissatisfied with the career his elders had chosen for him and the rigid course of training he must undergo to prepare for it," the boy makes his way to the emancipation of a Western ship. While there is scant evidence to support these suppositions about ʻŌpūkahaʻia's motivations, we might expect no more from a missionary society publication trying to appeal to contemporary Americans.[6] We might even anticipate this kind of treatment from Gavan Daws, an American writer who termed eighteenth-century Kānaka thieves and described the creation of a U.S. state of Hawaiʻi as the "sign and seal of Hawaiʻi's political maturity" rather than the consolidation of American colonialism. In his still-popular and influential 1968 history of Hawaiʻi, Daws terms ʻŌpūkahaʻia a "Hawaiian boy."[7] But we will be disappointed if we hope for much better from a leading American academic historian. A 2014 book by John Demos does much to understand ʻŌpūkahaʻia as a Kanaka, but then undercuts that effort by portraying an adult man as an infantilized "restless" teenager and using him as a player in a metaphor for the making of the United States. This account, too, does violence to ʻŌpūkahaʻia's life and to Hawaiian history (and even American history), twisting it to serve the purpose of creating an origin story to teach Americans "who we are as a people."[8]

ʻŌpūkahaʻia was no more an American than he was a "fifteen-year-old boy" drawn like a moth to the light of American learning and the

Christian faith. He was an adult Kanaka who, in 1807 at the age of about twenty (not fifteen), decided to embark overseas on the ship of a New England captain. He followed in the footsteps of Ka Wahine, Kaʻiana, and by then hundreds of other Kānaka who traveled to the great poles of the Pacific fur trade: northwestern North America and Macao. ʻŌpūkahaʻia spent six months working on a seal-hunting ship on the Aleutian "Seal Islands," sailed back through Hawaiʻi where the ship reprovisioned, then traveled straight to Macao. He then spent six months in Macao and Guangzhou (Canton), where the ship sold its sealskins and bought Chinese products—tea, cinnamon, silk, and "nankeen," a yellow Chinese cotton fabric in demand in the United States. After six months in China, the ship headed southwest around the Indian subcontinent, further south still around Africa's southern tip at the Cape of Good Hope, before heading northwest across the Atlantic, arriving in New York in 1809. ʻŌpūkahaʻia ended up in New England, where he worked for and boarded with American families and became increasingly interested in Christianity. He studied, learned to read and write and speak English well, and converted to Christianity. Like many Kānaka overseas, ʻŌpūkahaʻia grew ill and died from diseases to which he had no inherited immunities, but not before writing a short memoir that emphasized the story of his conversion. This memoir—or rather, an edited version of it surrounded by texts written by New England churchmen—became one of the major spurs for the movement to send American missionaries overseas, and first to Hawaiʻi.[9]

ʻŌpūkahaʻia—or Henry Obookiah, as Americans called him— wrote that memoir for an early nineteenth-century American audience and fulfilled what he expected they wanted: a tale that emphasized his transformation in New England into a Christian. This made it easier for Americans to read their fantasies into his life: with little explanation of his life in Hawaiʻi or his travels to the Aleutians and China, the real ʻŌpūkahaʻia became an imaginary Henry Obookiah, a blank slate for constructing stories of the glorious possibility of American Christian evangelization. To get beyond those fantasies to the historical reality of ʻŌpūkahaʻia and Hawaiʻi's early relationship to Christianity, we need to get closer to his life before he left Hawaiʻi in 1807.

On the opening page of the book he published under the title *Memoirs of Henry Obookiah*, American minister Edwin Dwight begins:

ʻŌpūkahaʻia. From Dwight, *Memoirs of Henry Obookiah,* 1819. Courtesy of the James Ford Bell Library, University of Minnesota.

Henry Obookiah was a native of Owhyhee, the most important of the Sandwich Islands. He was born about the year 1792. His parents ranked with the common people; but his mother was distantly related to the family of the king. Her name was Kummóoólah. The name of the father is unknown.

The first key thing to note here is the year: if "Henry Obookiah" was born in 1792, this would make him a fifteen-year-old adolescent when he left Hawaiʻi on an American ship. The second is the general vagueness about his social origins—common rank, a distant but undefined relation to royalty, and a father whose unknown name perhaps hinted at what Americans considered to be "illegitimacy." This fits with the general picture that Dwight and later American writers painted of "Obookiah": a dissatisfied youngster from an undistinguished family of the common people, who was orphaned in a war between unnamed chiefs, was taken in by an uncle who was a priest, and ran away from his uncle as a teenager to board a ship for adventure. All of these statements add up to present a certain image of a restless adolescent in rebellion against the restrictive "pagan" religion of his uncle who happened to stumble, in New England, on the saving light of the Christian gospel.

The real ʻŌpūkahaʻia could not correct this story about the fantasy Obookiah: he was already dead when it became popular. But the story did not end there, because of the way that Kānaka Maoli seized on the power of literacy to represent themselves and their own interests. Kānaka embraced literacy when missionaries came to teach it. But while missionary accounts emphasized the power of reading the scripture, there was also power in *writing*—including the power to present Kānaka from Hawaiian perspectives. This power of representation was magnified by the printing press, which created a mass medium in the form of widely circulated newspapers in Hawaiian. The first Hawaiian-language newspaper was published in 1834 on a press the American missionaries had brought from the United States. While that first newspaper was exclusively an organ of the mission, by the 1850s, newspapers were published weekly to report on news and the kingdom abroad. They circulated broadly in the islands, and their readership was largely Native Hawaiian. Until the 1860s, these newspapers were all under the editorial control of Americans and Europeans. In 1862, the first autonomous paper under Kanaka editorial control appeared. Native Hawaiian scholar Noenoe Silva has rightly argued that this marked a turning point in the history of Hawaiian politics as well as publishing.[10] But even before the appearance of that newspaper, some Kānaka were able to use the newspapers to offer distinctively Kanaka perspectives—even in missionary-controlled newspapers. S. W. Papaula did just that in 1865, when he published an account of ʻŌpūkahaʻiaʻs early years in a newspaper named *Ka Nupepa Kuokoa*. Two years later, Papaulaʻs newspaper story would serve as the basis for a translated and corrected Hawaiian-language version of *Memoirs of Henry Obookiah*. Papaula was well situated to write this account and have it published. He was a Christian minister and active in church affairs, a farmer, a representative to the Hawaiian Kingdom's parliament, and a contributor to the Hawaiian-language newspapers. He thus did not lack for legitimacy among the missionaries, but he also had access to information about ʻŌpūkahaʻiaʻs life in Hawaiʻi. He was a resident of Nāpoʻopoʻo in South Kona, where ʻŌpūkahaʻia lived for some years and where he had kin. Papaula squarely places ʻŌpūkahaʻia as a Kanaka in a Hawaiian context. Papaula opens his account with a genealogical statement that frames ʻŌpūkahaʻia within the genealogical conventions of Hawaiian narrative: "O Keau ka makuakane, o Komohoula ka makuahine. Hanau mai la na laua keia keiki, o Opukahaia" (Keau was the father, Komohoula was the mother. From these two was born this child, ʻŌpūkahaʻia).[11] Similarly, in the second paragraph, the newspaper version situates ʻŌpūkahaʻia in time and space in a manner that names his as a Hawaiian story: "Ua hanau o Opukahaia mamua ae o ke kaua ana o Kamehameha me Namakeha ma Kau. A ma Kau no hoi i Ninole kahi i hanau ai o Opukahaia" (ʻŌpūkahaʻia was born before the war between Kamehameha and Nāmakehā in Kaʻū. And Ninole in Kaʻū was the place where ʻŌpūkahaʻia was born).[12] What matters in this narrative are the specifics of site, of parentage, and of historical circumstance. Not only had Dwight gotten the year of ʻŌpūkahaʻiaʻs birth wrong (he was born in 1787, not 1792), but he had missed the whole point of situating the event in time: what mattered most was not the year per se, but that the timing and placement of his birth meant that his life would be marked by war. Papaula explains that he was born in Kaʻū before the outbreak of the war between Kamehameha and Nāmakehā, and goes on to explain that it was the outbreak of war that orphaned the boy, led him to flee Kaʻū, and ultimately ended with him going to live with an uncle. All of this links 'Ōpūkaha'ia's personal narrative to one of the most important events in Hawaiian history: the wars that brought Kamehameha to power as the first mō'ī (king) to rule the entire archipelago. Rather than appearing as inexplicable tragedies (or perhaps as tragedies needing no explanation due to the alleged savagery of the society into which he was born), the killing of his parents, his displacement, and his adoption by kin become legible to almost any Kanaka of the nineteenth century. They become not uncommon results of this time of war.

The difference between these openings suggests that the English-language version was cavalier with the facts and details of 'Ōpūkaha'ia's background because their accuracy was not really relevant to the purposes of the American evangelicals who wrote and read the book. To get that audience to pay for and pray for missionary enterprise to Hawai'i it was sufficient that they believe that an ignorant savage was brought to the light of Christian faith, and his tragic death was a call to bring the light of the gospel to his benighted homeland. But for Kanaka Maoli like Papaula, the specifics of Hawaiian history and geography mattered.

Moreover, Papaula uses specifics to show his readers that the 'Ōpūkaha'ia that boarded a ship to sail overseas was not a child but a man, and not an ignorant savage but a learned kahuna who continued to seek more strengthening learning overseas. After he was orphaned, 'Ōpūkaha'ia lived first in Ka'ū and then in Kohala before he was taken in by his uncle Kahua, his wife Hina, and their people, who lived in Nāpo'opo'o.[13] Kahua was one of the kāhuna at the heiau (temple) of Hikiau, a major temple complex in Kamehameha's home district of South Kona. He trained his nephew in the knowledge and the craft of the priesthood—the 'oihana kahuna. 'Ōpūkaha'ia cared little about farming, fishing, and other activities, Papaula reports: "O ke ao kahuna wale no kona makemake nui, a ua lilo ia he puni nana e malama mau, i na po kapu o ka Hainapule" (Kahuna training alone was his great interest, and he was consumed with observing and conducting the kapu nights of recitation of prayers).[14] Papaula so directly contradicts the suggestion in the English-language book that 'Ōpūkaha'ia was a boy or a youth when he left Hawai'i that one suspects he had read it. Papaula writes that long before his departure, the kahuna was already "he kanaka makua" (a mature person). With his uncle's training, he became a kahuna in his own right, building his own structure in the enclosure of Helehelekalani heiau at Hikiau. There he conducted the worship of the god to whom the heiau at Hikiau was principally dedicated: Lono, the god of growing things and fertility, but also Kūkaʻōhiʻalaka and Kūkāʻilimoku. The latter were two manifestations of Kū, a god of war, Kūkāʻilimoku being the specific god of war claimed by Kamehameha in his rise to power. Papaula emphasizes that ʻŌpūkahaʻia continued his quest for knowledge even at this point, saying that he continued frequently to "hooikaika" (make himself stronger) in the ʻoihana kahuna all the way up to the time of his departure for America.[15]

Moreover, Papaula uses two forms of moʻokūʻauhau (genealogy) in order to situate ʻŌpūkahaʻia and his voyage of discovery as emerging out of an illustrious Kanaka Maoli heritage, not cut off from it and not in teenage rebellion against it. First, rather than suggesting that he was a person of common stock with an unclearly named mother, an unnamed father, and tenuous ties to an unnamed king, we learn the names of family members and that they included relatively highly placed kāhuna. Papaula does not trace ʻŌpūkahaʻia's ancestry beyond his mother, father, and uncle, but by giving their names and the places they lived, he makes it possible for other Kānaka to trace his moʻokūʻauhau. The contemporary Kanaka Maoli writer Wayne H. Brumaghim uses Papaula as a source in explaining that ʻŌpūkahaʻia was "descended from the ʻĪ family of chiefs of Hilo, Hawaiʻi," had genealogical ties to renowned mōʻī of Maui, Oʻahu, and Hawaiʻi Island, and was distantly related to the nineteenth- and twentieth-century monarchs Davida Kalākaua and Liliʻuokalani.[16] Furthermore, Papaula situates ʻŌpūkahaʻia in a moʻo kahuna: a "genealogy of succession of priests."[17] He is tied to Kahua, then to Kahua's teacher, the illustrious kahuna Hewahewa. Hewahewa was the kahuna nui (high priest) of Hikiau heiau, and Kamehameha later named him to be his kahuna. More importantly, he followed in the priestly line founded by Pāʻao—the kahuna who brought the chiefly religion from Kahiki. Kānaka still emphasize continuity of training: a kumu hula (hula teacher) will commonly be discussed as the student of a particular kumu hula, who was a student of another kumu hula before her or him, and so on in a genealogy of teaching. Because Papaula's Kanaka readership were members of a nation who emphasize moʻokūʻauhau in this manner, these dual genealogies of descent and training signaled that ʻŌpūkahaʻia—and by extension his decision to travel overseas to explore the world—emerged from the center of Kanaka Maoli heritage and was tied to Kahiki.

Indeed, Papaula takes pains to make clear that that past and that heritage was still part of the Hawaiian present of 1865, when he published this story. ʻŌpūkahaʻia's aunt Hina still lived in South Kona, her hair gray and her skin wrinkled. More telling still was the vitality of ʻŌpūkahaʻia's heritage as a kahuna. At his small temple, Papaula recounts, ʻŌpūkahaʻia planted three coconut trees, perhaps because the coconut is a bodily form of the god Kū, to whom he dedicated much of his worship. Papaula takes the time to tell the reader that those coconut trees still bore fruit in 1865—a marker at the least of ʻŌpūkahaʻia's connection to that place, and perhaps even of the continued vitality of the Hawaiian religion in an allegedly Christian era. The past was not lost to Papaula, but tied to the present.

In fact, Papaula makes it possible for contemporary twenty-first-century Kānaka Maoli to see the continued presence of ʻŌpūkahaʻia's kin in Hawaiʻi, a way of situating his continued relevance for today. Brumaghim uses Papaula and other sources to trace the story of ʻŌpūkahaʻia's relations not only up to the first Kanaka born of Papa in the distant past, but up to contemporary Kānaka living in Hawaiʻi today. There could be no more powerful way of implicitly making a core argument in Kanaka Maoli society, studies, and politics: the Kānaka of the present are a continuity with their past. A person that Western histories look upon as a player in the story of the alleged rupture of the missionary era from the past is in fact a bridge that ties the present to the past in one narrative.

What was that alleged rupture? A story of travel and conversion, but one that, presented properly, reveals the way that Kānaka Maoli engaged Christianity sincerely but strategically, and with their eyes fully open. ʻŌpūkahaʻia, a man steeped in the esoteric knowledge of a religion and culture whose classical narratives recounted stories of great voyages and the movements of gods across oceans and lands to faraway places, was approached by one of the Haole on a merchant ship. This Haole had become the aikāne (the beloved companion, a term generally implying sexual intimacy) of the kahuna Hewahewa and asked ʻŌpūkahaʻia to join the voyage. A "mature person" of about twenty, he agreed and traveled to the Aleutian Islands, then Macao, and finally New England. Among the Haole, he continued to pursue the religious interests that had consumed him for years before his travel abroad. This meant delving deeply into the mysteries of the Christian religion and mastering its symbols and rituals. ʻŌpūkahaʻia did not move, as Demos argues, from "ritual to metaphor," from the mere formalism of what New England missionaries called "paganism" to meaningful Christian belief. The phrasing echoes centuries of Protestant Christian declarations that the austere worship of a redeemer Christ makes their religion different from, and better than, the ritualistic (without metaphor, without meaning) worship of Catholics, let alone "pagans" such as Kānaka Maoli. Hawaiian religion had its metaphors, and New England Protestantism had its rituals, and ʻŌpūkahaʻia was expert in both.[18] Indeed, scholars of New England Christianity going back to Edmund Morgan have identified the conversion narrative as central to that faith, and by shaping a life and then writing a narrative that so closely matched its forms, ʻŌpūkahaʻia demonstrated that he mastered Christianity's rituals of conversion.[19]

But even in New England, ʻŌpūkahaʻia did not cut himself off from Hawaiʻi and its Kānaka. Other Kanaka men also had boarded American ships and landed there, and ʻŌpūkahaʻia quickly became a central figure in a network that connected them. He was older than most of them. Some were laborers like Kenui/Kanui and Honoliʻi and, from what we can tell from the little there is written about them in English or Hawaiian, seem not to have had any particular religious training. Thomas Hopu was a sailor but had received religious training from his father, who intended him to dedicate himself to the worship of his god. There were probably differences in rank among them, but (as was true for Kaʻiana and Ka Wahine) there was the crucial common bond of Hawaiʻi between them.

Twenty of these young men became students at what we now know as the Foreign Mission School in Cornwall, Connecticut. ʻŌpūkahaʻia was the school's first pupil. The school was intended to train men from non-Christian nations to become missionaries to their own people. There they were immersed, in the middle of settler-colonial New England, in a cosmopolitan world of non-Western people in a settler colonial school: Hawaiians, Choctaws, Cherokees, Abenaki, natives of India and the Nuku Hiva archipelago, Chinese, and Malayans all convened there to study. What connections did these men from different places create among themselves, and what connections did they see that linked them? The Foreign Mission School was one of the crucial early nodes of communication among the disparate nations that Westerners had attempted to dominate. It was not the only such node; the Kānaka themselves had experienced two others, the diverse crews that labored on Western ships and the diverse population in port cities. These places, and especially the Foreign Mission School, point to a crucial lesson. Like Western ships, Western religion was a means of exploring and connecting to the world globally, not just the West. At the Foreign Mission School, Kānaka helped to create a context where, through a Western Christian institution, they could connect to learning about the world more broadly, and also connect to a diverse array of non-Europeans, including Pacific Islanders. Connection to other Pacific Islanders through connection to Christianity is worthy of note, because in the coming years, such connections became a mode of learning more about the globe and an important but underappreciated motif in the history of Kanaka engagement with Christianity. The Foreign Mission School was an important starting point for that Pacific connection, but it flourished in Hawai'i itself. The most famous of the missionaries to Hawai'i were the white Americans who arrived in 1820 and after, but those very missionaries recognized that it was Tahitians who were the most effective in first sparking widespread Kanaka interest in the religion. If anything, the story of the development of the American missions in Hawai'i has received disproportionate attention in Hawaiian history, but the role of the connection of Kānaka Maoli to other Pacific Islanders has been neglected.

THE LIGHT FROM TAHITI: TAHITIANS AS MISSIONARIES OF LITERACY AND CHRISTIANITY

As John Garrett has noted, in the two years after their arrival in Hawai'i, the American missionaries "found progress slow." Even with the assistance of Thomas Hopu (a friend of 'Ōpūkaha'ia's in New England) and other Kanaka converts, they made little headway in attracting the people to their religion. For help, they reached out to the London Missionary Society, a Protestant group that had enjoyed some success in Tahiti and the Society Islands. They were fortunate that, in addition to sending the English missionary William Ellis and his wife, the LMS sent nine converts from Tahiti and the Society Islands. It was a boon that Ellis could speak some Tahitian, and thus pick up Hawaiian more quickly.[20]

Tahitians were more valuable to the mission, however, as scholars

Malcolm Nāea Chun, Lilikalā Kameʻeleihiwa, John Garrett, Douglas D. Tzan, Marshall Sahlins, Dorothy Barrère, and others have recognized. A shared cultural background drew Kānaka to the Tahitians as a means to explore Christianity, the Pacific, and the world. Language was the most obvious advantage Tahitian missionaries enjoyed, as they could quickly make themselves understood by Kānaka: the Hawaiian language is closely related to Tahitian.[21] Just as important, the Tahitians came from places that were culturally akin to Hawaiʻi. Their homelands (like their languages) were not identical to what they encountered in Hawaiʻi, but the societies shared some elements of cosmology, gods, emphasis on genealogy, stories, and notions of sacralizing restriction (Tahitian *tapu* or Hawaiian kapu). Gender is one area where some shared cultural background made the Tahitians important vehicles through which Kanaka Maoli men and women could explore Christianity and, through it, the Pacific and the world.

Looking at one Tahitian couple, Naiomi (or Naomi) and her husband, Auna, demonstrates the way that engaging with Tahitians meant exploring the world through already existing Pacific connections. Auna was prestigious in his own society in ways that would have been immediately recognizable to Kānaka Maoli. The son of a priest (one historian says a high chief) from the island of Raʻiātea and designated by his father to become a priest, he was a "leading member" of the *arioi*, a select society of initiates skilled in esoteric knowledge, song, story, the arts, and eroticism. He was closely tied to Pōmare II, the second king of Tahiti, and an important warrior in that king's army. American missionaries and historians have expected men to be at the center of missionary work and privileged men generally in their stories. Because of this, they have noted Auna's exalted rank, learning, warrior background, and even his height, all characteristics that align well with Westerners' notions of what is prestigious among men. In contrast, they recorded almost nothing about Naiomi or the other Tahitian women who went to Hawaiʻi with the missionary party, even though all of the Tahitians traveled as married couples (as did the American couples, for fear that single missionaries would be more subject to sexual misbehavior). Most historians fail even to mention Naiomi, despite her early importance to the mission.[22]

The couple's backgrounds and kin mattered immediately upon their arrival in Hawaiʻi. By making a place for them in Hawaiʻi, Kānaka laid the path for Auna, Naiomi, and other Tahitian missionaries and teachers to build Christianity in the islands. Upon first landing on Hawai'i Island, Auna was eagerly welcomed by Kuakini, the chief who had been named governor of that island as Kamehameha and his son, Liholiho (Kamehameha II), built a consolidated authority over the archipelago. Kuakini immediately "manifested much attachment" to Auna, a fellow chief. He told him that other Tahitians were already on the island and took him to meet them. One was Toketa, whom Kuakini was already familiar with and who may already have entered his household.[23] A shared chiefly status created bonds between Auna and ali'i in Hawai'i. Because we don't know Naiomi's background, we can't say if she was an *arioi* initiate or if her status similarly made ties to the ali'i for her. But soon kinship relations were found that made a link to Hawai'i, in a way very familiar in Kanaka Maoli society: as the couple and the other missionaries proceeded to Honolulu, even before reaching shore, "we met a canoe, in which the wife of Auna recognized a brother, who had left the Society Islands on the *Bounty*, when the mutineers took possession of the ship." The man, Moe, was already a steward to Ke'eaumoku, the governor of Maui. To the English missionary, what was remarkable was the tie to the famous mutiny on the *Bounty*.[24] To the Kānaka, however, what mattered most was that layer upon layer of ties—language, chiefly social structure, cultural similarities, history, even kinship to people they already knew—made the Tahitian missionaries like Auna and Naiomi the perfect vehicle through which to explore a wider world, including Christianity.

Affinities between gender conventions may also have paved the way for Kānaka to embrace Christianity through Tahitians. Like Kānaka Maoli, Tahitian missionaries assigned separate but complementary roles to men and women in evangelizing, and this facilitated their evangelizing efforts toward Kānaka. Tahitian men like Auna presided over religious meetings of men, while Tahitian women like Kaamoku took charge of religious meetings of women.[25] American missionary wives also sometimes took responsibility for teaching Kanaka Maoli women and leading their activities, but American women came from a religious tradition that had often forbidden them from preaching to mixed congregations.[26] In contrast, women as well as men occupied positions of religious authority in Tahitian society. Kāhuna wahine (called "priestesses" by Westerners) were an established part of religious practice in

Hawai'i, and women as well as men were *arioi* in Tahiti.[27] In both places, Christianity would erode the place of women and exalt the power of men, replacing what was a tradition characterized by complementary roles for men and women with a hierarchy of men over women, and seeking to replace a cosmos shaped by akua (gods) both female and male with a universe created and ruled by a single, male deity.[28] But at this early point in the 1820s, the place of women in Tahitian religion and among the Tahitian missionaries may have laid the basis for the women and men of Hawai'i to perceive Tahitians as a means to explore the world through these Christians from Kahiki.

Moreover, Kahiki (or Tahiti as the new arrivals like Naiomi and Auna spelled it) enjoyed enormous cultural prestige in Hawaiian traditions. Kānaka knew that the priestly and chiefly religion of the heiau, the luakini (temple of sacrifice), and the kapu had been brought to them from Kahiki by Pā'ao centuries before.[29] That was the religion in which 'Ōpūkaha'ia and other kāhuna were trained. In 1819, that religious structure from Kahiki had been disrupted. It had been overthrown as what we could term a state religion (though a Hawaiian state was not yet truly formed) with the instigation of Ka'ahumanu (the kuhina nui, or regent, who held enormous power under the early reign of Kamehameha II), Keōpūolani, and Kapi'olani.[30] Historians have generally followed the example of the American missionaries in arguing that this created a religious vacuum that the American missionaries filled. Certainly, the Americans did gain a few early converts and chiefly friends, but the religion did not take off until a new religion came from Kahiki/Tahiti. It was in 1822, when *arioi* and other Tahitians arrived in Hawai'i, that Christianity gained momentum. Centuries before, Pā'ao had brought a new religion from Kahiki. When that religion was disrupted by ali'i who suspended the kapu, Kānaka looked again to Kahiki, which was present among them in the person of Auna, Naiomi, and other Tahitians.

At this point, Kānaka may have seen Christianity as a Tahitian religion as much as a Western religion. To embrace Christianity may have meant exploring and embracing Tahiti—and by extension the Pacific— at least as much as it meant adopting a Western import. Tahiti was both an object of exploration and a model, because of its affinities to Hawai'i and cultural prestige as the place of origin of the old religion. Kānaka constantly asked Tahitians about what things were like in their homelands: from the culinary (how did Tahitians prepare arrowroot? Auna showed them)³¹ to the religious (how did Tahitians convert to Christianity?).³² It was not only religious practices that Kānaka were adopting. In 1826, the missionary William Richards noted that Kānaka at Lāhainā had taken up the practice of hosting a feast with long speeches to initiate newly built houses, an idea that was "introduced here from the Society Islands."³³ Auna noted in his diary that when he spoke before gatherings on how Christianity had spread in Tahiti, Kānaka listened with interest.³⁴ American missionaries were conscious that Kānaka believed that Christianity could be useful to them in Hawaiʻi, because Tahitians suggested it had been useful in Tahiti (although, unfortunately, they did not name what Kānaka perceived the usefulness of the religion to be).³⁵ One of the most famous episodes in the early missionary history of Hawaiʻi was the mass burning of kiʻi (images) of the akua in 1822. The event cannot be understood outside of the way that Kānaka explored and engaged Christianity through Tahiti. Auna reported that Kaʻahumanu and Kaumualiʻi "talked to me a good deal about our burning the idols at Tahiti"—the apparent model for a conflagration that consumed one hundred and two kiʻi in Hawaiʻi on June 26, 1822.³⁶ He wrote that it made him think of the burning of idols in Papeete in Tahiti—and said the Kānaka were "following our example" in burning their kiʻi as Tahitians had done.³⁷

Moreover, Tahitians were crucial to introducing reading and writing to the Kānaka, which Kānaka embraced with famous vigor. Kānaka sought out the Tahitians as teachers broadly speaking, not just missionaries narrowly understood. Kānaka asked that Auna and Naomi stay in the islands to teach them rather than return to Tahiti.³⁸ Davida Malo, one of Hawaiʻi's most celebrated writers and one of the most important early converts, was heavily influenced by work with Tahitians. According to Kanaka historian Charles Kenn, it was inevitable that the Tahitians instructed Malo during his time in Kuakini's court—*before* Malo encountered Christian missionaries.³⁹ That influence continued after Malo embraced Christianity and moved to Maui and the Lāhaināluna seminary there. Malo worked closely with the Tahitian Taua in guiding and correcting missionary William Richards's translation of the Gospel of Matthew into Hawaiian.⁴⁰ Kānaka listened attentively to the preaching of Tahitian men and women. And they laid particular emphasis on learning reading and writing from them. Even before the Tahitian missionaries had arrived, the aliʻi Kuakini had turned to Toketa (also known as Toteta) to teach reading and writing to himself and people close to him. Toketa was not a missionary per se and probably not baptized when he arrived in Hawaiʻi. But he must have spent time around missionaries in Tahiti, because he understood reading and writing.[41] To Kuakini, this was valuable knowledge, and he sought it out from a Tahitian teacher. Other Kānaka did the same, and Tahitians became teacher/missionaries valued for their teaching of reading and writing.

KA PALAPALA: TEXTUALITY AND PERSPECTIVE

In the 1820s, there was an insatiable Hawaiian demand for ka palapala—a word literally meaning "a page with writing on it," but at the time connoting much more: literacy and literate knowledge as a whole. Aliʻi reached out to missionaries, and their entreaties make clear that ka palapala was as much a goal as Christianity—although, knowing their audience, the aliʻi emphasized religious motivations. In 1820, Kaumualiʻi, the mōʻī of Kauaʻi Island, sent a letter in English (perhaps penned by his son Humehume, who had been a student at the Cornwall Foreign Mission School in Connecticut, though the handwriting does not appear to be Humehume's) to the head of the American Board of Commissioners for Foreign Missions (ABCFM). Humehume thanked him for "giving my son learning" and encouraged him to send missionaries, who would presumably teach him to read, "the good book . . . the one that God gave us to read."[42] Similarly, his son Kealiʻiahonui emphasized that it was written knowledge that he valued when he wrote from Oʻahu to the ABCFM thanking them for "Ikohoouna ana mai nei ikatumu ao palapala ika olelo a Iesu Kraist" (sending a teacher here to teach the palapala of the word of Jesus Christ).[43] Kealiʻiahonui gave priority to ka palapala in part because of Christianity's scriptural orientation, but also in part because of Kanaka desire to access the outside world through text.

But foreign teachers were too few to teach the many Kānaka who wanted to learn ka palapala. As soon as many Kānaka learned to read and write from the Tahitians or Americans, they became teachers themselves. At first, classes were convened without schools themselves being formed. The missionaries were met with a tremendous wave of interest in ka palapala. Adults were at the center of this rapid embrace of ka palapala. By the late 1820s, one third of the populace was reported to be learning ka palapala. This form of education did not take place in schools, but rather in an intense and informal period of instruction, as newly literate Kānaka became the teachers to others who sought ka palapala. It is crucial to note that even at this early date, Kānaka Maoli themselves were coming to the fore of teaching in the islands, even though missionaries occupied some of the most visible positions in education.[44]

Kanaka students were using books in distinctly Kanaka ways that Americans and Europeans neither recognized nor fully understood. The politically involved teacher and writer J. H. Kānepuʻu, who grew up on Molokaʻi in the 1820s and 1830s, provided one of the rare portrayals of Hawaiʻi's common schools from the point of view of a Kanaka Maoli student in a semiautobiographical story he published in 1868. From it, we catch a glimpse of how teachers taught and students learned with the few textbooks they had access to. In the story, the teacher at the local school had few teaching materials available to him, but rather prepared his lesson from lessons he had memorized and that he passed on to his students through memorization as well. Kānepuʻu describes a basic geography lesson, in which the teacher wrote the names of "Na Aina" (countries) and "Na Kanaka" (peoples) on a stack of papers. Standing before the class, he would call out the names on these countries and peoples one by one, the students repeating after him. Repeating the lists, students memorized them just as they would lists of the months, and just as Kānaka had long memorized long and detailed genealogies. Classical Hawaiian education—the system of teaching in place prior to Western incursion and up until the early years of the nineteenth century—emphasized memorization in a process of group learning through chanting. This process was embedded in the culture of the lāhui Hawaiʻi (Hawaiian nation) and proved suitable for learning a wide variety of materials. Teachers like him (the main character's father, based perhaps on Kānepuʻu's own father) did not abandon this practice, but rather adapted it, integrating the use of text (the stack of written pages from which he worked, in the absence of scant textbooks) into the recitative practice of memorization.[45]

When textbooks became available, Kanaka teachers and students were able to incorporate their use in ways that merged with this method of learning through memorization, John Charlot's research reveals. Using books, Kānaka took Western knowledge and made it paʻanaʻau (literally "fixed in the intestines"), meaning memorized. This means of learning together through collective recitative memorization was particularly practical given that books were often in short supply. Americans were frequently puzzled and disapproving when they witnessed a group of Kānaka reading. "It is a very common thing," the missionary Lorrin Andrews wrote in 1834, "to see two, three, four, and sometimes as many as six persons, all reading out of one book at the same time." Kānaka laid the book "on the ground; or if in a house, in the middle of the room, at the center," then "prostrate themselves around as radii from that center . . . with their heads over the book." In unison, they read the book out loud. Through such choral reading, many Kānaka in the mid-nineteenth century memorized quantities of text and fact that astonished Westerners—and frustrated them, as they believed that memorization was a lower form of learning.[46] Moreover, from practice reading collectively in a circle around one book, many Kānaka learned to read texts fluently in ways that Haole considered upside-down or sideways. Seated in a circle, they needed this skill: "to some the book must be right end up, to others wrong end up, and others must read towards them, or from them, as the case may be."[47] The visual, aural, and educational effect of students in a circle reading and chanting a textbook strikingly reminds us of what Phillip H. Round writes about the American Indian engagement with the book: this was an intense engagement where Native people made uses of texts-as-objects (books) in ways that were as rooted in their own practice as they were strange to the settlers around them.[48]

Indeed, the Hawaiian noun ka palapala referred to above reflects the Kanaka appropriation of foreign text, written culture, literacy, and the knowledge encoded in them into one *thing* that could be put to use for Native ends via paʻanaʻau ʻana (memorization). As Noelani Goodyear-Kaʻōpua reminds us, when they incorporated literate knowledge, they did not just insert it into an unchanging block called a "culture."[49] In other words, cultures are illusory objects that cannot capture the dynamism of people and their history. But while culture is not a thing, Kānaka turned reading, writing, and texts into a thing, a noun called ka palapala. Having objectified ka palapala, Kānaka Maoli could use it for their own needs in a changing world. Furthermore, we can consider that the memorization of Western texts can also be a way of making that knowledge interior. There was no more powerful a way to turn ka palapala into something that could later be put to Native purposes than by making it paʻanaʻau.

THE MARKET FOR BOOKS

Kānaka did not only want the figurative palapala (literate learning); they also wanted books through which they could learn about the world. This brings us back to the vignette that opens this chapter, in which men paddled lumber twenty miles and built a home for a missionary in order to earn geography textbooks. Kānaka wanted books to read, and a recurrent motif in their relations with the Haole was their efforts to obtain books. Not all of these books were geography books per se, but all of them allowed Kānaka access to learning. William Richards, a missionary at Lāhainā on Maui, was especially interested in the commercial possibilities offered by the press the mission owned there. In 1826 he wrote, "Even the little books which we now have furnish us with all the vegitables, fowls, fresh fish and other little articles which we wish to purchase of the natives in Lahaina."[50] He repeatedly entreated his superiors in Boston to ship him paper to print on. Richards assured them, "Supplying us with paper will not add to our expense for our books are now [by] far the best articles of trade which we have."[51] In 1827, he wrote that he was convinced that with "10,000 copies of *any* small work" he could buy "the produce of the land." The mission, Richards noted, had recently had a bestseller with an edition of twenty-five thousand copies of a translation of the Sermon on the Mount, and the work had whetted an "appetite for more."[52] In fact, Lorrin Andrews noted disapprovingly that the Sermon on the Mount counted among the texts that "entire schools" had committed to memory so fully that it could not be used to test the students' reading skill.[53]

Native interest in this startlingly political sermon—in which Christ prophesied that the meek shall inherit the earth—might have spurred a more reflective and less business-oriented man than Richards to consider the cultural power of ka palapala for Kānaka:[54]

> Pomaikai lakou ke haahaa ka naau;
>   no ka mea, no lakou ke aupuni o ka lani.
> Pomaikai lakou ke u;
>   no ka mea, e loaa ia lakou ka olioli.
> Pomaikai ka poe akahai;
>   no ka mea, e lilo no lakou ka honua.
> Pomaikai lakou ke pololi, ke makewai a no ka maikai;
>   no ka mea, e hoomaonaia lakou.

Pomaikai lakou ke aloha aku;
no ka mea, e alohaia mai lakou.
Pomaikai lakou ke maemae ka naau;
no ka mea; e ike lakou i ke Akua.
Pomaikai lakou ke uwao;
no ka mea, e iia lakou he kamalii na ke Akua.
Pomaikai lakou ke hoomaauia mai no ka maikai;
no ka mea, no lakou ke aupuni o ka lani.

Blessed are the poor in spirit,
For theirs is the kingdom of heaven.
Blessed are those who mourn,
For they shall be comforted.
Blessed are the meek,
For they shall inherit the earth.
Blessed are those who hunger and thirst for righteousness,
For they shall be filled.
Blessed are the merciful,
For they shall obtain mercy.
Blessed are the pure in heart,
For they shall see God.
Blessed are the peacemakers,
For they shall be called sons of God.
Blessed are those who are persecuted for righteousness' sake,
For theirs is the kingdom of heaven.

The resemblance of the reiterative form of this passage to a Hawaiian helu (a listing, the same literary form that Kamakau used to enunciate Kaʻiana's exceptional learning) may explain part of what drew Kānaka to it, but its powerful message is inescapable.[55] This text, and the tremendous market for copies of it among Hawaiians, raises a number of questions. First, who might have been "the meek" in the minds of Hawaiian readers in 1827? There were undoubtedly many possible answers to that question depending on one's position in society, especially when we remember that "the meek" in this context might mean those who find themselves forced to accept the authority of others over them, whether or not they are truly meek. "The meek" could be impoverished and overworked makaʻāinana (commoners): aliʻi demands for makaʻāinana labor and obedience had grown as they pushed people to cut enormous amounts of sandalwood to sell into the China trade. This process likely worsened the effects of epidemics on makaʻāinana, as the resulting overwork made them more vulnerable to the diseases that were decimating the population, encouraging conversion to a religion that promised heavenly life after bodily death.[56] Or "the meek" could be aliʻi: individuals and families across the islands whom Kamehameha had forced to accept his authority as he rose to power over the whole archipelago. Or "the meek" could be Kānaka of all ranks: the past five decades had seen imperious Haole captains and missionaries arrive in their islands believing themselves to be godlike in their powers or carrying the light of the one true God to the benighted. To Kānaka in 1827, any of these and more could have been the meek that Jesus said would inherit the earth.

This raises a second question: What was this earth that the meek would inherit? What was it like? To learn about that earth, one of the key places that Kānaka could turn was literate knowledge: ka palapala. In differing ways, all of the first printed works in Hawaiian provided Kānaka means to learn more about that world. Texts that taught literacy opened the door to the others: religious texts that revealed secrets about a foreign religion, newspapers that contained essays full of hints about foreign countries, pedagogical pamphlets that offered glimpses of life in foreign places. And, most dramatically, literacy opened the door to the geography textbook *He Hoikehonua,* which was published by the missionary press in 1832. This was a book that Kānaka were prepared to labor for, to paddle heavy loads long distances for, and to build and thatch houses for, as they did for Ephraim Spaulding. That book was created to aid in the teaching of geography in schools to Kānaka Maoli. Its publication signaled the opening of a newly intense period of Kanaka study of geography. The inherently political nature of world geography in Hawaiʻi meant that this would also be a period of intense contestation over what lessons about global geography Kānaka would teach and learn in the classroom.

4
THE WORLD AND ALL THE THINGS UPON IT

GEOGRAPHY EDUCATION AND TEXTBOOKS IN HAWAI'I, 1831–1878

"The World and All the Things upon It"—the title of the work expressed the audacity of the Kanaka exploration of the world, and the content of the work expressed the political urgency of teaching geography in Hawai'i. The schoolteacher-intellectual J. H. Kānepu'u gave this title ("Ka Honua Nei a me na Mea a Pau Maluna Iho" in Hawaiian) to the serialized geography of the world that he published in the *Ka Lahui Hawaii* newspaper over six months in 1877. As befits the title, it was a substantial work, over thirty-two thousand words long. The title of the series proclaimed the expansiveness of the knowledge that Kānaka sought about the world. It proposed to gather all there was to know about the world and share it with Hawaiian readers. But when one reads the work, it is as if Kānepu'u was interrupted in his task. He does begin by presenting the geography of all the globe (its size, its oceans, its continents and rivers, and so forth) and at times returns to these topics. But Kānepu'u interrupts this global survey to propose a distinctly Hawaiian and ocean-centered perspective on the world, a perspective that countered the Western, colonialist, and land-centered perspective that dominated the textbooks used in schools in Hawai'i. Kānepu'u then moves on to address directly the most crucial issues of the time in Hawai'i, including the dispossession of Kānaka Maoli from the land by a growing plantation economy. He writes that Kānaka are increasingly landless, that land is commercialized, that Kānaka are forced to labor for wages rather than farm the food they need. Most of all, in the series Kānepu'u dwells on the danger that Americans would take over his country.[1]

The divergence between the title and the content of Kānepu'u's series is instructive. It reveals the way that the politics of the nineteenth century made it imperative for Kānaka to center Hawai'i in their studies of global geography. This did not mean that they abandoned the study of the globe or settled only for knowing their own homelands. In fact, Kānaka ambitiously pursued knowledge of "the world and everything upon it" all through the century. But because Americans in the Hawaiian educational system placed colonialist messages at the foundation of the curriculum, and because of the threat of domination by the United States, Kānaka found that in order to engage the world, they needed to defend Hawai'i and center it in their world geographies. This was not merely political necessity, however. As we have seen, perspectivalism (the notion that people could and should understand the world from their own perspectives) was a core notion in Hawaiian geographical thought. Thus it was both politically imperative and entirely in keeping with the heritage of Hawaiian geographical thought for Kānaka to place Hawai'i at the center of their world geography.

Kānepu'u's serialized work, like his career, also has much to teach us about schooling in nineteenth-century Hawai'i. The work laid claim for Kānaka to the roles of learning, teaching, and even producing knowledge about the world. Kānaka in the nineteenth century were students of world geography, teachers of world geography, and authors of works of world geography. Kānepu'u was all three. Kānepu'u was a student of geography: we have already encountered him through his semiautobiographical story of a childhood in Moloka'i when a schoolteacher father stood before the class, reading out loud the names of countries and peoples so that his students could learn them by heart.[2] Kānepu'u was a teacher of geography: he became a teacher at two public schools in Honolulu (one in Pālolo, the other in Waikīkī) in the 1860s and 1870s, when geography was a required topic in the curriculum and when schoolteachers taught all subjects.[3] And as his series on "The World and All the Things upon It" demonstrates, Kānepu'u was a writer of works of geography. When we think of Kānepu'u as a pupil, a teacher, and a politicized public intellectual writing on world geography, we bring together three roles that historians might often assume belong to different people, different eras, and different "races." Kānepu'u's career is, therefore, a striking reminder that Kānaka long occupied multiple roles in public education and in intellectual life in the Hawaiian Kingdom. In fact, Kanaka teachers took a leading role in answering back to colonialist texts that depicted Hawai'i as a minor nation confined to the benighted margins of the globe. This chapter is a close study of Hawaiian-language geography textbooks, the way geography was taught in the schools, and the work of Kanaka teachers such as Kānepu'u. It reveals generations of engagement by Kānaka in the struggle to define Hawai'i's place in the world. Nothing could be more important to a nation under cultural, economic, political, and military assault than that effort. Although Kānepu'u is not at the center of this chapter, his story and his vision of world geography as a political endeavor that could and should play a part in addressing the political needs of his people serve as a unifying motif for it. Kānaka dedicated their energies to the labor to define their place in the world from Kānepu'u's school days, through his own years teaching in Honolulu in the 1860s and 1870s, and to the perilous years of the end of the century, when U.S. occupation became a reality. In those efforts, we see a determination that geography (and schooling more broadly) could be a tool for the defense of Hawai'i's people and its national sovereignty.

A combination of cooperation and growing contestation between Kānaka and Haole shaped the history of education in Hawai'i from the 1820s to the 1870s—the years that stretched from Kānepu'u's childhood to adulthood. In a first phase from the missionaries' arrival in 1820 and into the 1830s, adult literacy was the focus. As Noelani Goodyear-Ka'ōpua has demonstrated, Kānaka and missionaries cooperated in the creation of a writing system, and mission schools became "points of access to the new skills of reading and writing." Enrollment peaked in 1832, when fifty-three thousand Kānaka studied in nine hundred schools and only fifty American missionary men and women were in the islands. As Goodyear-Ka'ōpua argues, that ratio makes it clear that already from this early moment, Kānaka were the bulk of the teachers in Hawai'i. Kānaka continued numerically to dominate teaching as the 1830s gave way to the 1840s and Hawaiian education arrived at a second phase: the creation of a national school system designed for "schooling children as proper national subjects." This larger goal was inseparable from the establishment of a constitutional monarchy in 1840 and was apparent in school attendance laws. By 1842, the law required an elementary education in reading, writing, geography, and arithmetic in order to marry or hold high office. Kānaka and Haole worked together on this national project, but in pursuit of what Goodyear-Kaʻōpua calls "competing projects of modernization and nation building." To the first superintendent of schools (the historian, scholar, and clergyman Davida Malo) and other Kanaka officials and administrators, the goal was to create an educated Hawaiian citizenry. In contrast, representatives of the "nexus of missionary and sugar business interests" envisioned a nation in the service of their business and the commercial economy. Accordingly, they worked to use the schools "as a way of sorting and segregating racialized citizen-subjects for an oppressive plantation society." Goodyear-Kaʻōpua has demonstrated that Kanaka administrators (notably Kekūanāoʻa, president of the board of education from 1860 to 1868) contested and resisted this effort: working to separate schools from church affiliation, promoting literacy in Hawaiian, and working to ensure that the common schools (free primary schools serving working-class Kanaka and Asian children) received adequate funding. This effort held off the hegemony of "white supremacist and assimilationist models of schooling" until the late 1870s and early 1880s. It received strong royal support in 1883, when King Davida Kalākaua forced the resignation of Charles Reed Bishop, a superintendent of schools who was shifting the school system to industrial education (preparing Kānaka for subservient status in the sugar-dominated economy) and working to replace Hawaiian with English as the language of instruction (preparing Kānaka for American linguistic, cultural, and political domination). But the ascendancy of the Haole missionary, commercial, and plantation forces meant that Bishop regained his position in 1887. Six years later, those same forces overthrew Queen Liliʻuokalani and declared a republic. In 1896, Hawaiian-medium instruction was effectively banned (and remained so until the 1980s). In addition to depriving children of schooling in their own ancestral language, it shut out most Kanaka teachers.[4]

This final effect is most important to this chapter, because evidence exists that Hawaiian adults had long played an essential countervailing and mediating role in the Hawaiian schools, protecting keiki maoli (Native children) from the worst of the increasingly colonial educational system in Hawaiʻi. Goodyear-Kaʻōpua has noted the efforts of Kanaka administrators and officials to hold off the missionary and sugar capitalist initiatives in education. Those initiatives would include publishing geography textbooks that taught Kānaka that they occupied an inferior and marginal place in world geography. However, a close examination of those same textbooks reveals fissures that left space for Kānaka to imagine alternative world geographies that centered Hawaiʻi and Hawaiian perspectives—and those fissures were most likely an artifact of Kānaka working on translating textbooks from English into Hawaiian. By the 1870s, that role as translator/mediator was threatened, as a new geography textbook was prepared by a Haole administrator with little or no apparent Hawaiian influence. Even then, however, there is evidence that Kanaka teachers interposed themselves between their students and this colonialist geography text, using it only selectively to teach lessons that gave them a sense of Hawaiʻi as an important place in the world in its own rights. Kanaka teachers remained essential to the resistance against colonialism even after the imposition of English-language instruction removed so many of them from the classroom. Teachers like Kānepuʻu were among the leaders of movements working in the late nineteenth and early twentieth centuries to defend Hawaiian sovereignty from Haole usurpation of power and annexation: to shape the place of Hawaiʻi in the world.

Geography textbooks, the way geography was taught, and the work of Kanaka teachers give us new insight into contestations over the politics of what Mark Rifkin calls "the imperial construction of space" in the nineteenth century.[5] From the time of the founding of Western-style schools in 1831 to the overthrow of the Hawaiian Kingdom in 1893, Kānaka Maoli and missionary educational authorities forwarded different geographies composed of imperial spaces of global hierarchy and national spaces of Hawaiian sovereignty embedded in an oceanic sea of islands.[6] Geography is understood in this chapter, as it was in nineteenth-century Hawaiʻi, capaciously, often including much that we might today consider history or politics.

This chapter argues that geographical education was a crucial site of contestation between missionary and planter-aligned Haole educationalists and Kanaka educationalists. Both groups sought to teach young Kānaka their place, both literally and figuratively, but they had very different visions of what that place should be. That is to say, missionaries who translated textbooks and established curricula intended geography to teach Kānaka the spatial organization of the globe, but also that they occupied a particular place in the global civilizational hierarchy of nations, people, and races. According to these missionary accounts, Hawaiʻi's place was neither at the bottom nor at the top of this civilizational hierarchy. It fell below the level of Europe, the United States, and (in some accounts) parts of East Asia, but Kānaka Maoli stood above the level of Africans, American Indians, South Pacific Islanders, and indigenous Australians. Moreover, it taught Kānaka that Hawai'i needed the guidance of missionaries and other Haole (that is, they needed colonial guidance) to move upward in the civilizational hierarchy. For all the colonial intent of this geographical schooling, Kānaka built several avenues of resistance to it, and in many cases, they built avenues of resistance *through* colonial schooling. Kanaka administrators and officials worked directly to shape the educational structure of the islands to serve a sovereign Hawaiian Kingdom. They were not, however, always successful in doing so, especially as missionary and planter hegemony was established in the years stretching from the 1870s forward and Haole administrators used education as another tool in what Jonathan Kamakawiwo'ole Osorio has rightly termed the "dismemberment" of the Hawaiian nation.[7] Yet less visible avenues of resistance to colonialism (and embrace of sovereign and decolonial learning) existed from the start of the educational system in the kingdom. Because Kānaka were essential to the preparation of educational materials and to teaching in classrooms across the archipelago, they had room to maneuver. They used this space to teach young people about the world from indigenous rather than colonial perspectives. Indeed, evidence exists that Kanaka writers and translators of textbooks and Kanaka schoolteachers taught geography in ways that diverged significantly from the ways the missionaries who ran the school system intended. This went beyond resistance to become what Scott Lyons terms a "Native signature of assent."[8] Kānaka Maoli embraced self-consciously "modern" modes of discourse (cartography and social scientific geography, in this case) for Native purposes.

GEOGRAPHY: REQUIRED READING FOR KANAKA STUDENTS

Geography was taught to students in Hawai'i's schools from the 1830s forward because geography was a priority not only for missionaries, but also for Kānaka as they struggled with pressures from colonial powers and Haole among them. In the face of those pressures, ali'i (chiefs) and other Kānaka demonstrated a commitment to teaching world geography to young people so that they could understand the world with which

Hawai'i would have to contend. From its beginnings in the 1830s, geography was part of the institutionalized system of education for Native teachers and students. The power of mapping and geographical description was one that was wielded against Kānaka, but also one that Kānaka seized for themselves. Critical cartographers recognize that maps are not neutral and objective representations of the world; they are instruments of power that represent the world in ways that reflect the agendas and interests of their makers. The same can be said of geography textbooks and atlases. In nineteenth-century Hawai'i, all of these kinds of geographical documents for school use reflected the Eurocentric, racist, and often colonialist agendas of white school authorities—agendas that included teaching Kānaka that they were lower than Europeans and Americans on civilizational hierarchies and therefore needed Haole intervention and guidance to survive, become modern, and thrive. But B. Kamanamaikalani Beamer and T. Kaeo Duarte make the essential point that although mapping was frequently a tool of colonialism, the nineteenth-century mapping of the internal boundaries of Hawai'i can be understood as an artifact of Kanaka agency. Ali'i dedicated energy and resources to the mapping of lands they worked to protect from colonialism. This same argument must also be extended to maps and works of descriptive geography that looked far beyond the shores of Hawai'i. Geography education was a site of contestation, but also a tool that Kānaka wielded in defense of their lāhui (nation).[9]

The first schools created in an educational surge in the 1820s had emphasized teaching literacy, mostly to adults. Beginning in the 1830s, missionaries and Kānaka established more enduring institutions with the schools at Lāhaināluna on Maui[10] and then at Hilo on Hawai'i. These boarding schools were intended to train Native men to become ministers and teachers. Later, a female seminary, meant to train Native women to be what the missionaries believed would be proper wives and assistants for male ministers and teachers, was established at Wailuku on Maui. Lāhaināluna, Hilo Boarding School, and the Wailuku Female Seminary formed the foundation of what the government would come to term in English the "select" schools. These "select" schools were soon joined by the Hale Kula no nā Keiki Ali'i (the Chiefs' Children's School, later known as the Royal School). Located first in the home of the missionary teachers who ran it and then in a separate building on the present site of the barracks of 'Iolani Palace in Honolulu, the Chiefs'

Children's School was the site of Western instruction for all of the mōʻī (monarchs) to follow Kamehameha III and many of the most important aliʻi nui (high chiefs) of the nineteenth century.[11]

Having created Western-style schools to train what the missionaries imagined would be Hawaiʻi's future preachers, teachers, political leaders, and the women who would marry them, the missionaries and the kingdom looked to the provisioning of schooling for the great majority of the populace. What emerged was a public school system of what was generally termed "na kula aupuni" (the government schools) in Hawaiian and "common schools" in English. The very design of this tiered system meant that far more Kānaka Maoli attended the common schools than the select schools. Common schools have received, however, less attention from historians than Lāhaināluna, the Royal School, the other select schools, and the Kamehameha Schools, which opened in 1887—perhaps because the select schools and the Kamehameha Schools trained well-known figures in Hawaiian history and also are better documented in archival and print records.

Geography had a place at the core of the plans for all of these schools. Lāhaināluna set the pattern. In June 1831, when the missionary members of the Sandwich Island mission met to plan this "High School for Teachers," they made an acquaintance with "the fundamental principles of geography" a requirement for admission, along with fluent reading and writing in the candidate's language (meaning Hawaiian for all the students of the time) and knowledge of "common arithmetic." To the school's founders, then, principles of geography were just as essential a foundation for further learning as reading, writing, and arithmetic. Nor did they consider the basic understanding of geography with which students were to enter to be sufficient for their future work as teachers or ministers. Geography joined religion and the "three r's" to make up the five subjects of instruction that the planners intended to teach at Lāhaināluna.[12] The same was true at the select schools that were founded later. The Wailuku Female Seminary, which the missionaries founded in 1836, started small but grew, with pupils averaging fifty per year after a decade. Although it emphasized teaching American domestic skills to its pupils, Minister of Public Instruction Richard Armstrong noted that it also trained them in "the common rudiments of education, including Mental and Written Arithmetic, Geography &c."[13] Similarly, although the Hilo Boarding School would eventually come to function as a manual training school, it also included geography as part of its broader curriculum. Founded in 1836, the school sent some of its students on for further training at Lāhaināluna. Its upper division included the geography that Lāhaināluna required for admission in its course of study, along with "mental and written arithmetic, algebra and music."[14] Geography, then, was a fixture at the Lāhaināluna, Wailuku, and Hilo schools, which were among the first founded and longest lived of the select schools. Reports on the curriculum of other select schools for Kānaka Maoli—including among others Makawao on Maui and Waiʻoli on Kauaʻi—suggest that administrators there also placed emphasis on teaching geography.[15]

The Chiefs' Children's School similarly trained its students in geography, as revealed by the Kanaka investment in teaching geography to young Kānaka. The sixteen male and female students who studied there in the mid-nineteenth century were among the most highly ranked individuals of their generation, among them the aliʻi nui Victoria Kamāmalu and Bernice Pauahi, and the children who would grow to become the mōʻī King Kamehameha IV, King Kamehameha V, Queen Emma, King Lunalilo, King Kalākaua, and Queen Liliʻuokalani. The missionaries wanted them to know English. This was in keeping with their larger goal of encouraging the leadership in Hawaiʻi to adapt itself to Western perspectives and Western power. It was, however, also in keeping with the urgent desire of aliʻi to train young leaders to preserve Hawaiʻi's own interests into the future—as Beamer has argued, "not to *Americanize* these keiki" but rather to "*internationalize* them."[16] Because this was a priority for the aliʻi, Kānaka Maoli placed education in the service of their own goals even in what would appear at first glance the most colonial of circumstances. Juliette Montague Cooke and her husband, Amos Starr Cooke, directed the efforts of the Chiefs' Children's School, but it is essential to remember that they did so at the behest of Kamehameha III. The Cookes dedicated much of their efforts to teaching their high-ranking students to adapt Western viewpoints in the classroom and Western domesticity in the schoolhouse's parlor.

Indeed, the parlor was an important site of education where royal students and American teachers gathered—but even there, the aliʻi parents took care to ensure that Kanaka perspectives remained central. In the 1840s, it featured what Juliette Montague Cooke described as "a splendid large map of the world"—a gift of Kamehameha III.[17] The gift of the king ensured that even in domestic settings, Hawaiʻi's young leaders would gain the global geographic perspective that they and the kingdom would need.

GEOGRAPHY IN THE COMMON SCHOOLS

Geography was also taught in the common schools, the schools that the vast majority of Hawaiians would attend and that were overwhelmingly staffed with Kanaka Maoli teachers. These schools were physically far humbler than the select schools, all of which were boarding schools. Select schools like Lāhaināluna and Hilo Boarding School boasted coral block or frame buildings to accommodate classrooms and housing for (mostly) missionary teachers and Kanaka Maoli students. In contrast, classes in the common schools met in what the president of the papa hoʻonaʻauao (board of education) reported in 1856 were "still nothing but miserable grass huts, without floors, windows or furniture, and some of them even in a great degree without thatch."[18] These were the schools in which the vast majority of Kanaka Maoli children were educated. In 1853, for example, 13,948 students were enrolled at 440 common schools—far more than the few hundred who were enrolled in all of the select schools combined. Common school students, like select students, received schooling in geography. In 1850, for example, 4,435 (28 percent) of the 15,620 students in the common schools received schooling in geography.[19] In 1853, the number was 4,229 (30 percent) out of 13,948.[20] In 1856, the number was 3,513 (35 percent) out of 10,076.[21] Through the middle of the nineteenth century, then, roughly a third of the students in common schools each year were receiving teaching in geography. The actual number of students exposed over the course of their schooling is far higher, because geography was being taught in certain grades and not in others, and some teachers emphasized geography more than others.

The teaching of geography in the common schools was hampered by a lack of funding in general and a lack of access to instructional materials in specific. In 1853, the minister of public instruction reported, "Of Atlases there are comparatively few in use, consequently the study of geography is in a great measure suspended."[22] In other words, a lack of texts disrupted the teaching of geography, and presumably limited what teachers could accomplish. Yet judging from the minister's own reports, the paucity of atlases did not put an actual end to the teaching of geography, even if it did slow it down. School reports continued to list the number of students studying the topic for years to come.

Artifacts of Educational Struggle: Atlases and Textbooks

Hawai'i's nineteenth-century geography textbooks are an essential archive: their pages are the hidden transcript of struggle over the power to define what the place of Hawai'i in the world should be. First of all, there is reason to believe that these books were more common in the better-provisioned select schools at Lāhaināluna and elsewhere where most of the teachers were taught. The content of the books may then have shaped the contents of the teaching that these teachers did, no matter what materials they had at hand once they were working in a common school. Secondly, we have already seen (in the last chapter) the intensity of the engagement with books through the processes of group reading that befuddled Haole and the extensive memorization that astonished them. Thus, though books may have been few at common schools, group reading meant that many Kānaka experienced them directly. Finally, the textbooks have the singular advantage of existing: textbooks have been preserved in libraries and archives, whereas vanishingly few examples of other potential sources about what happened in Hawai'i's classrooms (notably lesson plans and lectures) are available. Early atlases and textbooks therefore matter for our understanding of geographical education in nineteenth-century schools in Hawai'i— even though, as the second section of this chapter will argue, we cannot take for granted that what was printed in the book is what was taught in the classroom. In fact, textbooks reveal struggle over the content of geographical education that is evident in the texts themselves, in the ways they were used in classrooms, and in teachers' other writings and political activities. By transforming reading, writing, texts, and literate knowledge into an object called ka palapala (literacy and literate knowledge), Kānaka Maoli transformed foreign technologies and ideas into a thing that Native people could take up for their own purposes in a changing world. Geography textbooks and student atlases were one of the most important tools in that apparatus and became a site of struggle between Haole missionaries and Kanaka Maoli teachers.

When the minister of public instruction complained that teaching geography was hampered by a scarcity of atlases, the book he was referring to was almost certainly a slim volume titled *He Mau Palapala Aina a me na Niele e Pili Ana* (Maps with Questions Regarding Them), which was published at the missionary press at Lāhaināluna in 1840. The book featured ten pages of hand-colored maps, one page comparing the lengths of the world's greatest rivers, and twelve pages of questions to be used in map study—what nation could be found southwest of France, the approximate latitude of Lima, and so forth.[23]

Perhaps the most important thing to know about the *He Mau Palapala Aina* atlas is that it was Kānaka Maoli, not missionaries or other foreigners, who engraved the maps that illustrated it. The first atlas in Hawaiian representing the world in the Western cartographic code of image and word to Kānaka was made by Kānaka themselves. The maps bear their names: Kapehoni, S. P. Kalama, and others.[24] Nor was this the first time that Kānaka had made engraved world maps. In 1834, a fourteen-year-old boy named Kawailepolepo had engraved on copper plates a world map for printing on the missionaries' press.[25] The map and the atlas follow European and American models. Kawailepolepo worked from a map already present at the mission, and the Kānaka who engraved the atlas maps followed similar models. More important still, the documents follow European and American norms; they follow the visual vocabulary of Western cartography: north at the top, latitude and longitude lines, printed lines to separate nation from nation and land from water, and so on. All of these Western conventions would still have been novel to many Kānaka. Thus the printing of a map and the publishing of an atlas meant much more than the mere creation of pedagogical tools. It also meant the teaching of a particular kind of representational apparatus. The maps and atlases that Kānaka produced at Lāhaināluna were not the first ones in the islands—Western captains not infrequently provided Hawaiians with maps of the world. But they were the first ones made in the islands, and the first specifically made for Kānaka. What is more, they were produced by the labor of Native people. Pointing this out is not to imply that they worked entirely of their own initiative or under their own direction; clearly, missionaries were directing these efforts. But it is important to note that Kānaka were active participants in the first acts of Western-style world mapmaking in Hawai'i. Geographer Bernardo Michael has argued that maps are not simply the artifact of mapmakers, because "subaltern agents" "contest the simplistic representations of space and social relationships afforded by maps."[26] But here we have an even more complex and dynamic picture: Kānaka are not just the intended audience for maps and geography books; they are also coauthors of them. We must then be particularly attuned to what Michael terms their "cartographic agency."

He Mau Palapala Aina was not the first teaching tool for student use in geography in Hawai'i. It followed by eight years the 1832 publication of the first Hawaiian-language geography textbook, *He Hoikehonua, He Mea Ia e Hoakaka'i i ke Ano o ka Honua Nei, a me Na Mea Maluna Iho* (A Geography: A Description of the Nature of the World and of the Things upon It). Whereas the atlas, *He Mau Palapala Aina*, was a slim volume of maps and related questions, *He Hoikehonua* was a full-length geography textbook consisting of over two hundred pages illustrated with engravings. The missionary press produced 5,500 copies of the 1832 edition, 10,000 copies of the 1836 edition, and 10,000 copies of an expanded and more heavily illustrated 1845 edition—a total of 25,500 textbooks in thirteen years.[27] Rounding out the general geography textbooks circulating in Hawaiian in the mid-nineteenth century was *He Vahi Hoikehonua: He Mea Ia e Hoakaka'i ke Ano o ka Honua Nei* (A Small Geography: An Explanation of the Nature of the World). A slim volume of thirty-odd pages with neither maps nor illustrations, it was published by the Catholic missionary press for use in the Catholic schools.[28] The common schools in Hawai'i were divided into Protestant and Catholic schools, both of which received financing from local communities and governmental support. Given the greater number of Protestants, and the favor upon which the missionaries and government looked upon them, there were many more Protestant than Catholic schools. For this reason, and because the Catholic student atlas was so brief, this chapter emphasizes the teaching materials used in the Protestant schools.

He Hoikehonua and *He Mau Palapala Aina* were both closely based on English-language sources published in the United States. While the sources for the Catholic *He Vahi Hoikehonua* are not clear, portions of it resemble both *He Hoikehonua* and *He Mau Palapala Aina*. The principle source for the textbook *He Hoikehonua* was the most important American geography textbook of its day, William Channing Woodbridge's *Rudiments of Geography on a New Plan, Designed to Assist the Memory by Comparison and Classification*.[29] The title is important. Woodbridge was a professor of geography at Gallaudet, the innovative school for the

Title of atlas or textbook	Year first published	Major foreign-language source text	Notes
He Hoikehonua (A geography)	1832	Woodbridge, Universal Geography	World geography textbook, advanced level, published by the American Protestant missionary press
He Mau Palapala Aina a me na Niele e Pili Ana (Maps with questions regarding them)	1840	Olney, A New and Improved School Atlas	Student atlas published by the American Protestant missionary press
He Mau Palapala Aina a me na Niele no ka Hoikehonua (Maps with questions for geography)	1840	Hall, Child's Book of Geography	Student atlas published by the American Protestant missionary press
He Vahi Hoikehonua (A small geography)	1842	Perhaps Gobinot, Petite géographie élémentaire par demandes et réponses	Questions and answers about world geography, produced by the French Catholic missionary press
Ka Honua Nei (The world)	1873	Hall, Our World	World geography textbook, primary level, published by the American Protestant missionary press

Atlases and geography textbooks in Hawaiian schools in the nineteenth century.

deaf at Washington, D.C. Woodbridge was disgusted with most geography textbooks of the day, which he said merely presented "a mass of insulated facts, scarcely connected by any association but that of locality."[30] (Note that the pejorative term "insulated" preserves in its Latin etymology a scorn for islands—*insulae* in Latin—reminiscent of what Epeli Hauʻofa has termed "belittlement.")[31] Woodbridge's book favored (as the title put it) "comparison and classification." In brief, his textbook taught that one could classify nations into groups (no country was altogether unique) and compare them on a number of levels—including (as the titles of some editions of the work said) "degrees of civilization." In affecting the allegedly objective voice of the scientist, Woodbridge deployed a claim to the authority that science was coming to confer among Americans and Europeans in the nineteenth century.

The philosopher Thomas Nagel aptly captures the geographical power of this objective stance when he calls it the view from nowhere—a view that pretends to see the world from a stance outside of the world. Nagel criticizes the view on the grounds that "we and our personal perspectives belong to the world" and that denying this reality makes it impossible for allegedly objective descriptions of the world to truly describe it fully.[32] Perhaps even more important for understanding *He Hoikehonua*, to claim that one's own view is a view from nowhere (that is, to deny that one has a perspective) is to obscure and thereby enforce the power of the privileged scientist over the less-privileged "object of study." This is a particularly potent tool in colonialized spaces, as geographer Karen Piper has argued.[33]

This "view from nowhere" is, of course, diametrically opposed to the perspectivalism of Hawaiian geographical thought—a perspectivalism that named the north and south from the point of view of the sun, that described places as being in relative space such as ma uka and ma kai (toward the uplands or toward the sea), and so forth. This claim to geographic scientific objectivity thus made Woodbridge a useful source text for the missionaries in charge of preparing, printing, and distributing the *He Hoikehonua* textbook. In translation, his claim to authority became theirs and buttressed the other claims to power that they made (and believed): that their Christian faith and their Western civilization entitled them to authority over Kānaka. Remember that Woodbridge's textbook (and *He Hoikehonua*, the major Hawaiian-language geography textbook) was based on "the principles of comparison." To Woodbridge and the missionaries who drew upon them, western European nations could be empirically demonstrated to be the most civilized of nations. Although it was commonly referred to as Woodbridge's *Universal Geography*, the book presented a profoundly *Western* geography that obscured its colonialism behind the claim of universalism, a claim that was then translated—in the etymological sense of being "moved laterally"—to Hawai'i by the missionaries at Lāhaināluna.

COLONIALISM IN *HE HOIKEHONUA*

The Hawaiian-language text reveals how the missionaries intended geographical education to support their effort to establish their authority and the need of Kānaka Maoli for them. That effort is clear in the textbook's depiction of climatic zones and their correlation to human characteristics. Translating from Woodbridge, the text first introduces the notion of the globe and then explains that it is bisected by a pō'aiwaena (middle circle, or equator).[34] There lie two lines north and south of the equator: the "poai olu akau" and the "poai olu hema"—the north temperate circle and the south temperate circle (in English, the Tropics of Cancer and Capricorn). Between those two lines lies the "kaei wela" (hot zone) as the book translates "tropics." *He Hoikehonua* informs its readers that the climates and the people of the hot zone and the temperate zone could not be more different. In the tropics, the book claims, most of the time the weather is terribly hot, and the people suffer from "mai luku"—deadly diseases that kill many people. This is partly the result of climate, but the text also suggests that it is a result of the organic inferiority of tropical peoples: "Ua eleele ka nui o kanaka, ma ia kaei, a ua molowa, ikaika ole ka nui o lakou" (Most of the people are black [or dark] in this zone, and are lazy, and most of them are not strong).[35] For its part, the Catholic textbook *He Vahi Hoikehonua* nearly reproduces this wording, writing insistently: "Ua eleele ka nui o kanaka malaila, a he poe molova ka nui o kanaka malaila" (Most of the people there are black [or dark], and most people there are a lazy people).[36] The Protestant textbook, being longer, takes the time to elaborate on the point: "He ikaika no ka makemake, aka, ua pau koke ke aloha, aole kamau aku i na hana, a me ka imi naauao" (Their desire is strong, but their love is soon finished, they do not persevere in their efforts and in seeking na'auao [knowledge]).[37] Note that people of the "hot zone" are defined by what they supposedly lack (physical strength, perseverance in learning and working) rather than what they are—the very definition of a negative depiction. This is in keeping with the ways that Robert Berkhofer demonstrates that, over the centuries, white Europeans and Americans defined American Indians by the qualities they allegedly lacked. The missionaries brought this well-established Euro-American discourse to Hawaiʻi and attempted to teach it in their geography textbook. Having established (at least to its own apparent satisfaction) the innate inferiority of residents of the kāʻei wela, on the next page the textbook invites Kānaka Maoli to situate themselves in this lacking zone inhabited by lacking people: "Heaha ke kino o kanaka o ia kaei? Pehea ko lakou ano? . . . Owai la na moku o ka moana Pakifika ma ia kaei?" (What are the bodies of people of this zone like? What is their character? . . . Which islands of the Pacific Ocean are in this zone?).[38] In response to this final question, Kanaka students would find that their islands are located squarely in what the textbook described as a zone of disease, weakness, and sloth. Making clear the difference between the origins and nature of Europeans and those of Kānaka, the textbook describes the cooler climate zones as more moderate, invigorating, and healthful than the tropics. Indeed, both the Protestant and Catholic textbooks translate "temperate zone" as "kaei olu," which more literally means a belt (of land) that is cool, refreshing, and pleasant. *He Hoikehonua* pointedly asks, "O Europa anei kekahi ma ke kaei wela?" (Does Europe have any [lands] in the hot belt?).[39] The answer, clearly, was no. *He Hoikehonua* explains that the white people of the temperate zones were stronger than inhabitants of the tropical zone. They were enlightened in their literacy, skilled in their labors ("akamai lakou i na hana"), and knew how to do all kinds of things. Whether the Catholic textbook draws on the Protestant *He Hoikehonua* or merely on similar racial theories is not clear, but the point it makes is similar though more concise. In explaining why the zone is termed ʻolu (cool, refreshing, and pleasant), it suggests that is it not just the weather of the zone that is better. So are living things, including people. "Olu maikai kela hua ai keia hua ai; a ua nui ke kino o kanaka, he ikaika hoi a he keokeo ka ili" (All the fruits and seeds are pleasant; and people's bodies are big and strong, too, and their skin is white).[40] In both the Protestant *He Hoikehonua* and the Catholic *He Vahi Hoikehonua*, descriptions and questions encouraged Kānaka to situate themselves in space but also in a social hierarchy—distinctly below white people.

The system of global geography in *He Hoikehonua* situated Hawaiians in a broader hierarchy of degrees of civilization, not just in a binary system of difference with Europeans and their American descendants. As the Kanaka historian Kealani Cook has noted, the textbook emphasized to Kānaka Maoli that the world was inhabited by five lāhuikanaka. The word "lāhui" corresponds to several meanings in English (including nation, tribe, people, and others), whereas "kanaka" means person or people. Here the translators were conveying the word "race," the term used in the passage in Woodbridge's *Rudiments of Geography* that was the source of this passage in *He Hoikehonua*.[41] Following that text, *He Hoikehonua* presented Kanaka students with what white Americans were in the early nineteenth century declaring to be the state of the art in the scientific categorization of humanity. In this system, familiar up to the present day, there are five races on the earth, differing in the appearance of their bodies and named in reference to their place of origin. The races are "ko Europa poe kanaka" (Europe's people), "ko Asia poe kanaka" (Asia's people), "ko Amerika poe Inikini" (America's Indian people), "ko Malae poe kanaka" (Malaya's people), and "ko Aferika poe kanaka" (Africa's people). Large sections of the book are dedicated to giving Kanaka students the material to fit these races into the civilizational hierarchy. The socially determinative climatic zones described above were only the beginning of the ways the textbook constructed hierarchy through world geography. Students would also read about a three-part hierarchy of "Ko Kanaka Noho Ana" (People's Ways of Living). Some peoples were "naaupo" (ignorant). The first measure of this was that they did not know how to "malama i ka aina" (cultivate the earth). Describing them as lazy hunters or "he lawaia wale no" (just fisherfolk) who did not farm, *He Hoikehonua* reported "aole noho malie ma kahi hookahi, o ka hele wale no ko lakou makemake" (they do not live peacefully in one place [but rather] just go wherever they want). Most of all, they do not worship God. Having created the discursive category of the uncivilized, the text next summons the semicivilized into being, telling readers "he hapa ka ike o kekahi poe" (some people have partial knowledge). Naming Asians as part of this category, the text declared that they know how to work and some are literate, but they worship images and mistreat their women. Finally, the text reports on (and thus simultaneously generates as a category) civilized people: "He naauao kekahi poe" (Some people are naʻauao). These are the enlightened ones who are skilled in work, learned in ka palapala, knowledgeable in agriculture, peaceful in their good conduct under the law, and pious in their worship of the one true God.[42] In the context of the rest of the book, it is clear that here the text is referring to "ko Europa poe" (the European people).

Again, the purpose of teaching hōʻikehonua (geography) was, from the missionary perspective, to teach Kānaka their position in the world. The text made clear what the missionary educators thought was the appropriate position of Hawaiʻi vis-à-vis the United States and Europe. "No ko lakou naauao a me ko lakou pono, lanakila no lakou maluna o ko lakou poe enemi; no ka mea, o ka naauao ka mea e ikaika ai ko na aina" (Because of their enlightenment and their righteousness, they [white people] are victorious over their enemies, because enlightenment is the thing that makes the people in those countries strong).[43] The message to Kanaka students was clear: resisting American and European influence was futile. Resistance would make the Kānaka a "poe enemi" (enemy people) of the Haole, who were assured victory by their knowedge and righteousness. The fact that the translators resorted to a word of English origin, "enemi," is significant. The other Hawaiian words that come closest to corresponding to "enemy" are hoa paio, hoa kaua, and hoa pāonioni—meaning antagonist in a quarrel, antagonist in a war, or antagonist in a disagreement. The modifiers to "hoa" (which unmodified can simply mean companion, friend, partner, etc.) each suggest a situationally specific opposition—one is an antagonist in a particular circumstance.[44] In contrast, the use of the imported term "enemi" declared an implacable hostility between the Americans and Europeans on the one side and those who dared to oppose them on the other.

Haunted Maps: Centering Kanaka Perspectives in Maps and in Books

Kānaka Maoli could find themselves in Hawaiian-language geography texts in a number of surprising ways that subverted the geography textbook's message of Hawaiʻi's inferiority and marginality. As Piper notes, "colonial cartography... never quite suppresses alternative forms of territoriality, which continue to haunt the map."[45] The first and most telling example of this is in the organization of the 1840 atlas *(He Mau Palapala Aina a me na Niele e Pili Ana)* and the various editions of the *He Hoikehonua* textbook. The first two pages of the first atlas ever published in

Hawaiʻi present a map of the hemispheres, with the world centered on the Pacific Ocean. Hawaiʻi therefore is the landmass closest to the center of the image. The second map in the atlas, titled "Aina Moana," covers the entirety of the Pacific Ocean: north to the Bering Strait, southwest past Australia, and southeast to Tierra del Fuego. Again, this map is centered on Hawaiʻi. While it names the islands of the Pacific in great detail, it gives almost no information on the geography of the surrounding continents beyond the coasts. The third map in the atlas is a map of the principal Hawaiian Islands, showing the landmass of the islands but also internal political divisions (districts and towns). Only after these three maps, each centered on Hawaiʻi, does the atlas present a map that does not show Hawaiʻi: it is a map of the United States.

Given the visual vocabulary of the "view from nowhere" that these maps deploy, centering Hawaiʻi in the book's first maps is as close an approximation as possible to a perspectivalist view out from the shore of Hawaiʻi. All maps depend upon imagination. The question is, what imagining of the world will a map facilitate? No person could in fact ever have the kind of view that is presented in maps we are familiar with today: the flattening of the curve of the globe and the removal of the contours of the landscape make their view imaginary. But by placing Hawaiʻi at the center of its first visual representations of the world, the first maps in the *He Hoikehonua* textbook facilitated an imagining of the world as seen relative to Hawaiʻi.

Similarly, the linear organization of the *He Hoikehonua* textbook centers a Hawaiian perspective. The book opens with thirty-five pages of general and terminological discussion—latitude and longitude, waterfalls and volcanoes, the discussion of the races described above, and so forth. At that point, the book begins a region-by-region discussion of the world that is mostly focused on major nations. Tellingly, the first region discussed bears nearly the same title, "Na Ainamoana," as the map of the Pacific in the *He Mau Palapala Aina* atlas. This, of course, is the region of which Hawaiʻi is a part. Furthermore, following a one-page general discussion of the islands of the Pacific, there appear ten pages describing the Hawaiian Islands, including island-by-island accounts of the terrain and climate of each of the eight largest islands. This same organization is retained in all of the later editions of the *He Hoikehonua* textbook. Thus both the atlas and the geography textbook begin their portrayals of world geography by looking at Hawaiʻi.

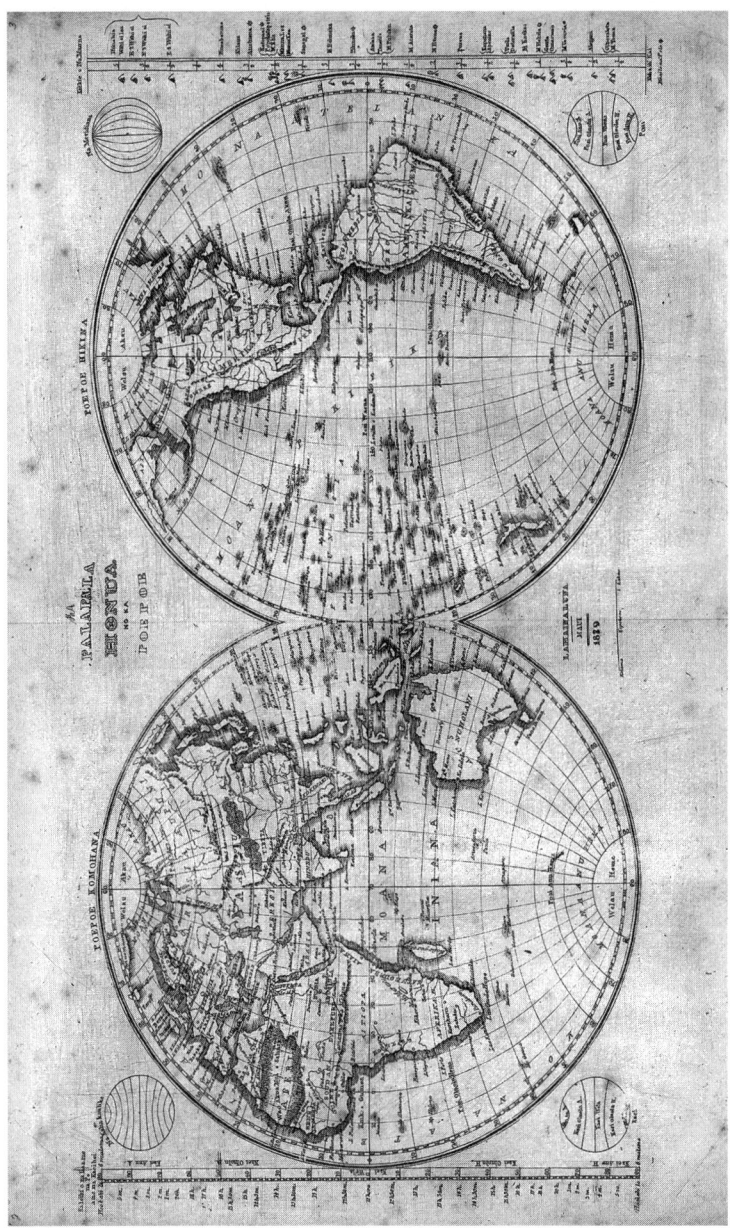

Map of the world in two hemispheres, with the Pacific at center and Hawaiʻi just right of center. Engraving by Kalama and Kapehoni in *He Mau Palapala Aina a me na Niele e Pili Ana,* 1840. Courtesy of the Library of Congress, Geography and Map Division.

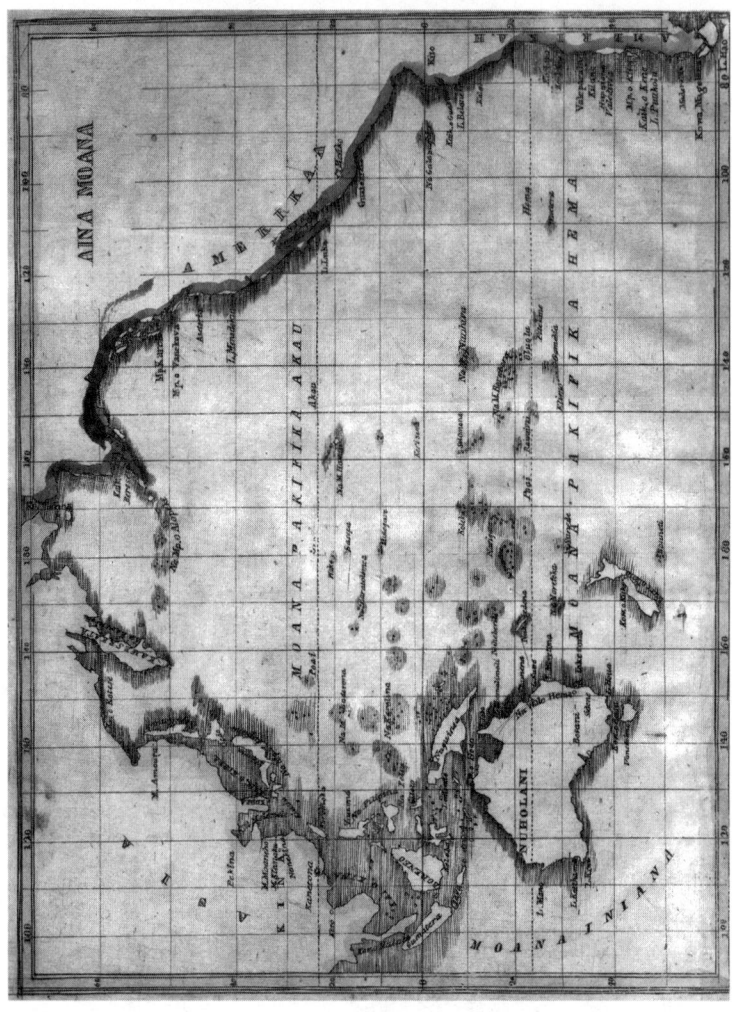

"Aina Moana," a map of the Pacific with Hawai'i at the center. Unsigned engraving from *He Mau Palapala Aina a me na Niele e Pili Ana*, 1840. Courtesy of the Hawaiian Mission Children's Society Library, Honolulu.

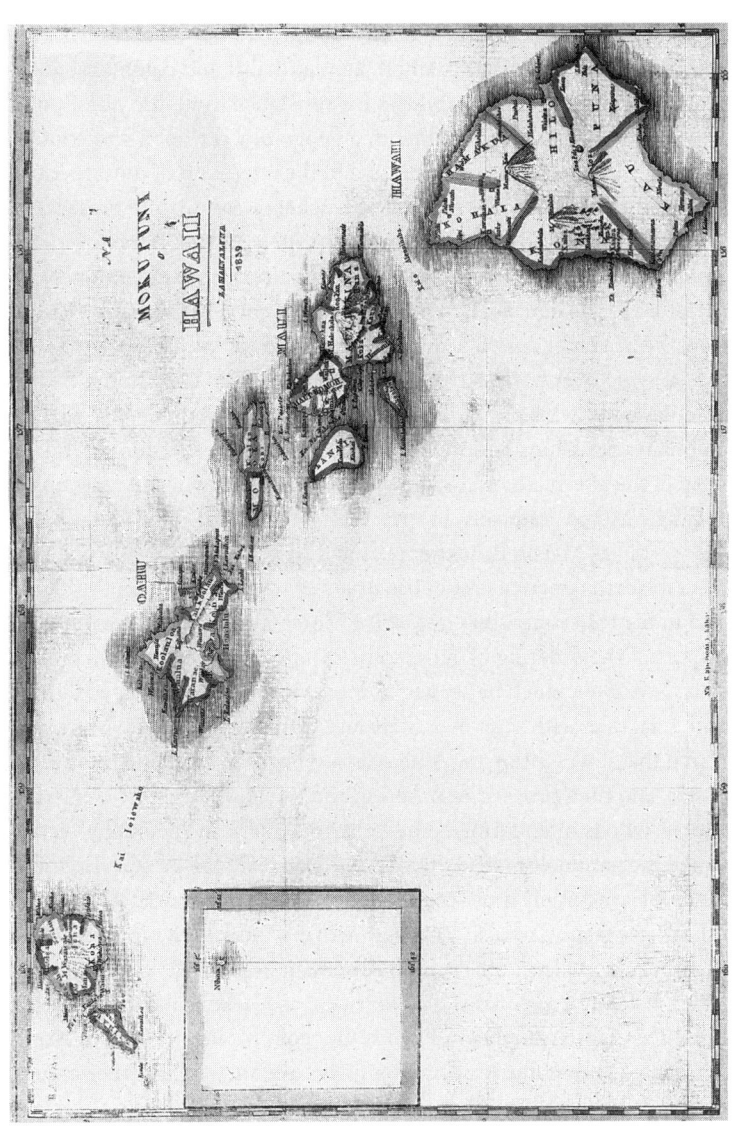

The Hawaiian Islands. Engraving by Kapehoni in *He Mau Palapala Aina a me na Niele e Pili Ana,* 1840. Courtesy of the Hawaiian Mission Children's Society Library, Honolulu.

This decision cannot be taken for granted, and the politics of it are complex: it both replicated American nationalist and colonialist geography and opened up possibilities for resistance to it. The question of what country would appear first in a geography textbook was contentious in the early American republic. In the aftermath of the American revolution, school boards and schoolteachers were eager to present the world from a self-consciously "American" perspective. Efforts to do so were made in any number of manners, from the celebration of American literature to the glorification of American history. In geography education, the placement of the United States in textbooks came to symbolize the struggle over perspective. The geography texts used in the British colonies had always presented Europe first, and then proceeded to their colonial possessions and the remainder of the non-Western world. In 1784, in the aftermath of U.S. independence, clergyman and geographer Jedidiah Morse "radically inverts the geographical representation of the world," as Martin Brückner puts it.[46] Morse's *Geography Made Easy* placed North America first in its survey of world regions. In this book, and in his 1789 *American Geography,* Morse made clear the nationalist import of the ordering of geography lessons. "Our young men, universally, have been much better acquainted with the Geography of Europe and Asia, than with their own State and Country." No more: Morse declared that it was fitting for Americans to begin by studying the United States, and then proceed with lessons on foreign lands.[47] In doing this, Morse, who is often termed "the father of American geography," established the nationalist standard for American textbooks. Later American textbooks, including those that served as sources for Hawaiian-language school geography books, followed Morse's model, giving the United States pride of place. This was an American imperialist standard in addition to being a nationalist one, for two reasons. First, it naturalized the presence of the United States as a settler colonial nation-state in North America—taking imperialism as a given. Second, as Brückner argues, early U.S. geography texts were teaching American students a "literacy for empire."[48] In their internal organization, by placing Hawai'i first, *He Mau Palapala Aina* and *He Hoikehonua* thus both reproduce the nationalist, colonialist, and imperialist organization that Morse established for the United States: placing the nation of their intended audience first.

Paradoxically, this imposition of the nationalist knowledge organization of the American texts centers a Hawaiian perspective in these books. In the *He Hoikehonua* textbook, the act of describing Hawai'i first means that all subsequent descriptions—of the United States, of Europe, of Asia and the remainder of the world—are viewed in a sense from a Hawaiian perspective. The apparatus of the text, numbered as to section and page to suggest a linear order of reading, directs the reader first to learn about Hawai'i, and only then to proceed from Hawai'i to learn about the rest of the world. Similarly, by opening with three maps centered on Hawai'i, the *He Mau Palapala Aina* atlas privileges a Hawaiian perspective on the globe—even while being a tool of missionary education. These were probably not the goals of the missionaries that oversaw the production of these important texts of geographical education. But arranged as they were in a way that mirrored (and thus reversed the vision of) American student atlases and geography textbooks, the Hawaiian-language books introduced a Hawaiian perspective that operated contrary to the explicitly racist and colonialist messages that appear throughout *He Hoikehonua*.

READING KA PALAPALA UPSIDE-DOWN: NATIVE PERSPECTIVES ON GEOGRAPHY TEXTS

Furthermore, the authors of the book chose to follow the conventions of Hawaiian grammar and Hawaiian discursive practice, in which one almost always speaks from a place—and this usage structurally preserves the perspectivalism at the heart of Kanaka geographic thought. In other words, the language itself subverts the authorial "view from nowhere" and asserts a view from Hawai'i. In certain passages, this is achieved through grammatical constructions. In the Hawaiian language, verbs are very commonly marked by the particles "mai" and "aku" that indicate where the speaker is located: "mai" following the verb indicates that the action happens in a direction toward the speaker, while "aku" following the verb indicates that it happens away from the speaker: thus "hele" means "move," "hele mai" means "come," and "hele aku" means "go away." Similarly, the particle "nei" after a place name indicates that it is where the speaker is located. "Nei" in this usage is a stylistic device rather than a grammatical necessity. It frequently indicates the affection of the speaker for the place mentioned, as in the phrase "Hawai'i nei," which is often heard even in English. Much the same effect is achieved by the mellifluous use of the word "kēia" (this) before a reference to a place, as in "keia pae ʻāina," meaning "this archipelago." Thus in *He Hoikehonua,* grammar and style literally reveal the writers' perspective from Hawaiʻi. That perspective is evident even in mundane sections, such as discussions of agriculture. The book describes the plants that grow "ma Hawaii nei" (*here* in Hawaiʻi), explains that taro is the traditional food "o keia pae aina" (of *this* archipelago), and discusses the crops that have "mai na aina e mai" (*come* from other countries).[49] Similar constructions name Hawaiʻi as part of ʻĀinamoana (Oceania) and the Pacific. The book informs readers that ʻĀinamoana is found in "ka Pakifika nei" (the Pacific *here*), that "ua nui na pele ma na aina nei" (volcanoes are plentiful in the countries *here*), and that Hawaiʻi is the biggest island "o keia aoao o ka moana" (of *this* side of the ocean). Note that all of these usages are choices: it would have been grammatically possible to simply write about plants that grow "ma Hawaii" (in Hawaii) or say that Hawaiʻi was the largest island "o ka aoao akau o ka Pakifika" (of the northern side of the Pacific). Such choices, however, would have neither grammatically made Hawaiʻi the point from which the world is viewed nor stylistically conveyed the affectionate warmth of "Hawaii nei" and "Pakifika nei." Instead, the creators of the book chose this wording, and in doing so inserted a perspective from Hawaiʻi into the book and the teaching of geography in Hawaiʻi.

This immediately raises the important question of whose agenda the textbook represents. Irrefutably, white American missionaries directed the preparation of *He Hoikehonua* and *He Mau Palapala Aina.* The textbooks that they were preparing were intended to serve the purpose of spreading American ideas and to operate in schools planned on an American design where Kānaka would be taught by teachers who had been taught by the missionaries. But while all of this is true, the artifacts of the books are themselves potent reminders that none of the educational enterprises in Hawaiʻi would have been possible without the labor and interest of Kānaka and the authorization and even support of aliʻi. In the case of the atlas and textbook, this means that Kānaka Maoli were essential to the project of translating European and American sources into Hawaiian-language texts and preparing engraving plates.

Missionary accounts of the means by which the Hawaiian-language Bible was translated suggest the ways that the text of *He Hoikehonua* may have been produced and hint at an important role for Kānaka in the production of it and the student atlas, *He Mau Palapala Aina*. By the 1830s, when they were preparing the Bible translation, a number of the missionaries had a usable knowledge of Hawaiian. They were not so fluent, however, that they could translate without constantly consulting with Kānaka who knew the subtleties and nuances of the language and its oral literature. To ensure that they were capturing what they considered to be the eternal truths of a sacred text, the missionaries drew upon their knowledge of Hebrew and Greek. But to make sure that these verities were faithfully reproduced in Hawaiian, the missionaries were dependent on Malo and other Kānaka who worked closely with the mission. It was to these Kānaka that the missionaries brought thorny issues for resolution. It was to these Kānaka that the missionaries gave drafts of the text for revision and correction. Although the missionaries presumably considered the stakes in translating a geography textbook or an atlas to be less high than in translating the Bible, they likely also depended on Kānaka to ensure the fitness of the translation.

It appears that this role for Kanaka guidance and editing introduced a subtle but powerful Hawaiian perspective into these tools of geographical education. Perhaps the best way to understand this is to return to John Charlot's sensitive portrayal of Kanaka engagement with books. Reading as a group, propped up on their elbows, arrayed like spokes, and facing the book at the hub, Kānaka read books in positions that Americans considered upside-down and sideways: after all, arrayed in a circle around the book, only one person could look at it "right side up." To Americans, this skill looked like a deficiency, but we would do better to interpret it as a literal manifestation of the ability of Kānaka to take the mission press's texts and bring an indigenous perspective to them. Kānaka working on the translation of books seem to have applied this same skill in the preparation of texts, and as we will see, teachers could adjust the use of texts to bring Kanaka perspectives to bear in geography lessons.

'ĀINAMOANA AND THE POWER OF COMPLEMENTARY OPPOSITIONS

More intriguing still is the term used in both the textbook and the atlas to describe the Pacific region: 'Āinamoana, a powerful term unknown in the language today.[50] 'Āina means "land" and moana means "ocean." Remembering that modifiers follow nouns in Hawaiian, we might translate ʻĀinamoana to mean "the lands of the sea," "the land from the sea," or even better, "sea-land." A number of things are remarkable about this term. First, it entirely bypasses foreign loan words. This is not "ka Pakipika," as "the Pacific" is transliterated in Hawaiian. Nor is it "Polunekia" or "Polunesia," as "Polynesia" is frequently rendered. The name the book and atlas give to the region are entirely indigenous in derivation. The term speaks powerfully to the sense of the embeddedness of Hawaiʻi in overlapping groupings that are variously called "Oceania," "Polynesia," and "the Pacific Islands" today. This point reminds us that although Hawaiʻi is currently subject to the United States (the political formation), it is not and never will be part of America (the continent). Rather, it is part of ʻĀinamoana. And if we look to comparative Pacific linguistics, the power of the term to remind us of Hawaiʻi's Pacific nature is even stronger. In Māori, the cognate of ʻāina is *kāinga*, meaning "home." *Kāinga Moana* would easily express the sense of the Pacific being an ocean home.[51] In Hawaiian also, ʻāina can mean homeland in a figurative sense. The term ʻĀinamoana thus maps Hawaiʻi and Kānaka Hawaiʻi definitively in a Pacific shared with other Pacific places and people.

The term ʻĀinamoana expresses the completeness of the region unto itself, rejecting the notion advanced elsewhere in *He Hoikehonua* that it is lacking and thus inferior. The term achieves this by uniting the two sides of one of the complimentary oppositions around which Kanaka cosmology is built: ka ʻāina and ka moana (the land and the ocean) but also ke ao and ka pō (the light and the darkness), ka luna and ka lalo (the higher and the lower), ka uka and ke kai (the uplands and the sea), ka wahine and ke kāne (the female and the male), and ke kapu and ka noa (that which is under sacralizing restriction and that which is free of sacralizing restriction). While people in many times and places organize their ideas through binaries, Stephanie Seto Levin has noted that in Hawaiian thought, "the procreative genesis of two complementary elements" first created and continues to create the world.[52] When elements of complementary opposition are in equilibrium, and when different elements of society are in balance, the situation is pono.[53] To these insights we can add that in Kanaka thought, these complementary oppositions contain completeness, encompassing the entire world. To unite the two elements of a binary—ʻāina and moana—into the name of one region was, therefore, to assert a sense of the region's sufficiency and totality. That assertion radically counters the racist depiction of the Pacific as a *lacking* region implicit, for example, when *He Hoikehonua* describes the tropics and its residents as lacking civilization, dedication to learning, ability to work, and so on. In the missionary text, "the tropics" may be lacking, but in Kanaka Maoli terms, ʻĀinamoana is complete.

It is complete, moreover, in a sense that is filled with power in Hawaiian thought, and marked by a marginal queerness in Western thought. In Western philosophy, binaries constitute nonoverlapping oppositions: to be male is to be not female; to be female is to be not male. Not so in Kanaka thought, which sees these two elements as complementary, each one containing within it elements that correspond to the other. In terms of gender, the difference between Western oppositional binaries and Kanaka complementary oppositions is stark. In the Western tradition, people who have not inhabited and performed either exclusively male or female positionalities in normative ways have conventionally been labeled as deviant, with this term suggesting that they are lacking by being outside the system. In contrast, in Kanaka society people who have lived and performed in ways that encompass both kāne and wahine were understood as quite powerful (and some still are, despite nearly two centuries of Christian influence in Hawaiʻi).[54] These complementary oppositions are generative. The world was made from the unification of ao and pō (day and night), after all. Thus to contain both kāne and wahine in one person is to embody tremendous power. Similarly, for a name to contain both ʻāina and moana imbues that place with all the generative power of the complementary opposition.

The origins of the term are uncertain: Was ʻĀinamoana a term used in Hawaiian for many years or centuries before it appeared in print in the 1830s? Or is it perhaps a neologism arrived at together by Kānaka and missionaries to translate "Oceania"? The answer is unsure. But whether the word was old or new, it is difficult to imagine an English speaker who was a newcomer to Hawaiian inventing the term. Its composition reveals a fluency with the structures of Hawaiian grammar and cosmology that suggests that ʻĀinamoana was almost certainly engraved into Hawaiian history by the Kānaka who assisted with the translations of the texts and preparations of the atlas. The use of ʻĀinamoana in the text and atlas encapsulates the complexities of the politics of geographical education in the period. In these books, the missionary colonialism of geographical education is unescapable. Yet the term reveals the ways Kānaka were able to bring opposing, indigenous perspectives into the very texts that aimed to teach them to accept an inferior and marginal place in the globe.

This power of Native perspectives is marked by its absence from a new student atlas that Minister of Public Instruction Richard Armstrong published around 1860. Whereas previous works seem to have been edited and translated by groups that included Kānaka, this work was Armstrong's. As is noted above, by the 1850s he reported that teaching in geography was hampered by the lack of atlases.[55] Kanaka Maoli teachers were eager to teach geography and were growing impatient with the failure of the ministry to get an atlas produced. Armstrong had endeavored to publish one. The method he employed, however, left less space for the sorts of Native perspectives that countered the colonialism of the earlier geography textbooks. Armstrong worked alone to prepare a list of Hawaiianized versions of English place names: Africa became "Aferika" and London became "Ladana," etc. Those names would replace English-language toponyms on maps drawn from a standard student atlas used in schools across the United States, William Augustus Mitchell's *Universal Atlas*.[56] Otherwise, the maps changed little. When Armstrong finally found a printer to work with, plates were printed in the United States and shipped to Honolulu. The result was a thirteen-map atlas bound with twelve pages of questions in Hawaiian that students were to answer by consulting the maps. Maps of the Hawaiian Islands were included, as were questions about them and about the Pacific. While this was clearly an adaptation for the Native teachers and students who would use the atlas, it largely was a simple transferral of Mitchell's atlas into Hawaiian. 'Āinamoana had no place on these maps or in these questions. Its inclusive embrace was replaced by a marginal space labeled "Polunesia" on the map titled "Poepoe Hikina" (Eastern Globe). Here, however, we have one hint of how Kanaka Maoli teachers (the teachers who had pressured Armstrong to provide them with an atlas) may have been able to provide their students with a Hawaiian perspective when using even so American an atlas. The map labeled "Eastern Globe" depicts what in English is called the Western Hemisphere. After all, from the perspective of Hawai'i, the Americas are to the east, not the west.[57]

As tantalizing as this map label is, it is unclear how the atlas was used by the Kanaka Maoli teachers. Armstrong himself makes clear they were eager for its production. Henry Whitney judged that there was enough popular demand for the atlas that he advertised in the Hawaiian-language newspapers that it was for sale in his Honolulu bookshop.[58]

Yet how would Native teachers, no matter how eager for an atlas, use an American school atlas transliterated into Hawaiian? The complexities of how Kanaka Maoli teachers brought Native perspectives to bear in using explicitly colonialist geography texts becomes clearer in the 1870s when a new textbook was produced.

KA HONUA NEI: MOVING HAWAI'I TO THE MARGINS OF THE WORLD IN THE 1870S

By the 1870s, Hawaiʻi's economy and its educational system had undergone significant changes. An increasing centralization of authority over schools had empowered white Americans in the kingdom's educational structure. Meanwhile, the agro-industrial production of cane sugar in American-owned plantations and mills had become the dominant force in the economy. The Department of Public Instruction served this new master with an emphasis on using schools to integrate Kānaka as proletarians in the emerging commercial and plantation economy.

It was at this time that educational authorities created a new geography textbook that left less space for Kanaka perspectives than *He Hoikehonua* and *He Mau Palapala Aina* had. In 1873, the board of education (which had replaced the Ministry of Public Instruction) published seven thousand copies of *Ka Honua Nei: Oia ka Buke Mua o ka Hoike Honua, no na Kamalii o na Kula Maoli o ke Aupuni* (The World: The First Book of Geography, for the Children of the Government's Native Schools).[59] It is a translation of an American textbook, *Our World, or, First Lessons in Geography, for Children,* by Mary L. Hall, first published in 1864 and revised and reprinted a number of times thereafter.[60] The translation was prepared by Harvey Rexford Hitchcock. Hitchcock was born in Lāhainā in 1835, only four years after instruction began at Lāhaināluna school. Named after his father, a missionary from Massachussetts who for a time worked at Lāhaināluna, Hitchcock's life and career was tied up with the development of Western-style schooling in Hawaiʻi. He would go on to serve as president of Lāhaināluna and inspector general of the board of education—in effect, the superintendent of the kingdom's school system.[61] Hitchcock was committed to making English the medium of teaching in all of the public schools. Toward that end, in 1887 he published (under contract from the board of education) a major English-to-Hawaiian dictionary that was explicitly intended to help Kānaka learn English.[62] This goal underscored what Hitchcock believed was the purpose of the Hawaiian schools: to prepare Kānaka to work in the service of an economy dominated by Americans—and perhaps even for annexation by the United States.

That American-centered vision was apparent in *Ka Honua Nei*. The new textbook hewed even more closely to its English-language source than did *He Hoikehonua*, as can be seen from its cover, first page, and maps. The cover reproduces the American image, retaining the English-language phrase "by Mary L. Hall." The first page similarly retains English in the plate. Because the text in the engraved plates for the images was retained, the maps in the book are precisely the same as the American version of the text, right down to the English-language labels in the book—unlike maps in the earlier textbooks and atlases, which used Hawaiian text. In the forward, Hitchcock explains that the maps and images were printed in Boston and were not translated, describing this as part of an effort to print the book more cheaply.[63] Yet retaining the English map and image labels is particularly powerful because it is just one of the ways in which *Ka Honua Nei* imposes an American perspective and American norms even more strongly than earlier textbooks and atlases. The cover features three images (see page 136). At the top of the page is a view of the globe, highlighting the North Atlantic and the North American coast. At bottom right is a schooner, its sails billowing as it leaves port. At bottom left, the cover displays a white teacher (probably male) showing a globe to three white children, a map hanging on the wall behind them. The group is clustered around a round table draped with cloth, suggesting the sort of domestic setting in which many privileged American children were educated. Together, the three images rightly inform the reader that the book will present the world from the perspective of white Americans and the United States. Looking out from their perspective, they become the center of the world. This message is reinforced by the woodcut on the first page of the book, showing a covered wagon pulled by an ox, a freshly felled tree, a farmhouse, and an American village in the background.

The organization and illustrations of the book emphasize Western perpectives and Western supremacy in a world organizational hierarchy, and it marginalizes Hawaiʻi. The book begins with a description of the globe as a whole and an introduction of terminology, then shows a map of the hemispheres. The map places the Atlantic Ocean at the center, thereby shunting Hawaiʻi to the edge of the world (see page 137). This is the opposite of the *He Mau Palapala Aina* atlas, which had centered

Hawai'i by placing the Pacific at the middle of the map. *Ka Honua Nei* goes on to describe the oceans, large island groups (including Hawai'i), mountains, rivers, and continents (with a map for each continent). It then provides country-by-country descriptions, a twenty-three-page island-by-island description of the Hawaiian Islands, and a list of English names for places and Hawaiian equivalents (Chattahoochie is "Katahuki," Mediterranean is "Waenahonua," etc.). The organization of the material in many ways centers Europe and the United States: the continent descriptions begin with North America; the United States is singled out for attention by getting its own full-page map; and the nation-by-nation descriptions begin with Greece, described as the font of Western civilization, and then proceed to cover the European nations. The illustrations and maps privilege Western colonialist and racist perspectives. Europe is illustrated with images of buildings bespeaking authority and civilization: a castle towering over the Rhine, ruins of an ancient temple in Greece, the Gothic spires of Westminster Cathedral in London. In contrast, the non-Western world is portrayed by savage nature and degraded humanity. South America is illustrated by an image of a jungle inhabited by monkeys and China by a picture of a group of men gathered around an outside cooking fire, two of them eating while sprawled on the ground. As for Africa, it merits no illustration beyond a map with the words "Uncivilized Negro Tribes" stamped in English across the middle of the continent. The one image that represents white and nonwhite people together points to the racial hierarchy that was a crucial lesson of the book. In the section on the West Indies, an engraving depicts four black men cutting sugarcane, which was gaining importance in Hawai'i at the time that *Ka Honua Nei* appeared (see page 138). A white man sits high on the wagon they are loading.[64] This image visually evokes the same message of global racial hierarchy that had been taught by the first Hawaiian-language geography textbook, *He Hoikehonua*, thirty years earlier. There had been a shift, however. *He Hoikehonua* had left discursive and visual space to examine the world from Kanaka perspectives. There was less space for that in *Ka Honua Nei*, with its more direct American voice and iconography. The image of the West Indian sugarcane plantation, moreover, suggests how the closing down of perspectives in the textbook corresponded to the racial hierarchy—white lunas (managers) over Kānaka and foreign Asian laborers—that structured the labor-hungry foreign-owned plantations that were growing at this time.

Cover of Hall and Hitchcock, *Ka Honua Nei*, 1873. Courtesy of the James Ford Bell Library, University of Minnesota.

The text reinforces these visual messages in describing a global hierarchy of nations and countries. At the opening of the section that describes a number of countries on each continent (more exhaustively for Europe and the Americas than for other regions), *Ka Honua Nei* follows the example of the English-language source text, Mary L. Hall's *Our World*, which introduces the terms "nation" and "country." To

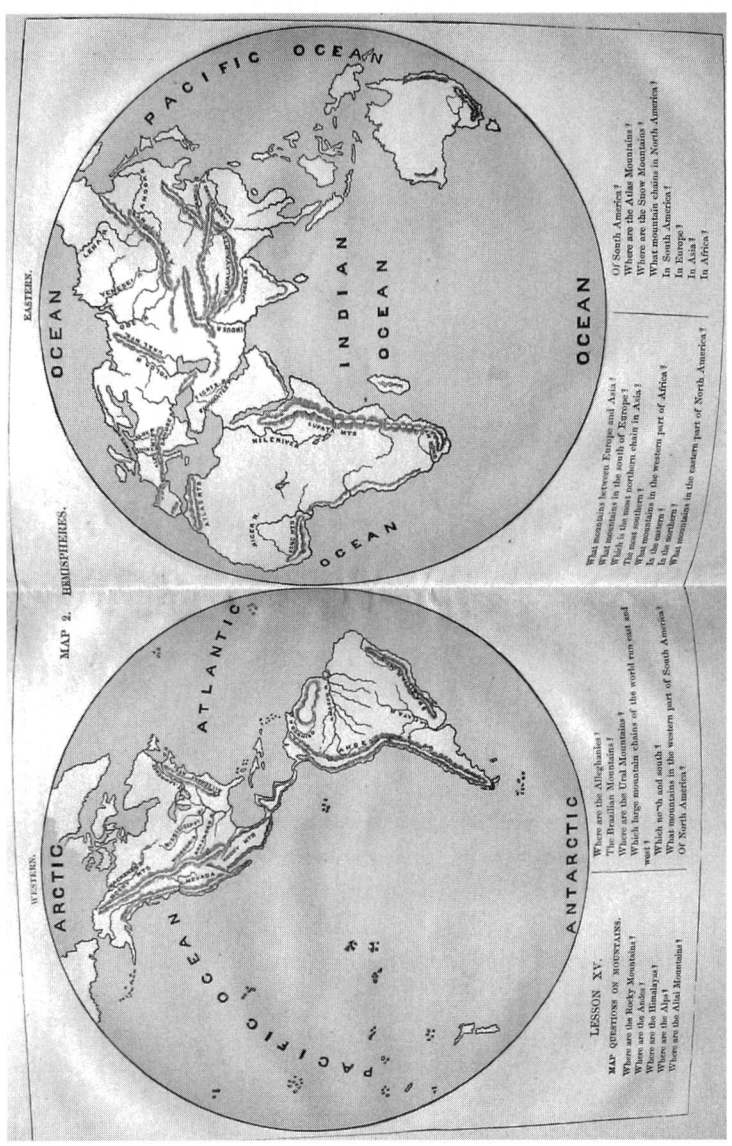

Map of the world in two hemispheres, with the Atlantic at center and Hawai'i moved to the left margin. From Hall and Hitchcock, *Ka Honua Nei*, 1873. Courtesy of the James Ford Bell Library, University of Minnesota.

Black men cut sugarcane while a white man supervises from a wagon. From Hall and Hitchcock, *Ka Honua Nei*, 1873. Courtesy of the James Ford Bell Library, University of Minnesota.

translate "nation," *Ka Honua Nei* uses the term "lahui," saying that the people who share a language, territory, and way of living form a lāhui. "Country" is translated as "aina," which also means "land" in Hawaiian. Closely following Hall's English-language text, *Ka Honua Nei* defines ʻāina in relation to lāhui: "O ka hapa o ka honua kahi i noho ai ka lahui, ua kapaia kona *aina,* a aohe kuleana o kekahi lahui e ma ua aina la" (The part of the earth where the nation [lāhui] lives is called its *country* [ʻāina], and no other country has a right in that land).⁶⁵ In one sense, the close translation of this portion of the American text left room for a notion of national political sovereignty that was in tension with the colonialist religious, cultural, and economic mission of men like Hitchcock in Hawaiʻi.

These notes of seeming enlightenment in the text are overwhelmed, however, by the dominant emphasis on categorizing the world into the civilized and the uncivilized. Immediately after having described a ʻāina as something like the sovereign country of a nation, the text says of nations: "he oi aku ka naauao o kekahi mamua o ko kehahi [*sic*: kekahi]. Mawaena o ka poe naaupo, aole buke, aohe luakini, aohe kula, aohe hale hana-lima; ua kau nui lakou i ka manao ana ma na mea ai, a me ke kaua ana. I ko lakou ike, he haiki loa, a ua pili i na hemahema o ka noho kino ana wale no" (some [nations] are more enlightened than others. Among the ignorant people, there are no books, there are no temples, there are no schools, there are no factories; they are stuck on thinking about food and warfare. Their knowledge is narrow and has clung to the needs of bodily life alone).⁶⁶ In this translation, *Ka Honua Nei* retains much of the wording of the original in *Our World*, but the final phrase in the Hawaiian text ("the needs of bodily life alone") evokes a particularly religious tone absent from the original: the missionaries suggest that ignorant people (non-Christians) neglect the life of the soul. Hall's English text had mentioned churches as one example of the ways that "savages, or ignorant people" lack the trappings and skills of what she calls "civilization." It did not, however, refer to the supposed lack of inner spiritual yearning at which Hitchcock's Hawaiian-language translation hints. When we remember that Hitchcock was the son and namesake of an American missionary, the particularly religious cast that he gives to the teaching of the geography of civilizational hierarchy is not surprising.

The image of the Pacific in *Ka Honua Nei* is markedly less compelling than the 'Āinamoana of *He Hoikehonua*, and this image establishes the basis for both civilizational hierarchy and the dependence of Hawai'i on whites for civilization. Describing the Pacific, the newer book writes that while it is calmer than the Atlantic, it is less frequently traveled, "no ka mea, he poe noho ma kahi hookahi ke poe kamaaina o ko laila mau kapakahakai" (because, the native people of its shores are a people who live each one in a separate place). Rather than a unified sea of islands of 'Āinamoana, these are isolated residents of specks of land occupied by inferior people, "naaupo a molowa" (ignorant and lazy), "ake ana e au mau ma ke kai, a e moe wale ma ua moena, ma kahi malumalu, a po ka la" (yearning always to swim in the sea, and just lie on mats, in a quiet place, all day long). The inferiority of other Pacific Islanders is both religious and racial; these are "he poe pegana" (a pagan people), but also "he ulaula ko lakou ili, a ane nele loa i ke kapa" (brown-skinned, and almost lacking in clothing).⁶⁷ Hawai'i occupies a superior position on the civilizational hierarchy than these people, but only thanks to the blessed presence of white missionaries, who are still more highly ranked. Fifty years before, the book claims, Hawai'i was a sorry place: "ua uhiia keia paemoku e ka pouli o ka naaupo a me ka hoomanakii" (it was stuck in the darkness of ignorance and idol worship). But over the past years, Hawai'i had made significant advancements toward civilization, as measured by the creation of schools. Hawaiʻi has benefited by the civilizing influence of the Americans, setting them above other Pacific Islanders. How fortunate for Hawaiʻi to have its level of civilization raised by "ka poe pono mai na aina Haole mai" (the righteous people from the Haole countries).[68]

This last phrase reinforces three interrelated crucial racist and colonialist messages in global geography as it was taught in these books. First, Kānaka Maoli were defined by their lack and insufficiency. Second, the world was organized into a civilizational hierarchy with whites at its pinnacle, Kānaka Maoli below them, and others (such as other, allegedly less fortunate, Pacific Islanders) lower still. Third, in order to advance in this hierarchy, Hawaiʻi and its people needed white people to teach and guide them. The voice of the missionary–sugar nexus's colonialism comes through loud and clear in *Ka Honua Nei*.

At the same time, the book ends with a countervailing note: a twenty-three-page chapter on the geography of Hawaiʻi, written specifically for the Hawaiian-language edition, that lays particular attention on the archipelago's economic geography. It was likely an invaluable resource to readers. It proceeds through the principal islands one by one, introducing readers to the physical landscape, climate, economic activities, political subdivisions, and towns and cities of the islands. The richness of the description lies in its multifarious detail. The Hāmākua coast of Hawaiʻi Island is a series of rugged valleys lush with vegetative richness and beauty. Maui is described district by district, its various internal divisions (moku, kalana, and ahupuaʻa) guiding the account. Unlike most accounts that focus only on the principal export crop of sugar and the importation of labor to produce it, the final section of *Ka Honua Nei* also notes products destined largely for the local market and the movements of the Native population in the islands. For example, it notes that on Molokaʻi, much salt is produced on the coast between Kawela and Pālāʻau, but that in recent years, the island's population had declined significantly due both to death but also to people moving to Oʻahu and elsewhere to find work. Readers learn that at Hāna, Maui, most people are farmers, and they plant tobacco, ʻawa (kava), sweet potato, and kalo (taro). With a level of detail matched by few outsider accounts of places like rural Maui, *Ka Honua Nei* notes:

> Kaulana o Makawao mamua aku nei no ka oihana mahiai huita; a o Kula hoi, no na uala maikai o laila. I keia wa, ua pau ke kanu

ana o ka huita ma Makawao, a ua lilo nui na kamaaina ma ke kanu ana i ke ko. Ke mau nei noo ke kanu ia ana o na uala ma Kula e kekahi poe, a o kekahi poe no hoi ke kanu nei i ke kurina, aka, o ka hana nui o ko Kula poe, oia ka hanai holoholona.

Makawao was formerly famous for wheat cultivation; and Kula also, for the good sweet potatoes there. The planting of wheat at Makawao has now come to an end, and the native people there have turned to planting sugarcane. In Kula, some people still plant sweet potatoes and some plant corn, but the great work of Kula's people is livestock raising.

The realities of the daily life of most rural Kānaka, and the local embeddedness of their lives, are better captured in this internally oriented description of Hawai'i's many places than in depictions that focus only on Honolulu.[69] It is essential to note, however, that the richest description is dedicated to commercial development in the islands. Where are deep harbors located? Which areas specialize in producing sugar, or livestock, or rice? Where are important business areas? Which areas have good roads that permit commercial intercourse, and which areas lack these roads? These are the questions to which the author returns time and time again.

The economic focus of the section is yet another example of how an educational apparatus (in this case, a textbook, though also the school system more broadly speaking) could serve the differing agendas of Haole administrators and Kanaka administrators and teachers. The chapter's commercial emphasis must be understood as part of the vision of a fully capitalist Hawai'i dominated by plantation commodity production. This depended on the transfer of lands, waters, and people (three elements that constituted the very substance of Hawai'i's ea, its life and sovereignty) to the service of plantations. The dispossession of Kanaka lands (which began with the 1848 Māhele [Division] and continued thereafter), the creation of a legal code and apparatus in the service of the plantations, the seizure of water that now irrigated commercial plantations for export crops rather than lo'i kalo (taro patches) and fields for Hawai'i's sustenance, the displacement of proletarianized Kānaka from rural communities to plantations and the urban commercial economy—all of these processes were accelerating in the 1870s. We must understand the privileging of Hawai'i's economic geography in the book as supporting this vision of what Hawai'i was. At the same time, however, this was one of the must sustained discussions of the geography of Hawai'i in print. It represented an opportunity to teach about the pae 'āina (archipelago) to its young people, which was also essential to adult Kānaka who worked toward a Hawaiian nation that was (as they understood it) modern, educated, and sovereign. Some of those Kānaka endorsed the commercial vision of the islands, others did not, but all would benefit from understanding the islands better, and understanding them in global context. The great-grandparents of children in the 1870s might have been taught the geography of Hawai'i though classically Hawaiian means. Now, the kingdom's schools joined families and other Kanaka spaces as important sites of geographical education. It was essential that they convey information about the pae 'āina. Yet all of this detailed knowledge about Hawai'i remains embedded in a text that was based on Hall's *Our World*, which described places like Hawai'i as needing enlightenment from the United States and portrayed Europe as the global focus of civilization. The historian cannot deny the power of the colonialist context.

SOVEREIGN PEDAGOGIES: NATIVE TEACHERS AND THE USE OF TEXTBOOKS

But if historians cannot deny this power, Kanaka Maoli classroom teachers could, and some resisted it. They did this in the same way that educators today might deal with a problematic textbook they have been given to use: they used only the parts of the book that supported the lesson they wanted to teach—in this case, sections that permitted students to regard the world from a sovereign Hawaiian perspective. In 1877, a three-member examining committee reported to the "luna kula o ka apana o Honolulu" (Honolulu district school director) on the functioning of schools in the district. In order to prepare the report, the committee, made up of B. W. Kawainui, Ab. Kalauli, and M. Kuaea (a Protestant minister), visited schools throughout the district and observed teachers and students at work. This included lessons on geography in classes that were using the *Ka Honua Nei* textbook. In praise of *Ka Honua Nei*, the committee noted, "Ma keia Buke e ike ai ke keiki i ke ano o ka Honua nei, a me ke ano o na Lahui Kanaka o ka Honua nei" (In this Book the child sees the nature of the Earth, and the nature of the Nations of the

Earth). Yet teachers were not having students simply read the book straight through. The committee noted, "Aole lakou i hoomaka ma ka haawina mua mai" (They did not begin with the first lesson). Instead, teachers had students jump first to the descriptions of the continents and then skip directly to the section on the Hawaiian Islands. The committee called out for particular approval "ka paanaau o na haawina e pili ana i na mokupuni o Hawaii nei" (the memorization of the lessons regarding the Hawaiian Islands).[70] But with gentle disapproval, the committee chided the teachers for using the books in a "kikokikoi" manner—skipping around the text rather than reading straight through.

We would do better to interpret it as classroom teachers' deploying a number of decolonial pedagogies. First, teachers chose to teach Native Hawaiian children first about the continents and then to induce them to learn in enormous detail a systematic geography of their own islands that was written from an internal perspective. By making this choice, teachers gave their students a wide sense of the world while cultivating in them indigenous Hawaiian perspectives deeply rooted in knowledge of their ʻāina. This would permit them to look at the world from Hawaiian perspectives and at Hawaiʻi from global perspectives. Moreover, the way they used the book meant they skipped many of its most corrosive sections, passages and images that promulgated a geography of racial and civilizational hierarchy. Even more than an act of Native resistance, this was an act of Native assent: an embrace of global geography and of new ways of knowing and teaching in the service of the Native present and future, while refusing to surrender the richness of that knowledge because of the racism and colonialism that infused the book.

The distance between the examining committee's wish that teachers would not "skip around" the text of *Ka Honua Nei* and the reality that teachers used the book for their own purposes is a yardstick to measure the political realities of common school educational administration in the kingdom in the 1870s. As Goodyear-Kaʻōpua makes clear, the growing power of the missionary–plantation nexus in the 1870s was tied especially to the central administration of the kingdom's schools. The examining committee referred to above may well have been an example of the efforts of the Haole in that administration to exert their power over the schools' largely Kanaka faculty. Yet the Hawaiian names of the committee members remind us that Kānaka Maoli also provided much of the staff of the educational apparatus, as well as the teachers. Even more important, the fact that the committee apparently did little more than rue the ways that teachers used *Ka Honua Nei* reminds us that the power of promissionary and procapitalist central authorities was limited.

NATIVE SCHOOLTEACHERS AND NATIVE POLITICS: POOHEA AND KĀNEPUʻU

To better understand the politics at play in the teaching of geography, it is worthwhile to turn briefly from the narrow consideration of geography and look more closely at Kanaka Maoli schoolteachers. They occupied the quasi totality of teaching positions in the kingdom's common schools in the early years of the school system and continued to be the majority even as the proportion of non-Hawaiian teachers grew in the 1880s and 1890s. Their own words reveal that they were able to find a way to live within the constraints of supervision while asserting a nationalist politics that resisted the colonialist intentions of that supervision. In fact, a close look at these teachers reveals the hazards in assuming that ties to the missionary-led educational system, or to the missionary enterprise itself, corresponded neatly to acquiescing to colonialism in Hawaiʻi. This complexity is visible in a group of documents that were generated by the same effort to centralize control that created the examining committee on schools in Honolulu. In the 1870s, the board of education periodically assigned essay topics to schoolteachers, perhaps as a part of accreditation or examination. Their answers, when placed in the light of what we know of their lives, demonstrate teachers' dedication to the defense of Hawaiʻi and the making of what they considered to be a modern nation. Some of their points—critiques of the hula, of "idolatry," of a love of material possessions over education, and so forth—could easily be mistaken for wholesale condemnations of Kanaka culture that would trouble many today who advocate for a stronger lāhui Hawaiʻi (Hawaiian nation).[71] It is important to remember that these appear to be assigned topics, however. Some teachers may have written their essays to play to what they knew to be Haole administrators' colonialist and racist prejudices. Other teachers may have used the essays to express sincerely held critiques of Kanaka practices such as hula, critiques based on their image of how Kānaka needed to change to survive and thrive as a modern nation. But in addition to critiques of Kānaka and their actions, some teachers also forthrightly pointed to the problem of Haole power, of Haole school administration, and of miseducation by a board of education that was showing no interest in educating an empowered Kanaka citizenry.

For example, J. W. Mahelona Poohea was a schoolteacher at Pelekunu, a tiny fishing and taro-farming village on the north shore of Molokaʻi.[72] Poohea was a member of the Ahahui Euanelio Hawaii (the Hawaiian Evangelical Association, or HEA), a group dedicated to proselytizing Christianity in Hawaiʻi and abroad. Moreover, Poohea was on the finance committee of his local church and served as a delegate to the HEA annual meeting in the 1880s. Despite his dedication to the missionary cause and his work in the missionary-affiliated school system, Poohea hardly approved of the colonial enterprise in Hawaiʻi. His own words in an 1872 board of education essay reveal that he rejected the presence of foreigners in Hawaiʻi and considered them to be a dangerous influence on Kanaka society. The assigned topic he wrote on was a pressing one for Kānaka Maoli: "He aha la na mea i pau ai ka make ana o keia Lahui Hawaii?" (What are the ways to bring an end to the dying of this Hawaiian Nation?). Hawaiʻi was suffering a calamitous demographic collapse of the Native population. In the 1870s, the Kanaka population had fallen to only about fifty thousand, a devastating figure when one remembers that historical demographers estimate the population at the time of European arrival at three hundred thousand to seven hundred thousand, and perhaps as much as a million. In his response to the question, Poohea struck a strongly antiforeign stance. First, foreigners spread terrible diseases among the people. The first course of action was therefore clear: "E hoi na kanaka o na Aina e i ko lakou wahi ponoi" (Send the people from other Countries back to their own places). Secondly, Poohea blamed the population decline on the Kānaka's desire for foreign wealth and goods: the desire for money and foreign clothing caused Kānaka to pursue excessive paid labor at sea and on land, encouraged prostitution, and "tempted children" (the last item being perhaps a reference to the prostitution of children and to the labor of boys on ships). It would be more healthful, he wrote, if they would return to fishing, farming, kapa (bark cloth) making, and other traditional pursuits. Third, Poohea blamed a disregard for law: people ignored the law of God and of the mōʻī. In this third diagnosis, one can see the merger of Christian and Hawaiian etiologies of disease. Disregard for God's law presumably incurred God's wrath, causing sickness and death. Disregard for the law of the mōʻī disrupted the proper relationship of makaʻāinana (commoner) to monarch, which in Hawaiian cosmology could disrupt pono (in this context meaning balance), resulting in illness. And finally, Poohea wrote, "O ka mea ai haole, oia kahi kumu e pau ai ka make nui ana o keia lahui" ([Avoiding] foreign food is one source by which to end the death of this people). The body of the Kānaka Maoli was suited to "na mea ai koekoe, a me na mea ai paakai, a me na mea ai momona loa" (cold food, salted food, and very fatty foods). Kānaka had changed their eating habits, however: "iloko o keia manawa ke hookomo pu nei me na mea ai pumehana a ka haole" (at the present time they are putting in the warm foods of the Haole). The result was that Haole food colonized the Hawaiian body, dispossessing Native food just as Haole people dispossessed Kanaka people: "i ke komo ana'ku o keia malihini iloko o ka opu, e kipaku mai ana kamaaina oloko o ka opu" (when these strangers enter the stomach they order the Native ones out). Poohea acknowledged that his theory was difficult to prove, yet to prevent further death among the Native population, he declared that the right course of action was obvious: "e hoopau loa ka ai ana ia mau mea ai" (the eating of these foods should stop entirely). Throughout Poohea's response, the danger that foreigners and their influence posed to Kānaka Hawaiʻi was explicit. Obviously, this Christian and teacher embraced instructing Native children in ka palapala and in Christianity, but he did not embrace Haole influence in the islands. In 1876, Poohea would be elected a representative (luna makaʻāinana) to the kingdom's legislature and would later work as an attorney in Honolulu and occupy a number of lower-tier government posts. Pelekunu, the village where Poohea once taught, has since been wiped from the map. The last inhabitant died in 1931, and in 1946 a tsunami swept away what remained of the buildings. But Poohea's essays remain to testify to the fact that even in rural villages like Pelekunu, Native teachers' politics contradicted the colonialist messages promulgated in textbooks like *Ka Honua Nei*.[73]

Similarly, some teachers emphasized that Kānaka Maoli had much to learn from Hawaiian-language sources and even perhaps Hawaiian knowledge, not just foreign sources. S. W. Kahoopii, a schoolteacher in Koauka, Waipiʻo Valley, on the island of Hawaiʻi, felt the need to open his essay by pointing out the incorrect premise of the question he was to answer: "Pehea la loaa ai i ka lahui Hawaii, ka waihona nui o ka naauao iloko o na buke Haole?" (How can the Hawaiian nation get the great treasury of knowledge in Haole books?). Before he could address the question, Kahoopii disputed it, writing "aole i manaoia aia iloko o na buke Haole wale no ka waihona nui o ka naauao; aka aia no iloko o na buke Hawaii kekahi waihona nui, o ka naauao a me ka ike noeau i-o maoli no hoi" (it is not thought that the great treasury of enlightenment is in Haole books only; there is a great treasury of enlightenment and truly wise knowledge in Hawaiian books, also). The rest of the essay, and the essays of other teachers who answered the question, make clear that they understood "Haole books" in this context to mean English-language books by Westerners. The passage is complex: it is not a simple rejection of all things Western or of Western ideas. In fact, Kahoopii emphasizes the great wisdom to be gained from the reading of Christian scriptures and the benefits of a strong command of reading English. But Kahoopii makes clear that Kānaka must read in Hawaiian sources, as well. First, he insists that in Hawaiian books one can find "ka ike noeau i-o maoli no." This is a difficult phrase to translate, because both "i-o" and "maoli" can be translated as "true and real." Yet "maoli" also refers to the Kānaka Maoli: Native Hawaiians, the true and real people of Hawai'i. Thus Kahoopii makes us see that "ka ike noeau i-o maoli no" can mean "truly true wise knowledge" or "truly wise indigenous knowledge." These texts included translations from English, but also the insights and ideas of Kānaka Maoli—insights that were also present in white-controlled newspapers. Kahoopii makes clear that Kānaka must learn from Kānaka—perfectly in keeping with the structure of the Hawaiian common schools where he taught, where Kānaka were the vast majority of teachers.[74]

Kanaka teachers could also be sharply critical of the kingdom's central educational administration in Honolulu, which was run by Haole and which was by the 1870s directing its efforts at training Kanaka workers for an American-dominated commercial economy, not Kanaka leaders for a sovereign nation. All of the monarchs defended their national sovereignty, but all also depended on a large number of foreign advisors and ministers. This was particularly true in education, as Western-style schools were originally an offshoot of the missionary enterprise. Thus when teachers or others critiqued administrative superiors in Honolulu, they leveled a critique at Haole administration. J. W. Kamoku, a schoolteacher in Waiehu, Maui, invoked the authority of Kamehameha III (Kauikeaouli) to declare that education was a fundamental value that school authorities were failing to uphold. In a paraphrase of the king's words, Kamoku wrote "o ko'u Aupuni, he Aupuni palapala" (my Kingdom is a Kingdom of literacy). Kamoku rebuked the kingdom's board of education (through which Haole were seeking to shape Hawai'i to their colonial vision) for failing to serve that royal vision of an educated and sovereign kingdom. He accused the board of education of ignoring requests for schools to be built where they were needed, paying Haole teachers more than Kānaka who did the same work, offering no specialized schools for the Kanaka poor, deaf, orphans, or those seeking vocational education, using Kanaka reform school students as contract labor in ways that exploited them rather than serving their educational needs, and building a system in which the most highly educated Kanaka graduates found themselves unable to find suitable employment and resorted to manual labor. In such a system, Kamoku suggested, was it any wonder that Kanaka parents did not always send their children to school? Kamoku's sophisticated anticolonial critique of the Haole-led educational system did not contradict his belief that Kānaka Maoli needed to abandon traditional beliefs and ways. In fact, he was committed to them doing so. His critique suggested, rather, that he believed that Kanaka Maoli teachers, not Haole administrators, were best fit to prepare Kanaka Maoli students to be modern subjects in a modern world.[75]

In that spirit, J. H. Kānepu'u offered to Kānaka the serialized work that is mentioned at the beginning of this chapter: "Ka Honua Nei a me na Mea a Pau Maluna Iho" (The World and All the Things upon It), a world geography that was centered on Kanaka perspectives and needs and stood up against the colonialist agenda of the educational system that we can see expressed in the *Ka Honua Nei* textbook. The beginning of the name of Kānepu'u's series, "Ka Honua Nei," gave it the power of a direct rebuttal of the textbook. The rest of the title ("a me na Mea a Pau Maluna Iho," meaning "and all the things upon it") referred back to the title of the earlier *He Hoikehonua* textbooks (*a me Na Mea Maluna Iho*, meaning "and the things upon it"). When Kānepu'u took the titles of the two major Hawaiian-language geography textbooks of his century and combined them to make his own title, he claimed the ground of world geography for himself and for Kānaka. Kānepu'u's series gives the reader insight into what it meant to teach global geography from an indigenous Hawaiian perspective in the 1870s. Indeed, it seems quite possible that the series published material from lectures and lesson plans that Kānepuʻu used in his classrooms in the Honolulu districts of Pālolo and Waikīkī.

From the first installment, the difference between Kānepuʻu's text and the *Ka Honua Nei* textbook is striking. After first calculating for the reader the circumference, diameter, and surface area of the globe, Kānepuʻu launches into its description—not with continents, not with nations, but with *oceans*. Not surprisingly, he begins with "ka Moana Pakipika," the Pacific Ocean. In an age of merchant power in Hawaiʻi, Kānepuʻu quantifies the enormous size and richness of the ocean, and its superiority to the Atlantic Ocean, in terms of dollars and cents. The Pacific is the largest ocean in the world, Kānepuʻu says, and if we were to estimate the value of all the underwater wealth that God has placed there, the sum would surpass a trillion dollars. The Atlantic, he explains, is second in size, and less blessed by divine beneficence: Kānepuʻu estimates that its underwater wealth does not surpass several hundred billion dollars.[76] In starting his world description with the oceans rather than the land, and by starting with the Pacific, Kānepuʻu puts forth a vision that privileges Hawaiʻi's place in the world. His monetary evaluation of the oceans would certainly have given pause to his fellow teacher J. W. Mahelona Poohea, whose essay objected to the glorification of wealth. Yet by using monetary value as a measure of the superiority of the Pacific (Hawaiʻi's ocean) over the Atlantic (implicitly, the ocean of the Americans and Europeans), Kānepuʻu bests the Haole merchants in Hawaiʻi with their own capitalist yardstick: money.

In subsequent installments of the series, Kānepuʻu continues to put forth a global geography that ties Hawaiʻi to the world in ways that center it in dynamic processes. In describing the continents ("na aina puniole"), he adds "na Ainamoana"—the plural form of the potent term for land and sea that is discussed earlier in this chapter. Unlike the earlier textbook *(He Hoikehonua)* and the student atlas *(He Mau Palapala Aina)*, he expands this term to include islands in all five oceans. He lays emphasis on the islands' placement, writing that God placed them here and there across the world when he established them. Thus rather than presenting Hawaiʻi and other islands as isolated places, Kānepuʻu presents them as occupying a divinely ordained place in a global order.[77]

Yet what most sharply differentiates his geography from that of the textbooks is that for Kānepuʻu, geography is a political act that can reveal injustice. It becomes clear as early as the third installment of the series that Kānepuʻu's geography is fundamentally a work of political economy in which the positionality and political beliefs of the author are made explicit—not obscured in a "view from nowhere" in the manner of nineteenth-century Western geography as exemplified by Woodbridge, Hall, and their translations into Hawaiian. In fact, Kānepuʻu makes the very calculation of the land surface and population of the Hawaiian Islands a political act. While the labor is burdensome, he writes, it is not in vain to present the information that he can gather from "na palapala aina Haole a me na buke Haole" (English-language maps and English-language books, which can also be translated as "white people maps and white people books") because Kanaka Maoli need that information. Moreover, too often they have not received the geographical instruction they need: "a aole no hoi i ao pinepine ia ma na kula maoli o Hawaii nei" (they have not frequently been taught in the Native schools here in Hawaiʻi). Publishing this in the newspaper, then, rectifies this problem and presents an opportunity for engaging Kanaka readers politically. Having calculated the land surface and population of the Hawaiian Islands, Kānepuʻu notes that this amounts to seventy-seven acres per person. He comments, "Aia la! Nui maoli na aina o ke kanaka hookahi, &c. Ina he ohana nui, aole e emi malalo o 1,000 eka aina no lakou" (There it is! There really is a lot of land for one person, etc. In the case of a big family, it would amount to no less than 1,000 acres for them all). But in Hawaiʻi in the 1870s, land was not distributed in that manner. When landownership was privatized thirty years before, many Kānaka Maoli were left with no lands, and many smallholders who received it found themselves quickly dispossessed. Kānepuʻu describes a situation in which whites have taken key positions in the administration of the government, leaving landless Kānaka no recourse and no explanation. Referring to the phrase "Hoʻoulu lāhui" (Increase the nation), the motto of King Davida Kalākaua, Kānepuʻu laments that if the Native population rebounded, the problem of landlessness would likely only worsen. "Mahea la e noho ai[?]" he asks—where would they live? He answers sardonically, "malalo aku paha o na konohiki ilipuakea kaumaha, a oluolu no hoi" (perhaps beneath the burdensome white-flower-skinned konohiki—pleasant indeed).[78] By calling white landowners and plantation operators "konohiki," after the headmen who administered land divisions and collected tribute for an aliʻi, Kānepuʻu likens capitalist burdens to the chiefly system that Haole derided as unjust. The injustice of the capitalist system built by the people he calls the "white-flower-skinned konohiki" is laid bare. Geography is a political act, and a powerful one, in Kānepuʻu's lesson.

Part of this political act of geography was to revise colonialist geographies. Kānepuʻu uses his close description of Hawaiʻi's economy, rivers, waterfalls, peaks and high points, capes and headlands, grasslands, interisland waters, towns and cities to correct "na hoike honua Haole" (which can be translated as "English-language geographies," "foreign geographies," or "white-people geographies") that neglect the archipelago.[79] His description goes beyond the one in the Hawaiʻi section of the *Ka Honua Nei* textbook. The series began with a broad purpose laid out in its title: "Ka Honua Nei a me na Mea a Pau Maluna Iho" (The World and All the Things upon It). As the series went on, Kānepuʻu seems to have decided that the most important contribution he could make in teaching world geography to Native Hawaiian readers would be to provide them with a detailed and politically sophisticated rendering of their own archipelago. While he does not ignore other places (Russia and Turkey receive particular attention), Hawaiian geography is the heart of the series. In fact, Kānepuʻu presents world geography in much the same manner as that described by the examining committee that visited schools in Honolulu in that same year, 1877: following a brief overview of the globe, he focuses on the detailed study of Hawaiʻi.

Moreover, Kānepuʻu's geography centers a Kanaka politics by deploying Native ways of knowing and teaching in its use of mele (songs) and moʻolelo (stories). At several points, when describing places, he references mele about them. Waiehu waterfall in Pelekunu, Molokaʻi; Puʻu Kānehoalani, a peak in Kualoa, Oʻahu; the Alakaʻi grassland on Oʻahu— Kānepuʻu describes these and other sites by referring the reader to mele.[80] Similarly, he describes Panaʻewa and Mahiki on Hawaiʻi Island as famous for the role those places played in "ka moolelo kaao o Hiiakaikapoliopele" (the legendary tale of Hiʻiakaikapoliopele).[81] Far from being simple references to familiar folklore, references to the oral and musical literature of Hawaiʻi were a powerful political act. In the previous decades, missionaries had sought to suppress Hawaiian story and song because they referred to the Hawaiʻi gods and often contained multiple layers of meaning, some of which were rich with eroticism. As Noenoe Silva has demonstrated, as part of their resistance to American colonialism, Native Hawaiians in the 1870s began to publish such stories and songs in opposition newspapers—most famously *Ka Hoku o ka Pakipika*, but also *Ka Lahui Hawaii*, the one where Kānepuʻuʻs series on geography appeared.[82] In his articles the reference to Puʻu Kānehoalani evokes Kāne (a god of whom Kānehoalani is one form) and Pele, god of the volcano. Kānehoalani was her brother (some accounts say her father). For their part, references to Panaʻewa and Mahiki evoke the Hiʻiakaikapoliopele story cycle, an epic that was a lightning rod for Christian disapproval and a pillar of Hawaiian tradition and Hawaiian resistance. The epic recounts (among other things) the migrations of deities, the origins of the hula, and the romantic rivalries of female gods. It describes a world in which, in the words of Charlot, "women have power, they have every human quality to a degree unlimited by gender ideals and stereotypes.... The passions of women for each other—both loving and hating, constructive and disruptive—are often the main motivations of the action. Those passions can be sexual, a clear reflection of the bisexuality common in classical Hawaiian life."[83] When Kānepuʻu deployed the reference to the Pele stories and the Hiʻiakaikapoliopele cycle in particular, he was doing two things. First, he was directly rejecting the missionary effort to suppress such stories and their references to the deities, hula, gender systems, and eroticisms the missionaries declared sinful. Second, Kānepuʻu was preserving Hawaiian ways of knowing geography. In these stories, knowledge about a place is also knowledge about the events in sacred stories that occurred there. These stories, moreover, reveal a world created by the generative relationship of complementary oppositions—here, the human and the divine and the male and the female. It is a world geography, then, that both shares a genealogical relationship with and can sustain a concept such as ʻĀinamoana—the extraordinary term for the islands of the Pacific that was put forth in the older *He Hoikehonua* textbook, the *He Mau Palapala Aina* student atlas, and (with a newly expanded meaning) Kānepuʻuʻs newspaper series on world geography.

Rather than internalizing and then teaching the colonialist lesson that Hawaiʻi was a small, backward place appropriate for colonial domination, Kanaka educators could use colonial educational establishments to teach sovereign lessons—a pattern we see in the career of Joseph Kahoʻoluhi Nāwahī at Hilo Boarding School. Born in 1842 in Puna, Hawaiʻi Island, Nāwahī studied at Hilo Boarding School before moving on to Lāhaināluna on Maui, probably completing his studies in the 1860s. As an exceptionally promising student, Nāwahī would undoubtedly have encountered the general education in geography offered at these schools and also the early textbooks *He Hoikehonua* and *He Mau Palapala Aina*. For all their complexity, their colonialist message about the inferiority of Hawaiʻi to Europe and the United States was inescapable. Trained in this colonialist global geography, Nāwahī returned to the Hilo Boarding School to work as an assistant teacher and eventually "managed" it (fulfilling the role of acting principal) in 1882.[84] And yet one could hardly claim that Nāwahī had accepted the colonialist message encountered in the books he had studied from and the one from which he perhaps taught at Hilo, *Ka Honua Nei*. Nāwahī became one of the leading lights of the struggle against American colonialism in Hawaiʻi. He served in the Hawaiian legislature from 1872 to 1892, was a member of the cabinet of the last sitting Hawaiian monarch, Queen Liliʻuokalani, and founded *Ke Aloha Aina*, one of the most visible of the newspapers to oppose annexation by the United States. Along with his wife, Emma Aima Nāwahī, Joseph Nāwahī founded the hui aloha ʻāina (patriotic associations) that pursued a petition campaign that opposed annexation and has inspired Kānaka Maoli in years since. Nāwahī may have been teacher and administrator in a school that was allied with elements that favored annexation to the United States, but he was nonetheless a force against it. Nāwahī, like other Kanaka teachers, was able to place himself between Kanaka children and the colonialist lessons of annexation-era Haole administrators.

ENGLISH-ONLY EDUCATION

Annexation in 1893 eroded Kanaka teachers' ability to perform this role, because banning Hawaiian-language instruction from the schools reduced the proportion of Kanaka teachers in the classroom. Already in 1890, thirteen out of seventeen public schools in Honolulu taught their classes in English. In 1896, three years following the overthrow of Queen Liliʻuokalani, the board of education instituted a ban on teaching in Hawaiian that would remain in effect until the 1980s. In 1892, 48 percent of teachers had been Native. By 1899, that proportion had fallen below 24 percent, and it remained roughly at that level for decades—and still (as Michelle Morgan has demonstrated) territorial administrators felt that the teaching corps was failing to "Americanize" children sufficiently.[85] And at a time when Hawaiian children needed Kanaka teachers more than ever to deal with the imposition of American occupation, colonialist English-language textbooks became the instruments of geography education in the occupied pae ʻāina. *Ka Honua Nei* was bad enough, but soon it was replaced by American textbooks in English: *Monteith's Elementary Geography*, also known as *Barnes' Geography*. The board of education first ordered the book in 1889 for use in English-language schools, and more copies were purchased as the entire school system was switched away from Hawaiian.[86]

The book was, if anything, more centered on American perspectives and more explicitly racist and colonialist than the Hawaiian-language geography books had been. Assuming that the reader was a resident of the United States ("the country we live in"), the book informed students that "the most enlightened and powerful nations in the world are in North America, Europe, China and Japan." Outside of those places, one likely was in a place where "heathens worship idols and do not acknowledge the true god"—"heathens" like the American Indian "savages who lived in huts or wigwams made from the skins of wild animals," or the people of India whom Monteith declared to be "very ignorant," or the residents of Africa's "Great Desert," which he summarized as a place of "unhealthy climate, savage tribes, and numerous wild animals," or the Aboriginals of Australia whom he summarily dismissed as "uncivilized and miserable." Kānaka would find they were part of the Malay race and lived in the "Torrid Zone," an allegedly energy-sapping region where climate and race explained the backwardness of its inhabitants. This new English-language geography was bound with a supplement on Hawaiʻi, and that did provide content on the islands. Like the Hawaiʻi section of *Ka Honua Nei*, however, it dwelt on commercial development and the advances foreigners and their rule had brought to the islands.[87] By 1893, the very geography of Hawaiʻi was being taught to young Kānaka in the language of their occupier, using a textbook that placed Hawaiʻi among the colonizable parts of the world. What is more, that book was largely being taught by settler teachers, not by Kānaka who could mediate between colonialist lessons in global geography and their Kanaka students' needs, as Kānepuʻu could.

This does not mean that Hawaiian resistance to colonialism or colonial visions of world geography came to an end in 1893: opposition to the illegal overthrow of the queen and the American occupation of Hawai'i continued, revealing that Kānaka did not swallow colonialist messages that they were in need of Haole rule. Even after annexation, Hawaiian thought (including geographical thought) was not "annexed." And yet there were real losses for Hawai'i in the teaching and learning of a colonialist form of global geography. One of the biggest of these losses was the teaching of a sense of 'Āinamoana, or of Hawai'i as part of Oceania. With the concept of 'Āinamoana fading, it was increasingly difficult to make Hawai'i look like a part of a larger oceanic world, and easier to mistake it for an island off the coast of California.

School training was not the only way that Kānaka learned about geography. From the late eighteenth century forward, Kānaka had experienced other places directly through their travels and labors throughout the Pacific and into North America. There they would encounter and experience racial systems of identification and hierarchy that they would have to navigate and into which they had to situate themselves.

5
HAWAIIAN INDIANS AND BLACK KANAKAS

RACIAL TRAJECTORIES OF DIASPORIC KANAKA LABORERS

In the Indian Cemetery in Chico, California, a granite marker reads:

John B. Azbill
1861–1932
Wailaki Yepim Maidu

Mele Kainuha Keaala Azbill
1864–1932
Kupuna Kaiana
Alii Nui o Maui
Hawaii Nei

Using a mix of Maidu and Hawaiian words and grammars, the stone declares that the graves hold the remains of John B. Azbill, a man of the Wailacki and Maidu peoples of California, and Mele Kainuha Keaala Azbill, whose ancestor was Kaʻiana, described here as the high chief of Maui Island in Hawaiʻi. The mingled American Indian–Native Hawaiian lineage that this gravestone represents has captured the imagination of a few white American and Kanaka Maoli writers since the 1950s. To tell their tale, they have depended in large part on information from Mele's descendants, especially information that her son, Henry Azbill, recounted and recorded in the 1950s and 1960s.[1]

What has made this story attractive to some white American writers is the perceived drama and romance of the Kānaka. In short publications and novelistic treatments, American authors have presented the story of this family as a glorious and surprising heritage: Ioane Keaala, a high-ranking aliʻi (chief), came from Hawaiʻi to California in the 1830s and married a Maidu American Indian woman named Su-My-Neh. Their daughter, Mele (or Mary), encountered Mōʻī (King) Kalākaua when he visited California, then traveled to Hawaiʻi, and served as a kāhili standard bearer for Liliʻuokalani, the last monarch of the Hawaiian Kingdom. Mele later returned to California and to the Maidu people. She became skilled in Maidu basketry and jewelry making, and her work is featured in museum collections in New York and California. The story of this woman's indigenous knowledge is remarkable in its own right. What is more, the aliʻi, the royalty, and the Kānaka add a Hawaiian exoticism to the story of allegedly "humble" Indians.[2]

While this emphasis on the Hawaiian is perhaps understandable, it prevents us from understanding these people's lives as they lived them. By emphasizing how Hawaiian these people were, the story misses the most distinctive fact of the lives of the many Kānaka on the West Coast: how *Indian* their lives became. They made their lives with American Indians, they linked their fates to American Indians, and they entered into American Indian communities. The story of Keaala and Mele would probably not be known if not for Mele's son Henry, who dedicated the years before his death in 1973 to the perpetuation of Maidu knowledge, culture, and identity.[3] Henry's efforts bear testimony to his ancestors' strong connections to American Indian people and communities.

On the Atlantic Coast of the United States from Maine to New York, Kānaka Maoli in the mid-nineteenth century lived among black people, made families with black people, and came to be seen as black people. This happened in smaller numbers and has garnered little of the celebration of Kanaka Maoli–American Indian connections on the West Coast, but the story followed very similar lines. In both places, male Kānaka landed on coasts as labor migrants and linked their lives to people who were by reason of race subject to colonization, enslavement, and exploitation.

This chapter traces this process on both sides of the American continent, and it emphasizes that this is a story about the Kanaka Maoli engagement with a notion that white Americans and other European-descent people were shaping in the nineteenth century: that race defined the contours of world geography. Race is after all in part a geographical concept. It had been part of the *He Hoikehonua* textbook that Kānaka began reading in 1832. The division of the world into races

Mele (Mary) Kainuha Keaala Azbill, undated photograph. Courtesy of the California State University, Chico, Meriam Library Special Collections.

Henry Azbill at the grave of John B. Azbill and Mele (Mary) Azbill, which he has decorated with lei, at Mechoopda Indian Cemetery, Chico, California, 1968. Courtesy of the California State University, Chico, Meriam Library Special Collections.

grew to dominance in the nineteenth and twentieth centuries. Indeed, although its roots were old, the notion that the world was divided into races was coming into a firmer and more scientized form in the nineteenth century. As Europeans and European-descent settlers shored up their power in the Americas, Africa, the Pacific, and the fringes of Asia, racial lines as Westerners understood them more and more predictably mapped lines between colonizer and colonized, enslaver and enslaved, dispossessor and dispossessed. When Kānaka engaged with that world, they had to engage with that system of race in some manner.

By and large, Kānaka Maoli in the United States engaged with this world of race by taking their place with the racialized and with the indigenous. I argue that in this coming together of Kānaka with American Indians and Kānaka with African Americans, we see Kanaka, American Indian, and African American agency—not just the impositions of a racist order from the outside that would group them together. I argue that gender, kinship, and genealogy (which is at the very core of Hawaiian studies) were crucial to this process. And I argue that in this coming together, we can see an earlier manifestation of a politics that is very present today in the notion of the indigenous as a category that connects Kānaka, American Indians, and others who have faced the onslaught of settler colonial impositions. Hokulani K. Aikau, in writing about Kānaka in diaspora, writes that "we maintain our indigeneity not only in relationship to home but also to the native peoples upon whose lands we dwell."[4] Indeed, Kānaka on the nineteenth-century West Coast made common cause with American Indian people, and even seem to have sharpened their dedication to the Hawaiian nation in the context of relations among indigenous people.

Social relations between Kānaka and African Americans and Native Americans did not occur only on land. Because so many Hawaiian, Native American, and African American men worked aboard ships (especially as part of the New England–based whaling fleet), many of the encounters between the groups must have occurred at sea. As work by Nancy Shoemaker has demonstrated, the history of relations among American Indians, Pacific Islanders, African Americans, and other racialized people at sea is important to understanding the indigenous histories of labor and racial formation.[5] This chapter, however, looks at relations on U.S. land to emphasize the importance of North American spaces in this story, and the importance of this story for understanding race in particular sites in the United States.

In a couple of well-known cases, aliʻi nui (high chiefs) and royal travelers in the nineteenth-century United States resisted racialization by refusing to be grouped with people of color. Resistance to racialization is captured in a famous story concerning Prince Alexander ʻIolani Liholiho. In 1845, when the prince and his brother Lot Kapuāiwa were traveling the United States by train, Liholiho confronted a white conductor who attempted to remove him from a train car. The conductor had "taken me for somebody's servant, just because I had a darker skin than he." The prince refused with outrage.[6] Five decades later, high-ranking Kānaka had learned how to circumvent the day-to-day impositions of racialization and segregation in the United States: Prince Kūhio Kalanianaʻole had traveled in the United States and learned the system of segregation. When Queen Liliʻuokalani brought a group of young people with her to Washington D.C. in 1899, he counseled the youths on how to avoid being shunted to the black sections of the train: converse with one another in Hawaiian rather than English, sit "where the white folks sit," and "tell them you're not Negroes." Lydia Aholo recalled that she and the other young people (Myra Kailipanio Heleluhe, Joseph Aea, and John ʻAimoku) who traveled with the queen used these techniques to make clear "we were not colored people."[7] In both of these episodes, we see Kanaka royals resisting the liabilities of racialization: Prince Liholiho fighting back against being subject to demeaning segregation, and Aholo and her mates pushing back against being considered black. These resistances correspond in many ways with the efforts of the kingdom itself to avoid racializations that cast it as inferior and colonizable.

Those trends, though important to nineteenth-century Hawaiian history, go in an opposite direction from what we see in the day-to-day lives of Kanaka laborers who dwelled long in the United States. Faced with racialization, they cast their lot with the racialized. And while there is an immediately identifiable politics of resistance in pushing back *against* racialization, this chapter explores the politics of identification and of kinship. This is a politics that we see in day-to-day acts of Kānaka, Indians, and black people making families and making homes and making lives together in the nineteenth-century United States.

The number of Kānaka in diaspora in the nineteenth century was astonishing. In 1845, Minister of the Interior Keoni Ana reported that about one in five men between the ages of fifteen and thirty were "wandering the oceans or in foreign lands." That is three thousand out of the kingdom's fifteen thousand men of that age. The men customarily traveled away for years: probably three thousand men were laboring overseas, but only 651 had embarked in the last year, suggesting that thousands were away for years at a time. Nor was this three thousand the total population of men who had left the islands. Keoni Ana noted that the cumulative number of men who had left and not returned could not be determined.[8] The gender dimension of this migration is essential to note: it was male because it was a labor migration that employed men. Men worked on ships across the Pacific and Atlantic, they labored in the Pacific Northwest in the fur trade, they provided much of the labor for the New England–based whaling fleet, in California they worked on ranches, and in California and Oregon they panned and mined for gold. In many cases, men who shipped overseas for labor stayed long after they left the ships. Kanaka men could be found in the hundreds in Papeete and in the dozens in Paita, Peru.[9]

Yet race as well as gender was at play in this labor migration, and in all these diverse sites one thing remained true: Kānaka lived in places where white Americans did not occupy a normative position, and Kānaka generally lived among dark-skinned people. This is a simple but important point. It corrects our distorted vision of the Kanaka Maoli past (and present) that constantly places it in relation to white people, especially white Americans. Just as the United States has established itself as an occupying power in Hawaiʻi, the United States (and also Canada) has occupied the center of attention in thinking about the experiences of Kānaka abroad. Historians have dwelt especially heavily on the experiences of ʻŌpūkahaʻia in the early nineteenth century. But the Kānaka who spread out across the globe inhabited sites where whites were rarely numerically dominant—including western North America.

We can consider, for example, the six Kānaka who lived in Fort Ross, California, eighty miles north of San Francisco Bay in what is now Sonoma County, in the 1820s. California at this time was Native American land, though Spain and Russia made claims to territory there. The 1820 and 1821 censuses of Fort Ross found six male Kānaka. Gerri was a sailor; Karya and Maktim were cowherds; no occupation is given for Kekʻkii, James Men'shoi, or Jack Fortunskii. Note the names: Men'shoi and Fortunskii are Russian, but Karya and Kekʻkii are perhaps Hawaiian names as transcribed by a Russian, and Jack and James may have taken their English names while working with Britons or Americans. These men's very names remind us that their overseas experiences involved much more than just Russian Fort Ross.[10]

More important, Fort Ross cannot be considered a Russian place even though it was an outpost of Russian power built by the Russian Empire. It was a mixed Native North American place in Pomo territory. In 1820 the fort had thirty-eight Russian men living among over two hundred people of Native North American origin, seventeen of whom were "Creoles" of mixed ancestry (eight male and nine female). Some of the Native residents came from California nations such as the Pomo, Chumash, and Miwok, but more came from the north: Aleuts, Kodiaks, Tlingits, and others. Moreover, all of the people who lived beyond the fort's palisades were Native. Kānaka were living in a Native American world with a European colonial incursion—and therefore a world with strong resonances to Hawai'i and its Western incursion.[11]

The Russians would not remain long in Northern California, but Kānaka, and their ties to Native American people, would take root there. In 1841, the Russian American Fur Company sold the fort to John Sutter from Switzerland. Sutter's workers (including Kānaka) all but disassembled the fort, transported the materials to what is now Sacramento, and built the stronghold that was known as Sutter's Fort.[12] Sutter arrived in California via Hawai'i, and ten Kānaka accompanied him—Harry, John Kapuu, Elena Kapuu, Manaili, Maintop, Manuiki (some of the preceding may be different names for the same person), and others. Only two were women, and there is little evidence that other Kanaka women came later. The nature of the arrangement between the Kānaka and Sutter, and the role of King Kamehameha IV in setting it up, is unclear. It may have been a three-year labor contract or indenture. In any case, at the end of the period, the Kānaka stayed on at Sutter's Fort.

Gender and sexual disparities of power in migration are vividly apparent here. First, male migration appears to have been the cause of female migration: these men were contracted to travel to California to labor, and those contracts resulted in their wives accompanying them. Second, as Kanaka Maoli historian Charles Kenn documents, Sutter wielded his economic, gender, and racial power to use Native American and Kanaka Maoli women as disposable sexual servants. Over a number of years, Sutter fathered several children by a Kanaka Maoli woman named Manuiki.[13] One white observer cast this as a consensual relationship, writing that Manuiki "favorably received" Sutter's "attentions."[14] "Consent," of course, must be understood as shaped by disparities of power. Sutter was the ruler of Sutter's Mill with enough power to make "consent" questionable. Could Manuiki have realistically refused Sutter? Eventually, Sutter's preferences turned to younger Indian girls. The same white observer wrote that he eventually discarded Manuiki and "kept a harem of young Indian girls in his ante room." He treated these girls or women as his racialized sexual servants, whose sexualized lack of power the observer conveyed by calling them by the orientalizing term "harem." Sutter enforced his privilege as the white lord of this domain, just as he had with Manuiki. When an Englishman had shown an interest in Manuiki, Sutter had "his Indians" (the wording again suggesting his power) bind the man with rope and imprison him in the fort's tower. When Sutter was done with Manuiki, he "gave" her to a man known in English as "Kanaka Harry."[15] Note that Manuiki and these unnamed "young Indian girls" were placed in the same position by virtue of their race and gender.

Race and gender—conjoined and inseparable—together created commonalities in the experiences of these girls and women and, by extension, in the experiences of Kānaka and American Indians generally. By tracing the lives of Kānaka in the area in the nineteenth century we can see that gender and race were fields through which Kānaka engaged with the Native people of the places they settled. The founding Kānaka at Sutter's Fort were joined by others over the years, until the discovery of gold and the resulting gold rush transformed the region. A number of the Kānaka joined the rush to find gold in the hills of Northern California, including Kānaka who had worked for Sutter.

First, though, a methodological point: other than the U.S. census, English-language sources from the time are of little use to this research project. The story I tell depends heavily on accounts in Hawaiian-language newspapers published in Honolulu in the nineteenth century, oral histories collected among California Indians in the 1960s, and the work of historians interested in Hawaiians abroad. Careful historical research on Kānaka in California, Oregon, Washington, and British Columbia by Charles Kenn, Jean Barman, Bruce McIntyre Watson, and others has made clear that a large number of Kānaka men lived with American Indian women (generally through nonformalized long-term cohabitation, but sometimes with the formal sanction of church and state) and had children with them. By and large, these studies have dedicated admirable effort to finding these Kanaka settlements and families and mining the sparse sources but have often been satisfied to demonstrate the presence of Kānaka in unexpected places (to borrow a phrase from Philip Deloria). They have generally spent less time considering what these connections might mean for understanding the larger questions of Kanaka Maoli and American Indian histories.[16]

A good place to explore those Kanaka–American Indian connections and their meaning is a site in the goldfields where a number of the Kānaka who had worked for Sutter searched for the precious metal. The 1860 U.S. manuscript census forms reveal that in El Dorado County was clustered a group of twenty-three men and one woman born in Hawai'i. But other than their birthplace, ages, and occupation (miner), the census tells us nothing of the lives of these Kānaka Maoli—not where their camp was located, not even a means to trace their identities, as they are listed under such Anglo-American names as Charles and Mary Aaron, Frank Harrison, and Thomas Boyd, names that they would have taken on after leaving Hawai'i and by which they were not known in Hawai'i. It is only by searching the Hawaiian-language newspapers published in Honolulu that we can discover their birth names and begin to understand their situation. *Ka Nupepa Kuokoa* of 1862 reported that these Kānaka Maoli (and one man named H. J. Ua from Mangaia in the Cook Islands, three thousand miles south of Hawai'i) had, just a few miles from Coloma, built very humble homes and planted vegetable gardens in which they cultivated Western crops. In the U.S. census, this appears to be an isolated enclave, mostly of bachelors. The Hawaiian-language newspapers, however, suggest that the Kānaka Maoli of El Dorado County were tied both to a network of Kanaka Maoli settlements in California and to local American Indian people.[17]

This can best be illustrated by the story of a young woman whose names were transcribed in the Hawaiian-language newspapers as Lakaakaa and Hitokane. She was a Concow woman born around 1844. The Concow (or Konkow or Konkau) are one part of the larger Maidu grouping of Native Californians. Her people's homeland, the Concow Valley, was in Butte County, a hundred miles north of El Dorado and twenty-five miles north of a place that white Americans called Oroville, meaning "Gold City."[18] To Concow people like Lakaakaa, this was not the gold country. It was home. They had not come from afar to seek fortune but had found themselves inundated by foreigners in search of a precious metal. Some Native people joined the search for gold, but in the 1860s many Concow still followed a seasonal round of activities— fishing, hunting, foraging for plant foods, and especially gathering and processing acorns. At least they did so to the extent possible; the gold seekers' diggings and flues and cattle disrupted the fish, plants, oaks, and game upon which Native Californians depended, and whites' violence threatened Indian people directly. It was dangerous for Native people to move about their homelands. It is no wonder that the white American trader Alonzo Delano reported that in the early 1850s, whenever groups of women and girls left a Concow village to gather foods, they brought "one or two men, to act as a kind of body-guard."[19]

The horror of settler brutality to Native Californians like Lakaakaa is almost unimaginable. Whites killed, raped, and enslaved California Indians with impunity during this time. The result: California's Native population fell by an estimated 80 percent between 1848 and 1860. In 1854, the federal government forced many of the Concow to relocate to Nome Lackee Reservation and the Nome Cult Farm, which would grow into the Round Valley Reservation where many Concow still live. Others were sent to the Mendocino Reservation, just south of modern-day Fort Bragg. In 1862, the Concow fled Round Valley for their homeland near the town of Chico. The next year, when Captain Augustus Starr brought them back to Round Valley, he left Chico with 461 Concow people. He arrived with only 277. The brutality of the forced march is still commemorated today.[20]

By 1854, Lakaakaa's family was so impoverished that, in search of food, she began to frequent the settlement of some Kanaka Maoli gold miners, some of whom had worked for Sutter. The site of the camp is probably at a place near Oroville that is remembered today as Kanaka Bar, near the mouth of Kanaka Creek, in the shadow of Kanaka Peak— place names attesting to the important presence of Kanaka Maoli miners in the region. She could communicate with the Kānaka Maoli, as they spoke the Concow language well, which suggests that relations between them and the Concow were ongoing. According to Thomas Gulick, a white missionary working in Hawaiʻi who visited this group of Kānaka in California, the Kānaka Maoli extended their practice of hānai adoption to Lakaakaa when she was ten. Hānai is a form of

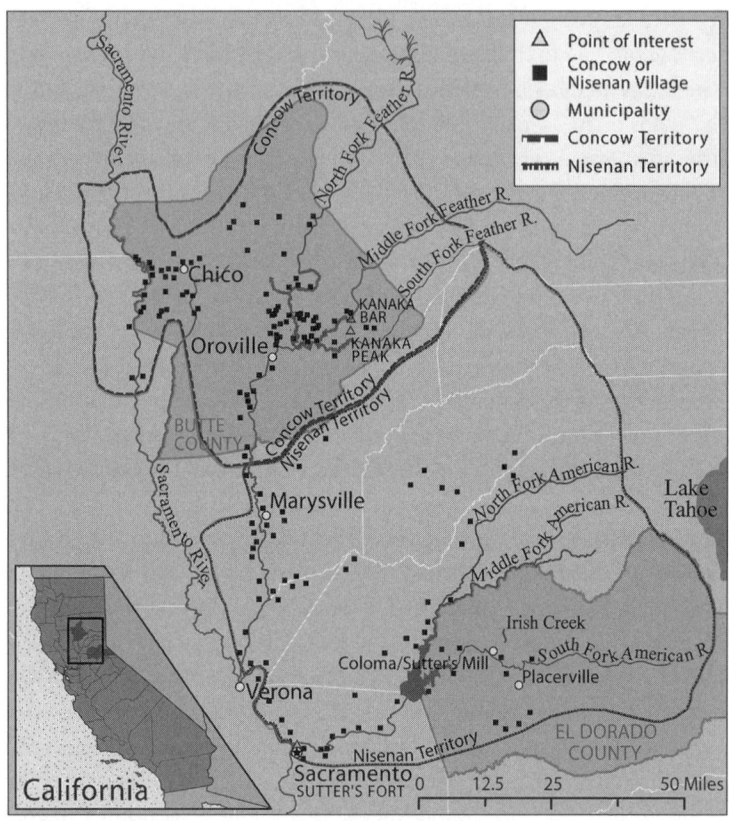

Sites of Kanaka Maoli history in Concow territory and Nisenan territory in northeastern California. Map by Matt Lindholm.

adoption that traditionally aims to strengthen bonds between families, rather than find homes for orphans. Through hānai, the Kānaka made Lakaakaa and her family into kin and gave the girl a new, Hawaiian name, Waiulili. Perhaps they intended to deepen the connection to her Concow family, or were motivated by sympathy for a starving child, or by a wish for female domestic labor. Their motivation was likely some mixture of these reasons. In any case, by 1857, Gulick reports that when she was about thirteen she "ua mareia" (was married) to a Kanaka Maoli named G. H. Kamakea.[21]

Here it is worth interrupting the narrative to consider the problematics of the misleading term "marry," which obscures the autonomous actions of the Kanaka men and American Indian women who formed these relationships. A number of historical accounts in English have done as Gulick did, and refer to Kanaka men and Indian women marrying. The word has the misleading connotation of Western and Christian conjugal norms of the time—lifetime opposite-sex pairing sanctioned by vows under the purview of church or state. Granted, this is a connotation only: "marriage" can in fact encompass many other meanings, from American common-law marriages to pairings endorsed by Native nations outside of American norms. Marriage was not a stable category. Nonetheless, in mid-nineteenth-century usages (especially those by a Christian missionary such as Gulick) the sense of sanctioned matrimony is present in the word "marriage." Yet there is little indication that the bulk of the pairings between Kanaka men and American Indian women on the West Coast were sanctioned by church, state, or covenant.[22] This becomes clearer when we look at the Hawaiian-language texts. For example, Pogue tells us that the Indian woman Kini (called Jennie or Jeanie in English-language sources) was Mahuka's "wahine" (woman), and elsewhere Mahuka is referred to as her "kane" (man). These are sometimes translated as "wife" and "husband" respectively, but that produces the same misperception as the word "marry." In fact, Pogue takes pains to differentiate cohabitation from marriage. He writes of Kini, "O ka wahine oia a Mahuka—Aole laua i mare i ka wa a'u i noho mai ai malaila" (She was Mahuka's wahine—they were not married at the time that I stayed there). It was only later, he writes, that they were married ("ua mareia") and now "oluolu ko laua noho pu ana" (live pleasantly together). As Pogue's final comment demonstrates, missionaries preferred formal marriage over cohabitation for Kānaka and American Indians.[23]

But cohabitation was the norm in these relationships, and a politically meaningful one: by following their own nonmarital ways of forming pairings, Kānaka and Native Californians were forging connections between their peoples through sovereign acts. Marriage as Pogue and other Anglo-Americans of his time understood it was an imported and foreign notion among Kānaka and among the Native people of Northern California. Historians sometimes impose this word and this concept on the pairings of Kānaka and Native Californians, but this is a colonial misreading. Note that in order to say "married" in Hawaiian, Pogue had to use an English loan-word, "mare." While Kānaka and

Jenny (Kini) Mahuka at center with Ellen Mahuka at left and Serrah Keaala at right, undated photograph. Courtesy of the California State University, Chico, Meriam Library Special Collections.

Native Californians like the Concow and Maidu certainly had practices in which opposite-sex couples created pairings, they were neither surrounded by the formalism of marriage nor covenanted nor assumed its lifetime status. Among the Concow and Maidu, a man interested in a woman would visit her family with gifts and express his will to live with her. If he was acceptable to the woman and her kin, he would simply stay. If the woman was uninterested in him, rejection was easy enough: "she would sit up all night."[24] In Hawai'i, established practice was similarly unformalized: couples that wished to form pairings did. And in both Northern California and Hawai'i, the dissolution of such a union was simple: either partner could dissolve the pairing at will. Of course, this sovereign act between individuals was rife with meaning beyond them individually. These were the social foundations that, as a later chapter will argue, were important for Kānaka coming to understand themselves as what we today would term an indigenous people.

The fact that most of these couples sought the sanction of neither church nor state suggests a resistance to the colonialism both of these peoples faced. Kānaka and American Indians initiated these pairings outside of the expectations of American missionaries—who had been active in Hawai'i for twenty years by the time the gold rush began, and who became active in Northern California as an effect of the gold rush. In Hawai'i, marriage was a practice introduced as part of a colonial project, and the American missionaries in California were trying to reshape Native forms of family making. Thus not only were Kānaka and American Indians making bonds between individuals and their communities, they were often intentionally doing so without entering into the kind of marriage that missionaries and the American state encouraged and sanctioned.

At the same time, American colonialism and the California it had shaped imposed constraints on the choices that Kānaka and Indians made. It would be unrealistic to expect any different. For American Indian women, dispossession by gold seekers (sanctioned and abetted by the American colonial state) had created the extreme poverty that made a coupling with a Kānaka gold miner a relatively attractive prospect. Remember that it was hunger that had first driven Lakaakaa (later known as Waiulili) toward the Hawaiians. For their part, Kanaka men had very few choices of female partners. There were few non-Indian women in the area of California where they worked. There was no large non-Native population in the area before the gold rush, and gold rush migrants from all the major sources of migration (the eastern states, China, Mexico, Chile, and Australia) were all but exclusively male. The small white female population, concentrated to the south in San Francisco, would certainly have been unavailable to Kanaka men, racialized as they were. The social bonds that Kānaka and American Indian women forged eluded colonialism but still had to act within conditions it had created.

The personal connection between individuals reveals how sensitive they were to the circumstances of history. Here we can return to Waiulili and Kamakea, who became a couple by 1857, when she was only about thirteen years old. Pairings at this age were not unknown in either Concow or Kanaka Maoli practice, and given the impoverishment of the Concow and the scarcity of prospective partners for the Kānaka Maoli, the pairing would have had advantages on both sides. The couple had two sons, Samuela and Kamakea Jr. After Kamakea Sr. died in 1859, Waiulili remarried. (Again, missionary Gulick uses the term "mare," or marry, though there is no evidence that the couple were formally married.) Her mate this time was another Kanaka Maoli named Edward Mahuka. With Mahuka, a native of Kohala on Hawaiʻi Island, Waiulili had a daughter, Rebeka. The couple left the Concow homeland, moving a hundred miles south, not far from Placerville, to another Kanaka Maoli settlement on Irish Creek, a stream named for yet another people active in the gold rush.[25] When they did so, they crossed an important territorial boundary, moving from the Concow homeland into the country of the Nisenan—although the whites among whom Waiulili moved may not have understood this. (Western settlers routinely ignored such boundaries.) To counter that obliteration, we can remember that Waiulili moved from the area of the Concow village of Čá-mpɨli to the vicinity of the Nisenan village of Koloma.[26]

The Kanaka Maoli settlement near Koloma (or Coloma, as white Californian mapmakers wrote it) is the hamlet that appears indistinctly in the U.S. census of El Dorado County for 1860. Lakaakaa had entered the Kanaka Maoli community through adoption and marriage and taken a Hawaiian name. She appears to have followed the Hawaiian practice of hānai adoption when she gave one child, Kamakea, to another man in the settlement, J. D. Kenao, so that he might raise him. In doing so, Waiulili was incorporated into the community via a Kanaka Maoli way of using adoptions to extend kinship bonds. (It is possible that Concow adoption and kinship practices were also in action here.) Waiulili was fluent in Hawaiian, and her mixed-ancestry children likely were also, as that was the language of the settlement in which they lived.[27]

A WEB OF RELATIONS AND OF MOVEMENT

Yet Waiulili and her children hardly cut themselves off from other Concow people. In 1862, for example, Waiulili's toddler daughter Rebeka spent time with other Concow and traveled with them in their subsistence work of acorn gathering and fishing. Meanwhile, Waiulili's mother had moved down to the mostly Kanaka Maoli settlement at Irish Creek to live with Waiulili and Mahuka.[28]

Here it is worth returning to the point that in the bonds that Kānaka and American Indians were making in California, we can see indigenous people autonomously remaking the geography they inhabited. In their movements, they evaded the imposed boundaries of colonialism to build and maintain the connections between them. When Waiulili, her mother, Lemaine, and her daughter Rebeka moved back and forth between Indian and Kanaka settlements, they repudiated efforts that began as early as 1851 to confine them to reservations.[29] In other words, by moving about California and living with Kānaka, one of the things they were doing was choosing *not* to move to the Round Valley Reservation, where the government wanted them. Agents of the Office of Indian Affairs there tried to make the reservation border a hard-and-fast line separating Indians on the inside from non-Indians on the outside. In accordance with the policy of the commissioner of Indian affairs, Round Valley Agent J. L. Burchard (a Methodist minister) tried to enforce a strict border-control policy that whites could enter only with his permission and Indians could leave only with a pass. He deemed Indians who left otherwise to be "escaping" and had them pursued. To Burchard, a strong border was a civilizing tool backed up by other disciplining tools, up to and including the lash. He unapologetically declared in 1875 that he and every previous agent at Round Valley whipped Indians, as they had no adequate facility in which to jail them. Yet as Burchard's comment itself reveals, the resort to the lash suggests the inability of Round Valley Reservation agents fully to restrict Indians, spatially or otherwise. As William J. Bauer demonstrates in his subtle analysis of the social meanings of the labor of Round Valley Reservation Indians, in this period and well into the twentieth century, they regularly moved beyond reservation boundaries for wage or subsistence work, and in the process maintained and even expanded spatially widespread social networks of Native people from a number of California tribes.[30] Waiulili's story illustrates this network from the perspective of off-reservation Indians. They used movement across boundaries to maintain the web of relations with other Native people, and indeed brought new people, such as Kānaka Maoli, into that network by making them kin.

Yet Waiulili was also immersed in her local context in El Dorado County, and in late 1861, Waiulili became a dedicated Christian during a Christian revival there.[31] Here it is worthy of note that Kānaka Maoli had experienced Protestant missionaries from New England beginning in 1820, and most had to varying degrees embraced the new religion. It seems that the miners with whom Waiulili lived were some of these Christians. Thus we have a Concow woman in the Nisenan country that is brought into Christian revival fervor partly through her entry into a Kanaka Maoli community in California, which had been converted by New England missionaries in Hawai'i. By any measure, this story exemplifies the way that Kānaka and American Indians created a cosmopolitan and complex world with the bonds they forged together.

In April 1862, smallpox swept through the mining settlement where Waiulili, Mahuka, and the others lived. In May, the disease killed four people there, including Waiulili and her birth child Kamakea. At the time of the epidemic, Theodore Gulick, a white missionary from Hawai'i, was visiting the settlement. He reported in a missionary-sponsored Hawaiian-language newspaper in Honolulu, "ua waiho aku makou malalo o ka honua, i ka hale kino lepo o Waiulili, me ka hoomana ana i ke Akua, a me ka manaolana, e ala hou ana oia" (we laid the earthly physical home of Waiulili beneath the earth, with worship to the Lord, and hope for her resurrection). Waiulili's mother, Lemaine, wanting a more visible and palpable source of comfort, declared: "Ina aole e laweia mai ia Rebeka i ikemaka ia, e make no kona kupunawahine" (If Rebeka is not brought so that I might see her, her grandmother will die). In time of crisis, the Concow grandmother summoned her Concow granddaughter, a half Kanaka Maoli girl with a name that marked her as the child of Christians. Yet Rebeka was a hundred miles to the north, in the Concow homeland, staying with other Indian people who were engaged in the summer salmon harvest. So Mahuka led Lemaine, a party of other Hawaiians, and the missionary up to Butte County to fetch his daughter, Rebeka.[32]

The tragedy was not over for Lemaine, however. A struggle over the custody of the now-motherless Rebeka ensued that reveals how American colonial and gender power hierarchies infused the marriages and other linkages that the Kanaka Maoli settlers and the Concow established between themselves. In most of Native California, and thus probably among the Concow, when the mother of a young child died, the maternal grandmother generally took the child to raise. Yet Mahuka wanted Rebeka. The missionary reported that Waiulili's mother, Lemaine, objected "me ke koi ikaika loa ana e waiho ia Rebeka me ia" (with the very strong demand that Rebeka be left with her). Rebuffed, she took action: "ua lawe o Lemaine ia Rebeka, me ka manao e holo loa aku a huna paha ia ia ma ka nahelehele" (Lemaine took Rebeka, with the plan of going far and maybe hiding her in the brush). The missionary interpreted this as an attempted abduction, chased down Lemaine, and turned Rebeka over to her father, Mahuka. The contest over Rebeka was surely deeply personal. Just as certainly, a struggle for a child between Concow and Kānaka Maoli, two peoples beset by the demographic collapses (such as smallpox epidemics and genocidal attacks) that followed contact with Western powers and settlers, was also a struggle for a future. Here, the intervention of a white American missionary favored Western patrilineal patterns and the Kānaka Maoli with whom he had longer-term ties.[33]

THE GEOGRAPHY OF KINSHIP

The movement of people through relationships of cohabitation, marriage, and hānai adoption knit a complex of relationships that connected people of the settlement to one another and to American Indian and Kanaka communities in California and Hawai'i. This geography of kinship comes into focus as we continue to trace the movements of Mahuka and Rebeka, who by 1867 had moved down to a mixed Kanaka Maoli and American Indian settlement at the confluence of the Feather and Sacramento Rivers. The hamlet was known in English as Vernon. Kānaka pronounced the name as Verenona, which is likely the origin of its name today, Verona.[34] Charles Kenn, who researched the Vernon settlement, reports that it was also termed "Puu Hawaii" in Hawaiian, which he translates as Hawaiian Refuge, and could also be translated as Hawaiian Hill. The Kānaka and American Indians there fished the river and sold their catch downstream to residents of Sacramento and upstream to miners in the gold country.[35]

Fishing was likely a more dependable living than gold mining and one they could turn to in the face of white violence. White miners pushed nonwhites out of the gold fields in the effort to monopolize its wealth, and tremendous violence by white settlers was devastating the Native American population of California.[36]

The fishing settlement demonstrated the common Pacific Coast pattern of Kanaka Maoli men and American Indian women forming families together, it showed the complex movement of individuals through multiple relationships in their lifetimes, and it suggests the spatially extensive web of connection that grew out of these relationships and the associated childbearing and childrearing. The Kanaka residents of Vernon included some men who had come to California with John Sutter (such as Mahuka) and others who seem to have arrived later. The Indian residents of Vernon included women (such as Kini, or Jennie) who had made partnerships with Kanaka men. Via the men, Vernon was tied to Kanaka communities in California and Hawaiʻi, and via the women, it was connected to Native settlements both on and off reservations. And in the persons of the children that Indian women bore by these men (such as Waiulili's and Kini's children), the place was tied to both communities. Another example of these connections: John Kapuu, from Hāmākua, Maui (who had traveled to California with Sutter), lived with a Maidu woman named Pamela Clenso at Vernon. Clenso outlived Kapuu and her next two male partners, both of whom were Kānaka: Richard Hakauila and Aihi Eel. By Hakauila, Pamela had a daughter, Lillie, who herself grew up to marry a Kanaka.[37] The multiple pairings of Kanaka men and American Indian women knit a dense web of connection between Kānaka and Northern Californians.

So did children: births connected the Kanaka and Native communities, as did the Kanaka Maoli practice of hānai adoption. For example, sources mention two children in relation to the name Kapuu. Hana was "he wahi kaikamahine" (a small girl) of "Kapuu ma" (Kapuu and his people). Harieka was a half-Indian and half-Kanaka child that had been hānai'd to the Kapuu family.[38] Notice that neither Hana nor Harieka is identified with just one male–female pairing. Instead, they are connections between families. Rebeka (daughter of the deceased Waiulili) was raised by the man who had fathered her, but her half brother, Kamakea, had been given in hānai to another Kanaka man to raise.

In case one were tempted to ascribe this likelihood to form pairings

John Kapuu and Pamela Clenso, undated photograph. From Kenn, "Descendents of Captain Sutter's Kanakas." Courtesy of the Conference of California Historical Societies.

between Kanaka men and American Indian woman simply to low status or low class, one can point to the counterexample of Ioane Keaala and Su-My-Neh. Because both were of elevated social status among their own people, their pairing may have helped to knit the alliance between Kānaka and Maidu in nineteenth-century California. Charles Kenn, who corresponded with Keaala's descendant Henry Azbill in the 1950s, declares that this aliʻi from Maui descended from the high-ranking Piʻilani line.[39] Descendants in California reported that he was a descendant of Kaʻiana, the great Kauaʻi chief whose travels took him to Macao, Zamboanga, and Nootka with John Meares. Keaala was another of the initial group of Kānaka to travel to California with Sutter and worked for him until he left California. While working on a riverboat on the Sacramento, he met and paired with a Maidu woman, Su-My-Neh, the daughter of a headman named Kulmeh.[40] Patrilineally derived authority mattered among the southern Maidu into which Su-My-Neh was born, and of course inherited authority was central to Kanaka society.[41] The pairing of Ioane Keaala and Su-My-Neh, then, brought together people who had distinguished parentage in their own societies. Indeed, it is plausible that their union expressed or even lay the foundation for an enduring alliance between the Maidu and the Kānaka in that region.

But whatever their status among their own people, in the settler colonial context of mid-nineteenth-century California, Kanaka men and Native women both found themselves scrambling for survival. This must have encouraged the formation of couples such as Kamakea and Waiulili, Mahuka and Waiulili, Mahuka and Kini, Ioane Keaala and Su-My-Neh, and Pamela Clenso and her three Kanaka partners. Those pairings and the children they produced and raised created and deepened the kinship bonds between communities. The building and maintenance of those bonds depended on the mobility of Kanaka men who left Hawaiʻi and the mobility of American Indian women who evaded reservation boundaries.

KANAKA RACIAL TRAJECTORIES: THE CASE OF EDWARD MAHUKA

That mobility evokes the ways that residents of these communities simultaneously evaded and lived within the boundaries of racialization that Americans were establishing in the region. They evaded those social boundaries in the same ways that they crossed spatial boundaries between Hawaiʻi and the United States or between reservation and nonreservation spaces: they transgressed categories that would have separated Indians from Kānaka. But they also lived within the boundaries of racialization to the extent that, as pairings of nonwhite people together, Kānaka and American Indians operated within the fundamental white/nonwhite binary that structured American society. The tendency to form Kanaka–Indian families was both an autonomous act and an act shaped by imposed racial categorization.

Looking at individuals over time reveals the complexity of this partly autonomous, partly imposed process. When we look at Kanaka individuals on the West Coast over time, we see a trajectory toward Indianness—but not an abandonment of Hawaiianness. Family-making between Kānaka and American Indian women was fundamental to this process. When Kanaka men paired with American Indian women in the middle of the nineteenth century, they embraced and entered into American Indian communities and seem to have taken on American Indian identities, all while maintaining Kanaka Maoli identities. In the 1850s, it was clear that Kānaka in California would be racialized, but they were assigned to many different racial positions

We can trace this trajectory toward Indianness through the changing racial classifications that state and federal officials mapped onto the Kanaka Maoli body of Edward Mahuka, partner to Waiulili and Jennie and father to Rebeka. Those categorizations stretch from the early days of the gold rush to his death at the end of the century. When the federal census enumerator (the person who actually took down census information) got to Mahuka in 1860, he threw up his hands. What race was Mahuka? The form allowed only three choices: white, black, and mulatto. Enumerators in California commonly ignored that rule and wrote in "Chi" (Chinese) and "Ind" (Indian). But the enumerator must have believed none of those fit Mahuka, so he just left the race column blank.

Mahuka was a racial cipher, as we can see from the many ways that state workers classified Kānaka born in Hawaiʻi. Between 1852 and 1860, Kānaka were classed as black, mulatto, white, and Indian—and given their age and their place of birth in Hawaiʻi, it is almost certain that most (though not all) of these Kānaka were entirely of Native Hawaiian ancestry. In Shasta County, the census enumerator marked Kakerku, Maka, Leon Black, Pakaha, and all the rest of the forty-five Kanaka miners in the county as black.[42] In Contra Costa County, a fifteen-year-old young woman named Hohano was termed a mulatto.[43] In Tuolumne County, the enumerator had not bothered to learn the Kānaka's names, listing all of them under the name "unknown." With the same lack of effort, he marked them as white by just continuing the ditto marks (") in the racial classification column that he had used for the white people above them on the list—white people whose names he recorded. The ditto marks (along with the lack of names) suggest that the racial classification of the Kānaka arose from his lack of dedication to his job more than a firm conviction that the men were white.[44] Meanwhile, mortality tables from Calaveras County declare that Kanaka gold-miners such as Kaunaea, John Wooster, and David Dorr were "Indian."[45] Given such confusion, it is not surprising that the enumerator just skipped the question on race when he got to Mahuka in 1860.

In 1870 and 1880, however, Mahuka was black, at least according to the census enumerators in Vernon, Sutter County. The lackadasical work of the census is visible here, too. Although all the Kānaka lived in one settlement and are listed in one grouping, the enumerator in 1870 marked the first five as white by merely continuing a line of ditto marks from the white people above them. He then seemingly realized his error and began marking the rest (including Mahuka) as black—including Mahuka's mixed-ancestry daughter, Rebeka (listed as Lipica Kapu). In the 1880 census, Mahuka, Rebeka, and the rest of the Kānaka remained black.[46]

But in the final official record concerning Mahuka, he appears as an Indian. On September 13, 1889, he died of a tumor. He was buried in the cemetery of the Maidu Rancheria at Chico (the same cemetary mentioned at the opening of this chapter, where the remains of John B. Azbill and Mele Kainuha Keaala Azbill are interred). In the state's record of death, his race is given as "Indian" although his place of birth is still recorded as "Sandwich Islands." Edward Mahuka had entered the United States public record as a racial cipher, lived much of his life under the label "black," but in death, he was buried in Indian land and named an Indian.

Edward Mahuka's ancestry had obviously not actually changed, but there was a certain truth in the state's record: the trajectory of his life was toward the American Indian community. The strongest reminder of that trajectory can be seen in the location of the grave where he lay and in the identity of the woman who lay beside him: Jane Mahuka (whom we have encountered as Jennie, Jeanie, and Kini). At some time after his first wife, Waiulili, had died, Mahuka and Jane began to live together. Later, they were formally married.[47] Mahuka had already partnered with one Concow Maidu woman, Waiulili, and then partnered with another, Jane, with whom he had two children, Ellen and Albert. Like

Mahuka, Jane's race slips and slides from one colonial racial category to another. In the 1880 census, she is recorded as a black woman born in the Sandwich Islands—in other words, as a wahine Kanaka Maoli, a Hawaiian woman.[48] Yet Jane's death record declares her, too, an Indian woman, and she was buried next to Mahuka in the Indian cemetery at the Chico rancheria.

Was Jane a black woman, a wahine Kanaka Maoli, or an Indian woman—or was she, also, a racial cipher whose race must be understood situationally? Apparently, just as whites thought that Kanaka men took on the Indianness of the Native Californians among whom they lived, Native Californian women could take on the birthplaces and ascribed races of their Kanaka partners or husbands. If we seek a certainty based on her ancestry, an excellent source is a report on the Kānaka of Vernon that was written by J. F. Pogue, a white missionary working in Hawai'i who visited the community in 1868. With no apparent doubt, he described Jane as "he wahine Inikini" (an Indian woman) in an article published in a Hawaiian-language newspaper in Honolulu. Given that Pogue was familiar with Kānaka, spoke Hawaiian, and spent several days in Vernon, his judgment is more trustworthy than that of a census enumerator who passed quickly through the community and did not speak Hawaiian. Charles Kenn, who communicated with the descendants of the Vernon Kānaka in the 1950s, describes her as "an Indian woman of Wintu stock." Still, she lived in a settlement of Kānaka, married a Kanaka, and probably spoke the Hawaiian language (the lingua franca in other majority-Hawaiian settlements with Native American residents). Her ties to Hawaiians were so strong that an outsider like a census enumerator took her to be Hawaiian—and another census enumerator who considered Kānaka to be black, took her to be black. She had cast her lot with Kānaka, and thus took from them the instability of their racial categorization. But one thing remained constant for both Jane and Mahuka: they were not described as white.

Dark skin was a contemptable racial stain in the eyes of the whites around them, according to J. F. Pogue, and that stain made them black. In a report in a Hawaiian-language newspaper in Honolulu, he wrote: "Hoowahawaha aku la kekahi poe Haole ia lakou" (Some Haole people treat them with contempt). He explained this was not due to their conduct: "Aole no ko [sic] lakou hana pono ana, aole hoi no ka hana kolohe" (Not because of their good actions, not for their bad actions). Rather it was for their race: "no ko lakou ili-ulaula no" (because of their brown skin). 'Ula'ula describes what are in European color classifications a range of hues, and sometimes the term was used to describe American Indians' skin color. But Pogue makes clear that in this case, the racial referent is black. He reports that in the eyes of the Haole in California, "he poe lakou i ano like me ka poe Negero" (they are a people like the Negro people).[49] They were dark, they were racialized, and the racial positions they could take (or would have mapped onto them) were delimited by that fact. Kānaka would not be considered as white, their numbers were too small to be recorded as their own race, and thus they came to be marked as some other nonwhite race.

But which one? Gender and sexuality as well as color shaped the outcome because family making through opposite sex pairings, childbearing, and childrearing were crucial to making the enduring link between Kānaka and Native Americans. In other contexts, being seen as "people like the Negro people" might have resulted in California Kānaka increasingly becoming part of African American settlements and communities. This does not appear to have happened for a number of reasons. First, despite the hardship they faced, American Indian communities had a visible presence in Northern California in the 1850s and 1860s that African American settlements did not. More important, like other gold rush settlements, African American settlements in California were heavily male. Because gender (in the guise of heterosexual pairings) was so crucial to the building of Kanaka Maoli–American Indian bonds, there was less likelihood in mid-nineteenth-century California that Kānaka would tend to enter African American communities. The same can be said of West Coast Chinese communities, which were almost exclusively male. This meant that there were apparently only a few West Coast children of mixed Kanaka Maoli–Chinese descent, like Ioane Amiuna, who lived in a Kanaka Maoli goldmining settlement in Jacksonville County, Oregon, in 1868. Californios, the long-time resident Mexican population of the region, had normal gender ratios. But young women from their community tended to marry either Californio or white American men, so heterosexual bonds did not draw Kānaka into their fold. Gender ratios of racialized groups and sexuality—specifically opposite-sex pairing—created the conditions in which bonds between Kānaka and American Indians were built.

Note that it is quite likely that male–male sexual intimacies and even

Mele (Mary) Kainuha Keaala Azbill with Henry Azbill in a cradleboard, about 1900. The girl at right may be Henry's sister, Cora. Courtesy of the California State University, Chico, Meriam Library Special Collections.

enduring partnerships arose between Kanaka Maoli men and men of other races in the region at this time. As we have seen in the aikāne (same-sex romantic companion) relationships mentioned earlier in the book, Kānaka accepted such practices, and Christianity threatened them but hardly wiped out male–male relations. But these relations are not as clearly documented as marriages and mixed-ancestry children. Moreover, because they were nonreproductive, they did not create lineages like the Keaala-Azbill line that was mentioned at the opening of this chapter.

In the context of mid-nineteenth-century California, then, the most common racial trajectory for Kānaka was toward Indianness. Mahuka had paired with at least two women, one certainly American Indian and the other probably so, and fathered at least three children by them. He lived with Concow Maidu and Wintu people and spoke Concow Maidu. He made families with American Indian people, and he lay with them in death. This does not mean that he was Indian, but it does mean that he entered California as a racial cipher and ended his days there as a man who had cast his lot with American Indians.

ENDURING TIES: KANAKA IDENTIFICATION AND KANAKA NATIONALISM

But even as Mahuka built enduring ties to the Native people of his new land, he remained tied to his homeland and to a Hawaiian national project. Mahuka is suprisingly well documented in the public record in Hawaiʻi because he maintained his connections to Hawaiʻi for decades after he left. As Noenoe Silva has argued, debate in Hawaiian-language newspapers and the publication of newspapers under indigenous control were central to Kanaka resistance to American colonialism in Hawaiʻi. A crucial organization in this was the Ahahui Hoopuka Nupepa Kuikawa o Honolulu—the Special Newspaper Publishing Association of Honolulu. And not only did Mahuka subscribe to Hawaiian-language newspapers, not only did he write to them, in December 1861 Mahuka became a founder of a Coloma, California–based group called the Kaikaina o ka Ahahui Hoopuka Nupepa Ku i ka Wa o Honolulu—the Younger Sibling of the Special Newspaper Publishing Association of Honolulu.[50] Mahuka was the luna hoomalu (presiding officer), and he and the rest of the fourteen members of the organization (which included at least one woman, "Mrs. Mary") sent their moral and financial support for the building of a newspaper autonomous of white missionary and business oversight. This was an indigenous nationalist act. Hawaiʻi, and a Hawaiian national project, continued to be important to Mahuka even as he built a life that tied him to American Indians.

There was a connection in the ways that Mahuka built ties to indigenous people in California and to the indigenous national project in Hawaiʻi. The making of Kanaka Maoli–Indian ties was not foreordained for Mahuka or any of the many other Kānaka along the Pacific Coast. It was not a given that Kānaka would perceive and deepen a connection to American Indian people in 1850 or 1860 just because we term both peoples "indigenous" today. Yet many built lives with and entered into American Indian communities and became kūpuna (ancestors) to American Indian people today. The Hawaiian historian Charles Kenn traces Mahuka's American Indian descendants into the twentieth century, and Jean Barman and Bruce Watson trace the American Indian descendants of Kānaka into the twenty-first.[51] What we must remember is that founding these lineages took choices. Mahuka may have begun his time in North America as a racial cipher, but it took choices by Kānaka and by American Indians for them to make a life together. In fact, they are ancestors of our present, in which we see both Kānaka and American Indians as sharing a common bond as "indigenous."

ATLANTIC COAST: BLACKNESS, "PEOPLE OF COLOR," AND THE QUESTION OF CONTINUITY

When we shift our gaze to the other side of the continent, to the Atlantic Coast, we see that Kānaka also lived among racialized and indigenous people there, in this case people termed "black" by white Americans. As on the Pacific Coast, Kānaka Maoli in eastern towns and cities mostly lived among or in close proximity to other people who by dint of their race were subject to colonization and slavery. Rarely did they live with white Americans. This flies in the face of famous stories about Kānaka in the nineteenth-century United States. As we have seen in previous chapters, the story of a Kanaka convert named ʻŌpūkahaʻia was made famous by New England missionary societies. He lived with white families and embraced Christianity in their homes. His conversion was a spur to New England Christians to initiate and support the missionary effort in Hawaiʻi. But ʻŌpūkahaʻia's famous experience among whites was far from the norm.

The norm for East Coast Kānaka was to live among black people and to be categorized as black. A search of 1850 U.S. manuscript census returns for the eastern United States for people born in the "Sandwich Islands" (as Americans of the time termed Hawaiʻi) reveals dozens of men of working age.[52] Most of these men were mariners or maritime laborers who lived in maritime communities—towns like Nantucket, Massachusetts, or Portsmouth, Maine. The census also located sailors who were in ships at port—for example, the Kānaka aboard ship in the port of New London, Connecticut. Wherever they lived in New England, Kānaka Maoli mostly made their lives among black people. They worked with them on ships. They lived in the same neighborhoods as black people. More often than not they lived in black people's homes. In fact, Kānaka were normally categorized as "black" or "mulatto" in mid-nineteenth-century New England census returns. To understand this better, it makes sense first to look at the living circumstances of these men and then to think about what being black among black people meant for Kānaka in New England in the mid-nineteenth century. Nantucket Island, just outside Boston, was a base used by many Kanaka whalers between voyages. Because fishing and whaling loomed so large in the history of the island, probably the largest concentration of Kānaka Maoli on the Atlantic Coast in the mid-nineteenth century resided there. Fishing was a core part of the subsistence of the island's indigenous Wampanoag. As European settlers occupied the island in the seventeenth and eighteenth centuries, it became a center of commercial fishing and then, in the nineteenth century, a global center of whaling. Native Americans and African Americans were important laborers—and, in the person of Paul Cuffee, even occasional captains—to this whaling fleet. Because Nantucket was second only to New Bedford, Connecticut, as a New England home to the fleet that sailed the Pacific and operated out of Hawaiʻi, many Kanaka sailors and whalers passed through Nantucket. Between voyages, Nantucket became a base for some of them.[53]

Nantucket was also the home of a large black population, and it was with them that Kānaka found a home. Many black men were whalers, but unlike the Kānaka, their population was not entirely male. Many black women and their families took in boarders. In Nantucket, Kanaka lodged in a neighborhood referred to sometimes as Newtown, and sometimes by names that expressed the racialized status of its inhabitants: Negro Town or New Guinea. Newtown was founded by descendants of the first African and African American slaves that whites brought to Nantucket in the eighteenth century. After emancipation became effective in Massachusetts in 1783, these slaves became the leading families of Newtown. By 1822, Kānaka were living among these families—the Barneys, Maxceys, Whippeys, and others. Seven Kānaka were attending Sunday school with African Americans. The population of Kānaka, and other Pacific Islanders (who also lodged in Newtown), continued to grow. In 1825, the local newspaper reported "more than fifty natives of the South Sea Islands [were] employed on board whaleships belonging to this port" and that many were "now on the island."[54]

On shore, the boardinghouse was "the most dominant institution in mariners' lives." One place to find them and other Pacific Islanders in Newtown was William Whippey's "Canacka Boarding House." The name of the residence (a variant spelling of Kanaka), its location, and its management points to the way that Pacific Islanders from different places came together in a place identified with blackness and its black residents. Whippey was "probably half Maori," according to Frances Kartunnen. His life demonstrated the intimacy between Pacific Islanders and people of African descent: he was married to an African American woman and had taken on a surname shared by many African Americans in Newtown. His rooming house made a home for "Kanakas," a term that included all Pacific Islanders in the English of this time. Elsewhere on the island, and in the nearby whaling center of New Bedford, boardinghouses in black neighborhoods were commonly where Kānaka lived. Jack Mowee (Maui), Joseph Fife, and Joseph Henry lived next to Guy Mills, listed as black and born in New York. Other Kānaka like Henry Mowee, Jim Kanaka, Bradley Robinson, and Levi H. Springer were listed as close neighbors to black men from the West Indies. In Nantucket and the rest of New England, Kānaka boarded in the homes of or near to black people.[55]

This residential pattern raises the question of whether sexual intimacies, and possibly kinship ties, connected Kānaka to people called black. But gender imbalances discouraged the creation of heterosexual pairings, and the sailor's transient life discouraged the making of lasting pairings (whether same-sex or opposite-sex) on shore. Distance from Hawai'i meant that any such pairing was unlikely to be recorded in documents that exist in Hawai'i. Kanaka men commonly lived with American-born black men, but it is difficult to determine whether these were more than simple housemates. As for opposite-sex couples, only two including a Kanaka appear in the census records from 1850. John Swain, from Maui, lived with Mary Swain (listed as black) in New Bedford. James Perry, a sailor from Hawai'i, lived with Mary Perry (listed as mulatto) in Portsmouth, New Hampshire.[56] Remember, these men were almost all sailors. As sailors in port between voyages that lasted one, two, or three years, they were highly transient and unlikely to create lasting pairings of the kind we saw among the more settled mining and fishing population in California. In fact, none of the Kānaka in Nantucket in 1850 can be readily found on the census there just ten years later, in 1860. Nor were there as many records in Hawai'i of the type we have for Mahuka—letters written from Hawaiians or reports on their status in the Hawaiian-language newspapers. Mail communication with

Hawai'i was likely less frequent than it was in California, so fewer documents in Hawai'i can shed light on relations of Kānaka with the black people with whom they lived.

WHAT DOES "BLACK" MEAN?

Kānaka lived with black people in New England, but being "black" in New England in 1850 did not mean that one was necessarily of African descent. Many were, and Nantucket was famous for its large population of free people of African descent, some born in the United States, others born in the Caribbean. But the categorization of "black" (and occasionally "mulatto") also applied to a broad range of people on the island who lived in Newtown: men born in Hawai'i, Java, Sāmoa, Tonga, Tahiti, and other Pacific Islands. (One New England whaler of the 1850s even remembered Kamehameha V as a "a big, black Kanaka.") Moreover, American Indian people in New England and the mid-Atlantic states had long shared the categorization of "people of color" with African Americans and were often termed black. Like Kānaka Maoli they often shared spaces together with African Americans. Many families were composed of American Indian women and African American men and their children. Whites (such as the census takers upon whose records historians often depend) often ignored that people of mixed African-indigenous descent often belonged to and identified with American Indian communities. Thus records simply categorize them as "Negro" or "mulatto."[57] Therefore, because Kānaka were termed as black, when one talks about relations between Kānaka and black people, one is in fact discussing the relations *among* black people—some of whom were almost certainly American Indians.

Along the New England coast and further south, Kānaka Maoli were listed as black and lived with others who were labeled in the same way. Significantly, in communities with an overwhelming white majority and only a few black households, Kānaka nearly always lived in blacks' homes and neighborhoods. Southold, a village on the tip of Long Island in New York that was closely integrated with New England's maritime economy, did not have a large black population, but it was with the village's few black residents that Kānaka found their home. Four Kānaka lived there: Benjamin (age twenty-five), Fred (twenty-one), Ebenezer (twenty-six), and John (sixteen). All are listed as black. Their surnames do not appear in the census; the Southold census gave that privilege only to "heads of household." They are listed in a house with six other people, all of who are listed as black, and the census listing raises questions about the relation between them. Sydney and Hannah Hallock (both U.S.-born, and thus not Kānaka) are listed immediately before Amaret, a baby. Given that this matches the convention in the census for listing parents and children, Amaret was likely their baby. Two other women, Mayet and Lyser (again, both U.S.-born) are listed, immediately preceding baby Amelia. Were any of the Kānaka Amelia's father? The source makes it impossible to know for certain but bears witness to the intimacy of domestic spaces shared by Kānaka and others who were called black.[58]

Some Kānaka did become fathers and ancestors of "people of color" in the region, but the transience of sailors' lives meant that connections to communities of color in New England looked different from the long-term residential ties that were growing in California. In 1863, one New Bedford resident made clear that Kānaka were among the groups from which what he called "colored people" in the area got their ancestry. He wrote that the "staple" was African but that foundation was deeply mixed:

> With the staple of the African there have been mingled the blood of Massasoit [Wampanoag] and Uncas [Mohegan]. Fayal [the Azores] has sent its contribution, and the feeble Cannacker [Kanaka] and the sturdy and warlike Marquese [Marquesan] have added to the diversity.[59]

The transience of sailor life makes it difficult to know much about Kānaka who fathered children by women termed black while they were on land, but this document tells us that such children did exist. Furthermore, the census does not reveal the relationship between a woman in New Bedford or Nantucket or Portsmouth with a Kanaka who was away at sea when the census enumerator came to the door. Kānaka thus fathered some children, leaving behind people of Kanaka ancestry in the region, but the transience of sailor Kānaka on the East Coast made the history of their relationship to black people there different from the connections that Kanaka miners and fishermen built with American Indians.

We need to think about how the population of color had developed to understand why it included such a variety of origins. By the time that settlers in the British colonies fought for independence from the Crown, the making of families between American Indians and African Americans was already the foundation of the category of "people of color" in the region. As African-descent and American Indian people interacted and intermarried, we see the same combination of self-identification and ascribed categorization from without as I have referred to in the African American and Kanaka Maoli relationship. This is not to say that American Indian people on the eastern seaboard called themselves "black." But it does suggest that there was a real social proximity of Native people to people of African descent behind the census categorization. At a second stage, as the whaling industry brought people from around the Atlantic and Pacific to New England, racial ascription from white Americans lumped Kānaka from Hawai'i, other Pacific Islanders, Azoreans, and others into the already mixed Native–African groupings of "people of color" and "black." By midcentury, Kānaka in the East had followed their own racial trajectory, toward blackness.

The social geography of the world that Kānaka were exploring included experiencing racial categorization as something lived together with other dark-skinned people as well as imposed by light-skinned whites. In the Kanaka Maoli diasporic experience of racial construction in the mid-nineteenth-century United States, despite the differences among Kānaka and American Indians and African Americans and Fijians and others, all of these could situationally belong to one category. In the 1850 census, that category is called "black," and sometimes its fuzzy boundaries of appearance and of social belonging are indicated with the term "mulatto." Today we might choose a different term than "black." We might instead term all of these as people whose racialization made them variously liable to colonization, enslavement, and labor exploitation.

This is even underscored by the very exceptional circumstances of the few Kānaka who lived in the homes of people who were called white in the census. Joseph Banks, a forty-two-year-old Kanaka laborer who was categorized as a mulatto, lived in Queens, New York. He shared lodgings with a man who was listed as white in a neighborhood (present-day Elmhurst) that was overwhelmingly white in 1850. But when we look at the birthplaces of Banks's lodging mate and his neighbors, an important ambiguity arises: they are largely Irish born. As Noel Ignatiev has demonstrated, the Irish in America dwelled at this time at the fringes of whiteness. They were racialized as different and Other, and their homeland was subject to British colonialism. Already in 1850 and in coming years, they were violently distancing themselves from African Americans and other nonwhites like Banks, and in the process claiming whiteness as their own.[60] Kānaka inhabited a world of shifting racial categorization. Unlike Banks, most did not live among whitening people like the Irish. In their geographical proximity and their domestic intimacy with other black people, we see Kānaka taking their place on the other side of a great global racial divide: with people of color.

THE LIABILITY OF KANAKA BLACKNESS AND OF KANAKA INDIANNESS

What did it mean for Kānaka Hawaiʻi to be black in the eastern United States in 1850? That depended especially on who was terming or treating Kānaka as black. When white Americans assigned the classification, blackness was a marker of racial inferiority. When the *Nantucket Enquirer* published an article describing Hawaiʻi for its readers in 1833, it informed its readers that to know how "highly amusing" meetings of Kānaka were for a white person to look on, all they had to do was "imagine a motley group of darkies."[61] But being called black—whatever one's Kanaka, African, American Indian, or other ancestry—carried liabilities in New England that were not amusing at all. Living in places like Portsmouth or Nantucket or Freehold gave Kānaka up-close and personal lessons on the meaning and power of race in American society. Black roommates, neighbors, Sunday school classmates, and fellow workers were subject to racial oppression, and (after the passage of the Fugitive Slave Act in 1850) they lived in real danger of being captured, enslaved, and sent to the South. On the other hand, when black people took Kanaka sailors into their homes and communities, they made a space for them in this new place. Thus there was inclusion in blackness, and there was a liability to blackness.

But then again, to return to California, there was a liability to Indianness, as well: overwhelming and officially sanctioned violence and the fact that one could be dispossessed at will. In the 1960s, Henry Azbill told the white ethnographer Dorothy Hill many stories about his great-grandfather Ioane Keaala's life among Indian people. Many have been reprinted and retold. To outsiders, they may appear romantic. To the present generation of Kānaka Maoli, they may affirm a connection to indigenous people on the American continent. But this is a connection that came at a steep cost. One story has apparently never made it out of the archives and into the public conversation. In 1966, Azbill explained: "Whenever Keaʻaʻla tried to establish himself, and worked some land, a white man would file on it." These repeated dispossessions and dislocations "drove him crazy." According to Azbill, in the early 1870s, driven to despair or fury or perhaps some combination of the two, Keaala "killed his wife ... and committed suicide." The children were split up, the boys going to live with "uncle Mahuka" in Vernon, the girls going to Chico.[62]

Keaala had lived in California for three decades and knew what it meant to become a member of an Indian community and to accept racialization. White mobs had been attacking Indians and running off nonwhites since the height of the gold rush: Chinese, Indians, Mexicans, Chileans—all were liable to dispossession by whites. Yet Keaala fled the liabilities of neither Indianness nor racialization—and would have lacked the resources to do so if he tried. Instead, like most Kānaka living and working far to the east in New England, he accepted the liability of racialization. Like many Kānaka living and working on the West Coast, he accepted the liability of Indianness. Keaala entered into the Indian community and cast his lot with them. That lot included being dispossessed of his land, just like any other Indian. It was part of what he took on when he made his trajectory toward Indianness. In the end it was too much to bear.

This gives us another way to look at Kānaka Maoli on the East Coast of the United States. Their most easily documentable contact with other black people was residential proximity—its own kind of intimacy—and this was in part an effect of segregation in port towns and cities. This is the repressive side of their residential closeness. But we must remember, as we did when thinking about the bonds between Kānaka Maoli and American Indians on the West Coast, that racial designation is also a means of shared identification. In other words, the making of Kānaka into black was an imposition from whites on the outside, but African Americans and Kānaka also acted to build the connections that bound them *together*.

We have already seen that Edward Mahuka and other Kānaka in Coloma, California, subscribed to Hawaiian-language newspapers, and even supported the important movement to found the first newspaper under Kanaka Maoli editorial control in 1862. They did that even as they enmeshed themselves further in Native Northern Californian life. Making new connections to Indians and deepening connections to home politics were not separate processes, let alone contradictory ones. Rather, they suggest that local connections to Indians helped them to imagine themselves as part of a free and sovereign Hawai'i. Local connections and homeland connections tied them to a broader world of indigenous people, a world we are very much interested in today. And in California as in New England, Kānaka built connections with racialized people. Racialization was something they could not always evade.

This made their lives different from royal travelers in the nineteenth-century United States. The famous episode in Hawaiian history (mentioned near the opening of this chapter) concerning Princes Alexander 'Iolani Liholiho and Lot Kapuāiwa's expulsion from first-class train seats is noteworthy in Liholiho's diary in part because it is one of few examples of explicitly stated racism in it. Due to their high status and wealth and their being accompanied by a white American teacher, it is likely that most of the racism they faced in America was more subtle. Denied the cushioning benefit of elite status and wealth, the working-class Kānaka who lived as black under white American racism daily undoubtedly gained a deep appreciation for what racism meant.

Kānaka who returned to Hawai'i brought those lessons home with them and were able to share the lesson there. Such lessons, moreover, could transfer across boundaries of rank and gender in everyday conversation. In February 1884, Queen Emma (the widow of Alexander Liholiho, who reigned as Kamehameha IV) hired a Kanaka plumber and musician named Poka to install gutters and a water tank on her home. She apparently knew him well and (in an act that points both to her graciousness and suggests that ali'i and maka'āinana [commoners] of this time were not always as distant from one another as we imagine) invited him to eat an early afternoon "breakfast" with her. He had been to California not long before, where he had worked with "the band boys"—presumably the Royal Hawaiian Band. He gave her a generally unfavorable opinion of the visit. The group was unable to find lodging, the queen wrote, on account "of their being slighted for their color (dark)." Emma was quite familiar with Haole and fluent in Anglo-American culture; her hānai father was an Englishman, and she had been educated by American missionaries at the Royal School. Her station, however, likely protected her from the most egregious racial abuse. Thus while racism such as Poka recounted was not a surprise to her, it was freshly brought to her attention by a working-class Kanaka. Kānaka Maoli who traveled to the United States gained clear lessons about what race meant in America and the world and what side of the racial line they fell on, and they brought those lessons back to Hawai'i with them.[63]

On both sides of the continent, Kānaka came together with racialized people, lived among them, and cast their lot with them. For all the liabilities that this entailed, most Kānaka who stayed on the continent for extended periods seem to have made this choice. Unlike Prince Liholiho on the train in Virginia, they do not seem to have vigorously insisted that they should not suffer the indignities and injustices that racialization entailed. Instead, they made their lives among American Indian and African American people.

But to the extent possible, they seem not to have abandoned their homeland or their ties to the Pacific. For Kānaka in New England, the nearness of other Kānaka eased the isolation that distance from Hawai'i caused. Those who did not return to Hawai'i could maintain local ties to other Kānaka and other Pacific Islanders—as 'Ōpūkaha'ia did with his friends, and as sailors in Whippey's Canacka Boarding House did on Nantucket. For Kānaka in California, this was easier, especially for those who worked in Northern California sites like the gold country or Vernon. The economic capital of their region, San Francisco, was in constant communication with Honolulu via the ships that sailed back and forth between them. We have already seen that Edward Mahuka and other Kānaka in Coloma, California, subscribed to Hawaiian-language newspapers, and even supported the important movement to found the first newspaper under Kanaka Maoli editorial control in 1862. Kānaka in the nineteenth century entered into American Indian communities, living and dying with them, but they also kept their ties to Hawai'i and other Pacific Islanders when possible. And for many Kānaka in these same years, religion—both Christianity and non-Christian Hawaiian religion—was a crucial way of imagining relationships that placed Hawai'i at the center of a global geography of sacred power, not at its periphery.

6
BONE OF OUR BONE

THE GEOGRAPHY OF SACRED POWER, 1850S–1870S

In October 1860, Iosepa Opunui, a member of the Kanaka community in the California gold-mining town of Coloma, wrote to the *Hoku Loa* newspaper with tears of joy in his eyes. He wrote that he was deeply moved by the recent issue of that newspaper that had reported extensively on the work of Kanaka missionaries living thousands of miles from home in the Nuku Hiva archipelago and Micronesia. He praised the missionary-affiliated newspaper (to which he subscribed): it was an aid in the work to "hoomalamalama" (enlighten) those far-off islands and "ko makou one hanau hoi" (our birthplace as well). Moved by this effort at Christian enlightenment, Opunui and his neighbors founded the Ahahui Misionari, Coloma, Kalaponi (Missionary Society, Coloma, California) and dedicated prayers and dollars to support the Honolulu-based Ahahui Euanelio Hawaii (Hawaiian Evangelical Association). Opunui's letter supported their efforts as a whole but paid particular attention to Nuku Hiva, calling on all Hawaiians in California to send financial support for the missionary efforts among the people there. He explained it meant so much to him that Kānaka were missioning to Nuku Hivans: "No ka mea, o lakou no ka iwi o ko makou mau iwi, a me ka io o ko makou nei mau io, nolaila ko'u waimaka helelei" (Because they are indeed bone of our bone and flesh of our flesh—and thus my falling tears).[1]

There is an interesting grammatical uncertainty here: When Opunui said "they" were bone of Hawaiians' bone and flesh of Hawaiians' flesh, did he mean the Kanaka missionaries to Nuku Hiva, or did he mean the native people of Nuku Hiva? He might have meant either or both. The missionaries he is talking about were Native Hawaiian preachers and ordained ministers that the Hawaiian Evangelical Association sent out to win converts for Protestant Christianity in the Nuku Hiva archipelago.

As Kānaka Maoli, the missionaries could clearly be "bone of our bone" to Opunui, and thus deserving of support. But so, too, could the people of Nuku Hiva—*if* Opunui imagined them as connected to him in this profound way. This reading of Opunui's letter suggests that he felt a sense of identification with the Nuku Hivans and therefore a special responsibility to sponsor missionaries to them. Note that in Opunui's wording, that kinship to Nuku Hivans *preceded* what he hoped would be their embrace of Christianity. American missionaries and Hawaiian church members frequently referred to church members as hoahānau, or cousins. In this usage, the kinship came from Christianity. But in Opunui's letter, the hoahānau kinship to Nuku Hiva may have been older, deriving not from conversion, but from bone and flesh kinship ties between Kānaka and Nuku Hivans. After all, Opunui was writing in response to an article that had asked, "Ko oukou mau hoahanau ponoi lakou, aole anei? Ua like pu ko lakou helehelena me oukou, ua ano like ka lakou olelo me ka olelo Hawaii. Me he ohana hookahi la" (They are your own cousins, are they not? They look like you, and their language is kind of like the Hawaiian language. Like a single family).[2] The only difference between Hawai'i and Nuku Hiva? "Ua loaa i ko Hawaii nei i ka malamalama e ola'i. Ke noho nei ko Nuuhiva iloko o ka pouli, a me ka malu o ka make" (The people of Hawai'i received the light of life. The people of Nuku Hiva are living in darkness and the shadow of death).[3] The physical and linguistic similarities of Nuku Hivans to Hawaiians proved they were kin and made Hawaiians responsible to enlighten their benighted cousins to the south by bringing them the Christian gospel.

It should be noted that Americans (including missionaries) played an important role in affirming the Pacific bonds between Pacific Islanders, especially the closely related groups that they termed "Polynesians." This runs counter to the present-day spirit of asserting this kinship in a "sea of islands," a kinship that resonates powerfully with a decolonial sense that Pacific identities precede, endure, and will persevere despite the belittling carving up of the ocean into so many allegedly small, isolated, and powerless spots of land. Yet note that the above passage that points to Kanaka–Nuku Hivan kinship and similarity is in the second person—"They are your own cousins, are they not? They look like you." The voice suggests that the writer was a non-Hawaiian, probably an American (given their leadership at *Ka Hoku Loa*, which published the piece). The piece points to kinship and also emphasizes Kānaka's allegedly superior level of enlightenment in regards to other Pacific Islanders. In this piece, a Haole voice encourages a Kanaka to identify with other islanders, but also to feel superior to them. But the sense of identification was not a Haole invention, as Opunui's words make very clear. Even the biblical phrase he references—"bone of my bone"—emphasized how this call for Christian evangelization was deeply rooted in Hawaiian beliefs. Although it is drawn from Genesis, "ka iwi o ko makou mau iwi" (bone of our bone) also evokes the powerful and positive association that bones held for Hawaiians. People guarded, respected, and treasured the bones of their dead. The protection of iwi kūpuna (ancestral bones) is still of the highest importance to Kānaka today. And most important for understanding a phrase like "iwi o ko makou mau iwi," Hawaiians refer to bones in a range of potent expressions that have iwi as a root. A kulāiwi is one's homeland. Iwikuamoʻo, which means backbone, is a term for a close relative to whom one is tied by a bond of loyalty.[4] Practices and words such as these express the meaning of bones for Hawaiians, whose very term for themselves as indigenous people—ʻŌiwi—invokes the power of iwi.

This chapter uses Opunui's brief letter to open up a window on the ways that some Kānaka Maoli at home and in diaspora in the middle of the nineteenth century thought about their country as having a very important place in what we can call a geography of mana, meaning sacred power. In Kanaka geography, the knowledge that the islands, like the people, were descendants of the akua (gods), and the knowledge of the action of the akua in particular sites in the islands, meant that the islands were imbued with mana. But beginning in the 1820s, American missionaries had taught Kānaka that theirs was a land of darkness in need of Christian light. Missionary descriptions of world geography and Hawaiian geography tended to portray the lands of Hawaiʻi as desacralized, even profane. But some Kanaka Christians and intellectuals used both Christian and non-Christian Hawaiian means to meet this challenge head on. Opunui's letter, for example, invokes a notion, common by midcentury among Kānaka and studied in depth by the historian Kealani Cook, that Hawaiʻi was a land of Christian enlightenment that could and should bring the light of the gospel to lands in darkness, especially other Pacific Islanders to whom Hawaiʻi was tied by bonds of kinship.[5] This notion placed Hawaiʻi as a site of enlightenment in the global geography and regional geography of spiritual power. With this notion, Kānaka resisted colonialist geography, but they also replicated and participated in its structures. We hear colonialist echoes in Opunui's celebration of Hawaiians bringing enlightenment to their benighted cousins to the south. Kanaka Congregationalists thus both resisted and replicated the colonialist geography of sacred power.

These ideas coexisted in the complex cultural context of the time with another notion that also contradicted the image of Hawai'i as a land of spiritual darkness. The nineteenth century saw Kānaka research, publish, and disseminate mo'olelo (stories) that affirmed that mana imbued Hawaiian places. Nineteenth-century Kānaka maintained the long-standing forms of Hawaiian geography that was embedded in these stories—especially the knowledge of the migrations that brought akua to the islands, the places they were from, and the knowledge of the wahi pana (storied places) of the archipelago. This kind of assertion of Hawai'i's place in a geography of spiritual power obviously existed in tension with the notion of Hawai'i as a paragon of Christian enlightenment. Yet these apparently contradictory initiatives were deeply related, and overlapping groups of Kānaka pursued them. Both initiatives posited that Hawai'i was a space of spiritual power, not a land of darkness or spiritual need. Both responded to Western lessons in geography that restricted Hawai'i to the benighted periphery of the world, dependent on the Haole for spiritual illumination. Both positioned Hawai'i as a central site in the geography of sacred power.

SACREDNESS, POWER, AND PLACE

Cook demonstrates that the belief that Hawai'i was a source of enlightenment for other Pacific Islander nations was linked to a dynamic that was at the core of Hawaiian cultural, religious, and political history in the nineteenth century. Between the 1840s and the 1860s, he argues, "the devout core of Native Hawaiian Congregationalism embraced a Protestant vision of the world as separated between the mālamalama, the light[,] and the pōuli, the dark." The "greatest threat [to] the lāhui" (the Hawaiian national community) lay in "the remnants and reminders of their own na'aupō, unenlightened, past." One means of differentiating themselves from that supposedly dark past was to evangelize to other Pacific Islanders, whom these devout Congregationalist Kānaka figured as "archaic versions of Hawaiians, of 'how we used to be.'" This simultaneously distanced Kānaka from the pōuli (darkness) and "strengthen[ed] their status as a lāhui naʻauao [enlightened nation], placing them on an equal theological and cultural footing with the empires."[6] This chapter emphasizes that this missionary initiative was one of a range of ways that Kānaka responded to American missionaries' attempts to desacralize Hawaiʻi's landscape and place in the world.

To understand the challenge of the Christian desacralization of Hawaiʻi's geography and the response to it, it is important to discuss three key concepts: kapu, mana, and wahi pana. Kapu is a cognate of the Tahitian word *tapu,* which gave rise to the English word "taboo." In part for this reason, English-speakers might assume that "a kapu" simply means "a prohibition." In fact, the meaning of kapu is much more expansive. Mary Kawena Pukui and Samuel Elbert note that kapu can mean a "taboo" or "prohibition," but can also mean a "special privilege or exemption from ordinary taboo." These terms seem contradictory because they are based on Western categories. The idea that joins prohibition and exemption from prohibition in a single word is more fundamental and appears as the definition goes on: "sacredness; prohibited, forbidden; sacred, holy, consecrated … sanctify." When something is declared to be kapu or is made kapu, it is surrounded by particular rules of conduct that correspond to its sacred status. A kapu, then, is not necessarily a marker of what is forbidden because it is wrong (an immoral act, a disgusting food, etc.) but a marker of what is restricted because it is special (acts that can only be performed at certain times or by certain people, foods that can only be eaten at certain times or by certain people, etc.).[7] Kapu was a central structure in the religious and political order of Hawaiian society, and as that society underwent rapid change in the decades after the arrival of Europeans, kapu came under pressure. In 1820, when the first American missionaries disembarked in Hawaiʻi, they learned that they had come at a moment of great transformation in Hawaiʻi that is remembered as "the overthrow of the kapu." That moment came out of a long and complex process. For decades, foreigners had flouted many of the islands' kapu, and apparently not suffered the wrath of the aliʻi (chief) or the gods for their actions. Some Kānaka had similarly broken important kapu, most famously the requirement that men and women eat separately, called the ʻai kapu (the kapu eating). It appears that they may have begun doing so when they were among foreigners, aboard foreign ships at harbor, and out of sight of the kāhuna (priests). Moreover, rules about kapu eating were lifted in the period of mourning for Kamehameha, who died in May 1819, as was customary following the death of a mōʻī (monarch). The change was limited only to the chiefs, did not extend throughout the land, and could be assumed to be temporary. But as the months went past, enormously powerful women, Kaʻahumanu (the kuhina nui, or regent, who ruled with the young King Liholiho), Keōpūolani (Liholiho's mother, by genealogy the highest-ranking individual in the islands), and the powerful and influential Kapiʻolani, seem to have come to oppose the ʻai kapu on a more permanent basis. They did not return to eating separately from men, and at Kailua they quietly ate in mixed company. Under their influence, Liholiho broke a number of key kapu related to gender. He ate dog meat that was reserved only for female aliʻi, entered a house that was noa (unrestricted) to them but forbidden to him, and touched items that were kapu to him. In a culminating moment in early November 1819, Liholiho made an overt act of sitting down to eat with women in the midst of a well-attended feast. There was an uproar. Within a week, Kānaka had laid waste to heiau (temples) around the island, and soon around the archipelago—despite the firm opposition to the overthrow of the kapu system by other high-ranking aliʻi, notably Kekuaokalani. As historian Lilikalā Kameʻeleihiwa has argued, breaking the ʻai kapu was an act of enormous political importance.[8] In optimistic moments, the missionaries thought that the breaking of kapu (those concerning eating and others as well) meant that they were entering into a spiritual void that they could quickly fill with Christianity. But the missionaries were very much mistaken when they allowed themselves to think this meant the islands were a spiritual tabula rasa on which they could build a Christian church, and historians have been mistaken to believe that the missionaries arrived in a "spiritual vacuum."[9]

The specifics of certain kapu were in question, but Hawaiian practices and beliefs—especially those tied to places—remained vital. Central to these practices and beliefs was the idea of mana, a Hawaiian word that is often translated as "spiritual power." Mana is both power itself and "the aura of power" and its political dimension is inescapable: it carried (indeed was) "an inherent quality of command and leadership." But mana was not mere human power. It "owed its primary origin to the gods." Mana imbued both people and places and things—"*mana* could be emitted from a rock, the bones of the dead, the medicine that cures or the potion that kills." Some people, especially leading aliʻi, were suffused with enormous mana.[10] So too were certain places, many of them the sites that are referred to as wahi pana. Wahi pana is sometimes figured in English as "sacred sites." A better translation would be "celebrated, noted, or legendary places." The term contains in it the notion that these sites are socially and discursively constructed. The telling of stories about a place—especially stories that narrate the deeds of akua and other beings there—is often what makes that place a wahi pana.[11] Moreover, many wahi pana are of the utmost political importance. In the ideational structure of early nineteenth-century Hawaiʻi, the sacred could not be separated from the political, and thus the mana of wahi pana would be described in English as both spiritual and political power. For example, no wahi pana is more literally exalted than the summit of Mauna Kea—the highest peak in the islands of the Pacific. The famous song "Hānau a Hua Kalani" recounts that this summit is a child born of the mating of the akua Wākea and Papa and that Mauna Kea is an honored ancestor of the Hawaiian people. But this kapu site, this site bound by sacralizing rules, is also of great political importance. "Hānau a Hua Kalani" was composed to celebrate the birth of Kauikeaouli in 1813. Kauikeaouli was the son of Kamehameha I and his highest-ranking wife, Keōpūolani. He grew to reign as Kamehameha III.

The song (as transcribed and translated by Mary Kawena Pukui and Alfons L. Korn) opens by narrating a difficult birth:

O hānau a hua Kalani
O hoʻonā kū i luna,
O momoe o maʻule ka piko.
O kolokolo ia pō ke ēwe,
O mulea, o malahia ka nalu, ke aʻa.
O hoʻonā kū o ka malama,
O kaʻahē a ke ʻīloli,
O hoʻowiliwili e hānau Kalani.
ʻO ia hoʻi, ʻo Kalani, hānau Kalani.

The chiefess gave birth,
she bore in labor above,
she lay in a faint, a weakness at the navel.
The afterbirth stirred at the roots, crept in darkness,
in waves of pain came the bitter bile of the child.

This was a month of travail,
of gasping labor,
a writhing to deliver the chief.
He is this chief, born of a chiefess.[12]

As this is a birth chant for Kauikeaouli, we know that the words tell us of the long labors of Keōpūolani to deliver the child. Indeed, the very difficulty of the birth ties this song to Kauikeaouli: he was stillborn following a terrible labor, and only breathed and lived following the prayerful intercession of the kahuna Kamaloʻihi. But the following stanza makes clear that this song chants of more than just one birth:

O hānau ka honua, a mole ka honua.
O kolokolo ke aʻa, ka weli o ka honua.
O lani weli ka honua, o lani ʻiʻī.
O holo pū ka mole, o ʻuʻina ke aʻa,
O hale ka pou lewa ka honua.
O pali nuʻu ka honua, ākea ka honua,
O honua kū, o honua noho ka honua,
O honua lewa, o honua paʻa, ka honua,
Ka honua ilalo, ilalo nuʻu ka honua.
O honua a Kea, na Kea ka honua.
O honua a Papa, na Papa ka honua.
ʻO ka hiapo honua a Papa i hānau.
ʻO ia hoʻi, ʻo ka honua, hānau ka honua.

Born was the earth, rooted the earth.
The root crept forth, rootlets of the earth.
Royal rootlets spread their way through the earth to hold firm.
Down too went the taproot, creaking
like the mainpost of a house, and the earth moved.
Cliffs rose upon the earth, the earth lay widespread:
a standing earth, a sitting earth was the earth,
a swaying earth, a solid earth was the earth.
The earth lay below, from below the earth rose.
The earth was Kea's, to Kea belonged the earth.
The earth was Papa's, to Papa belonged the earth,
the earthly firstborn borne by Papa.
He is the earth, the earth that was born.[13]

This stanza chants the birth of a land, born to the akua mother, Papa (or Papahānaumoku in its fuller form), and sired by the akua father, Kea (or Wākea, as he is more generally known). As the "Mele a Pakui" explains (see chapter 1), these were the parents of a number of the Hawaiian Islands. By placing the birth of the chief next to the birth of the land, and by similarly emphasizing their drama, the birth chant for Kauikeaouli evokes an identification between the newborn chief and the land, his royal mother and the earth-mother god Papahānaumoku (roughly, Papa-who-births-islands), and his royal father and the sky-father god, Wākea. Note that these identifications sacralized the aliʻi and also the land, and their sacredness is bound up together in kinship. The chant then goes on through a series of stanzas revealing that the mana of the cosmos is tied up in the chief's birth. Each stanza narrates one birth: of the night (a reference to the meaning of Kauikeaouli, "Placed in the Dark Sky"), of the island of Hawaiʻi, of the cloud (another reference to Kauikeaouli's name), of the mountain (an allusion to Mauna Kea, which bears the name of its father-god, Kea), of the sun, and of the ocean. All of the cosmos and its mana are united in this sacred birth.[14]

Although it is cosmic, this mana is neither placeless nor impersonal: it is tied to Mauna Kea and Kauikeaouli. In terms that can refer to both the mountain and the man, the song chants:

O hānau ka mauna a Kea,
ʻŌpuʻu aʻe ka mauna a Kea.
ʻO Wākea ke kāne, ʻo Papa, ʻo Walinuʻu ka wahine. . . .
Hānau ka mauna, he keiki mauna na Kea . . .
ʻOia hoʻi hā, o ka mauna, hānau ka mauna.

Born was Kea's mountain,[15]
the mountain of Kea budded forth.
Wākea was the husband, Papa Walinuʻu was the wife. . . .
born was the mountain, a mountain-son of Kea . . . .
He was this mountain's growing, this chief: so was the mountain
   born.[16]

These verses give the name of Mauna Kea in its fuller form: "ka mauna a Kea," "the mountain of Kea," with the grammatical particle "a" (of) indicating Wākea's fatherhood of the mountain. Yet the rest of the mele (song) reminds us that Wākea is also the father of Hāloa, who is the ancestor of all Kānaka. The mountain's meaning is tied up not only in its height, but also in its kinship with the gods and the Kānaka and its identification with the newborn high chief. The mana of this kapu place and the mana of this kapu infant are cosmic, but they are also tied to a specific Hawaiian site.

Mauna Kea is just one example of a truth in the Hawaiian order of things: just as kapu, mana, and wahi pana were concepts that were tightly intertwined, the spiritual, the political, and the geographical were bound together in a way that legitimated Hawaiians' sense of the specialness of their place in the world. Kānaka understood Hawai'i to be an archipelago born from and inhabited by gods. (These ideas are still potent today. The defense of Hawai'i and its sacred sites against destructive "development"—such as the 2015 struggle against plans to build a nineteen-story telescope on the summit of Mauna Kea—still merges Hawaiian ideas of the spiritual, political, and geographical. Indeed, the 1813 birth song "Hānau a Hua Kalani," discussed above, has been a touchstone in that movement.) As Kānaka moved through their islands, they experienced a landscape that reminded them that they and their islands (who were their kin through their mutual ancestors Papa and Wākea) were born of the akua and that the akua continued to be present and act there.[17]

THE MISSIONARY EFFORT TO DESACRALIZE THE HAWAIIAN LANDSCAPE

For those who aimed to Christianize or colonize Hawai'i, the Kānaka sense of the mana of their geography was a challenge to overcome. To succeed in their efforts at evangelizing Christianity and Westernization more broadly, the missionaries would have to confront this sense of sacred landscape. We can catch a glimpse of the way that missionaries did this by looking at Western-style schooling on geography and such geography texts as *He Hoikehonua* (see chapter 4).[18] Although it was intended as a general geography textbook, it also featured content that was specifically religious. The 1836 edition of *He Hoikehonua* demonstrates how the missionaries used academic geography in the effort to undercut Hawaiians' sense that their archipelago was unique and sacred.

He Hoikehonua divides the world into four great parts: the "pegana" (pagan), the "Mahometa" (Mohommedan), the "Iudaio" (Jewish), and the "Kristiano." Of the "pagan," it says:

> Oia no ka poe pule i na 'kua wahahee—i ka la, i ka mahina, i na Hoku, i na muliwai, i na holoholona nui, i na ia o ke kai, i na manu o ka lewa, i na kii i kalaiia, a me na mea liilii e ae he nui loa.

These are the people that worship false gods—the sun, the moon, the stars, rivers, large animals, the fish of the sea, the birds of the skies, graven images, and many different small things.

Contrast this to the depiction of Christians as high-minded people: "Ua manao lakou e ola ana ka uhane o na kanaka i ka wa pau ole, ke paulele lakou maluna o Iesu" (They think that the soul will live forever, [and] they have faith in Jesus). The book evinced its own faith that the superiority of Christianity would overwhelm other religions. "Ke i mai nei ka olelo a ke Akua, e hiki mai ana ka manawa e lilo ai na kanaka a pau, i poe Kristiano" (The word of God speaks forth, and the day will arrive when all people will become Christians).[19]

To convince Hawaiian students that they belonged to just one confused, "pagan" people among many, the Hawaiian textbook deploys what William C. Woodbridge, the author of the American original, declared in his title to be a central project of his book: to teach "comparison and classification" in geography.[20] The book is replete with examples of other benighted lands to which Hawaiians could compare themselves and with whom they might classify themselves. Examples are global, but particular emphasis is placed on the ignorance of Pacific nations from Nuku Hiva to Australia.[21] Woodbridge's system of classification (as adopted by the missionaries in the textbook they printed and used to teach teachers) told Hawaiians they were inferior, and his system of comparison served to tell Hawaiians that they were not unique.

No example demonstrates this more powerfully than the section on Hawaiʻi Island, which contains the following passage on volcanoes:

> I ka wa mamua, ua manaoia e ke kanaka, he Akua o pele, a hoomana lakou ia ia. Aole lakou e ike, aohe no i lohe, ua nui

na pele ma na aina e. I keia manawa, ua maopopo o Iehova ke Akua, nana i hana i na pele a pau, a me na mea a pau. Aohe Akua e ae; oia, hookahi wale no.

In earlier times, people believed volcano was a God, and it was worshipped by them. They did not know and had not heard that there were many volcanoes in other lands. Now they know that Jehovah is God, [and] it was He who made all the volcanoes, and all things. There is no other God, only one.[22]

This passage further underscores the spiritual dimension of the lesson through careful wording and capitalization choices. One might expect the text to state that "ua manaoia e ke kanaka, he Akua o *ka* pele"—the people believed that *the* volcano was a God. Instead, the text reads that "ua manaoia e ke kanaka, he Akua o pele"—people believed volcano/pele was a God. The missionary text exploits the fact that pele (meaning volcano) and Pele (the akua of the volcano and of fire) are the same word. A geographical text becomes a religious attack—an example of the ways that the missionaries' global geography sought to desacralize and marginalize Hawaiʻi by making Hawaiian geography (and the whole social, cultural, and political system of which it was a part) one mistaken belief among many mistaken beliefs of benighted people around the globe.

Note that the American missionaries did not seek to replace the Hawaiian geography of the sacred (wahi pana and connection with Hawaiʻi's many akua) with a secular geography. After all, they believed that their own god's power was active in the world. Instead, the missionaries forwarded two different conceptions of the spatial nature of what they would understand to be God's power. First, they implied that that power flowed from a place they called "ka ʻĀina Hemolele" (the Holy Land). This is the sense of sacred spaces that we encounter in a textbook titled *O ka Hoikehonua no ka Palapala Hemolele* (The Geography of the Holy Scripture), first published in 1835, then expanded and republished only three years later in an edition of two thousand copies. Biblical geography would remain a topic on which future ministers would continue to be trained and tested for decades to come.[23] The biblical geography textbook is the counterpart of books that are still published today in most of the languages of the world, informing readers about the location and nature of places mentioned in the Hebrew Bible and

Christian scriptures, and even (in the case of the Hawaiian textbook) on esoteric topics such as the possible location of the Garden of Eden.[24] Yet the book can be used to represent a particularly important message in Christianity as the missionaries in Hawai'i preached it: the site of sacred history was a place called the Holy Land, and it was far away—most emphatically, not in Hawai'i. *O ka Hoikehonua no ka Palapala Hemolele* described that place as one whose name had changed many times due to holy events that were recorded in scripture. It had been "Kanaana" (Canaan), then "Ka Aina i Oleloia" (the Promised Land), then "Ka Aina o ka Israela" (the Land of Israel), then "Ka Aina a Iudaia" (the Land of Judea), and then "Palesetina" (Palestine). That was not its most descriptive name, however:

> Ua kapaia hoi, o ka Aina Hemolele, no na mea nui i hanaia malaila a haiia mai maloko o ka olelo a ke Akua; oia hoi ka hanau ana, a me ka noho ana, a me ka hana mana ana, a me ka eha ana, a me ka make ana o Iesu Kristo, ko kakou mea e ola'i.

> It was even called the Holy Land, because of the great things that were done there and proclaimed in the word of God—the birth, life, miraculous work, suffering, and death of Jesus Christ, our savior.[25]

If Palestine was the land that was holy, what could Hawai'i be but profane? This countered the Hawaiian sense of a landscape of *wahi pana*, a landscape with sites infused with *mana*.

This brings us to the second geography of spatial power taught by the missionaries: the division of the world into the 'āina na'auao (enlightened lands) and the 'āina na'aupō (ignorant, benighted lands), as described by Kealani Cook. These terms can be loosely translated as "enlightened countries" (enlightened in the sense of brought into the light of Christ) and "ignorant countries" (benighted in their sinfulness and ignorance of Christ). The Hawaiian terminology is literally visceral because whereas English names the heart or brain as the seat of thought and of spiritual states, Hawaiian locates them in the na'au, or intestines. "'Āina na'auao" can therefore be translated literally as "daylight-intestine lands" and "'āina na'aupō" as "night-intestine lands." These terms evoke the depth of feeling associated with belief in enlightenment or benightedness—conditions that were far more than intellectual. In this dualistic geography of illumination, Europeans and Americans came from enlightened lands to bring Hawaiians Christ's salvation. This geography of light and dark differed from the idea of sacred geography as presented in the book on Bible geography. That book presented Palestine as the Holy Land. In contrast, in the geography of light and dark, the source of enlightenment was not Bethlehem or Jerusalem, but the Western nations (the 'āina na'auao).[26] The geography of the light and dark and the geography of Holy Scripture did, however, share a crucial similarity: both posited that the source of holiness was distant from Hawai'i.

Note that a third possible Christian geography of sacred power would have run counter to both of these ideas that distanced the sacred from Hawai'i: a geography emphasizing the presence of the Christians' god among his faithful. Indeed, one can sense that idea in the words of one of the first Kanaka Christians, Thomas Hopu, who had traveled with 'Ōpūkaha'ia to New England, converted there, and then returned to Hawai'i. In 1828, Hopu wrote from Hawai'i to Edwin Welles Dwight, a friend and mentor from his days in New England. Writing of the relationship between Kānaka and Americans, Hopu stated (in English) that "as two peoples, we met together with the love of Jesus in our hearts."[27] One can question the accuracy of Hopu's description of the encounter between Kānaka and Americans, but the logic of his statement evokes a famous passage from the book of Matthew: "For where two or three are gathered together in my name, there am I in the midst of them."[28] If Kānaka and Americans met "as two peoples ... with the love of Jesus in [their] hearts," Christ had to be present in the place of that encounter—whether it was New England or Hawai'i did not matter. Hopu's letter was just one convert's statement, but it points to the way that Kānaka could, using Christian premises, challenge the idea that they lived in a landscape of spiritual darkness. If Christ was among them, how could Hawai'i be na'aupō? This Christian geography was implicit in the actions of many Hawaiian Christians in the nineteenth century.

RESITUATING HAWAI'I THROUGH DENOMINATIONAL CHOICE: THE CATHOLIC, EPISCOPAL, AND MORMON CHURCHES

For many Hawaiians, Christianity would be at the center of response to colonialism. By the middle of the nineteenth century, most Hawaiians were Christian to some degree or another. Of course, numbers are suspect and degrees of devotion are unsure, but in 1853, when the Hawaiian Kingdom had just over 71,000 citizens, it counted 56,840 of them as Protestant (80 percent) and 11,401 as Catholic (16 percent), leaving just 4 percent non-Christian. About 30 percent of the kingdom's citizens were official members of a Congregational church.[29] Measured in terms of numbers of converts, the American Congregationalist missionaries had clearly achieved remarkable results. Still, as the presence of over eleven thousand Catholics in the islands suggests, Kānaka did make choices about how they would affiliate themselves with Christianity. Congregationalism was strongly associated with the American missionary establishment, and Kānaka who chafed at American influence in their islands could look to other denominations, giving a political dimension to the religious question of Christian denominational choice. Options included the Roman Catholic, Episcopal, and Mormon churches.

French Catholic missionaries had first arrived in 1827, but Kaʻahumanu and Kamehameha III (both of whom were loyal to the Congregationalist Church) formally banned the priests beginning in 1829. It was only in 1839 that Kamehameha III extended toleration to Catholicism, and over the coming years, Catholics continued to suffer incarceration and abuse.[30] Growth was slow, but a sizable minority of Catholic Hawaiians did emerge—as we have seen, 16 percent of the population in 1853. It is noteworthy that the Catholic Church in Hawaiʻi seems to have been gaining members in the 1860s. As Cook notes, "In 1861, Father Lyman of Hilo [a Congregationalist clergyman] reported that the Catholics, or Papists as he insisted on calling them, were starting to draw significant support away from the Congregationalists."[31] The 1860s were a time when the politics of culture in Hawaiʻi were at a boiling point, with much of the conflict playing out on newspaper pages. And while Catholicism remained a minority religion in the islands, it did boast two of the most important Kanaka writers of the cultural ferment of the 1860s: Kepelino Keauokalani (who was raised a Catholic) and Samuel Manaiākalani Kamakau (who converted to Catholicism as an adult).

These two Catholics were among the most important figures in the preservation of polytheistic Hawaiian sacred traditions, histories, and geographies of sacred power. Kepelino, a grandson of Kamehameha I, collected and published hundreds of pages of histories, songs, stories, prayers, genealogies, and information on practices from farming to fishing. For his part, Kamakau was (as we have seen) one of the most important Kanaka historians of all time. In part because it was available in English translation, these men's work (along with that of Davida Malo and John Papa 'Ī'ī) long provided the bulk of the documentary basis for the historical study of Hawai'i in exclusively indigenous times and in the first decades of relations with outsiders.

While Kepelino was raised by Catholic parents, it may seem surprising that Kamakau embraced Catholicism. Not only was he a Congregationalist for much of his life (and thus familiar with the anti-Catholicism of the New England missionaries), Kamakau was a proud Hawaiian, resistant to the designs of Western empires. France had threatened war in 1839 to force Kamehameha to declare that Catholicism would be tolerated. France had actually invaded Honolulu in 1849, in part over the imprisonment of Catholics and in part over trade issues.[32] One might expect Kamakau, then, to resist a denomination that was associated with French imperialism. But France did not represent the same threat to Hawai'i's sovereignty that the United States did in the 1860s, when he converted.

Kamakau appreciated many elements of Catholic doctrine and practice, some of which he explicitly likened to elements of Hawaiian religion. Indeed, Kamakau simultaneously distanced himself from Hawaiian religion and suggested that it bore the imprint of Catholicism. Kamakau made these points by indicating a number of features common to Catholic worship and heiau practice. He defended prayer before images of the saints, rejecting Protestants' claims that it constituted idol worship. He likened it to the use of ki'i (images, in this case carved images) of the gods in Hawaiian religion. Kamakau declared that in both cases, images were mere likenesses that helped the faithful to reflect on the divine—not actually objects of worship. He cited communal and simultaneous prayer by the congregation as a practice shared between Catholics and the Hawaiian religion, and even the fact that, he claimed, in both religions the congregation prayed with their right hand pointing toward the heavens. Kamakau took particular care in defending Catholics from charges of polytheism, declaring that believing in the sacred power of angels, or praying to Mary, did not constitute the invention of false gods, but merely asking a being to intercede with the Christian god on one's behalf. Here we can see the nuance and even ambiguity of Kamakau's efforts to defend Catholicism while likening

Samuel Manaiākalani Kamakau. Crayon portrait drawn from an original in the collection of Mrs. B. L. Kamakau of Nāpoʻopoʻo, Kona, Hawaiʻi. Courtesy of the Bishop Museum, Honolulu. www.bishopmuseum.org.

it to Hawaiian religion, because he pulls back from valorizing Hawaiian religion. He explicitly disavowed what he called "ka hoomana pegana" (the pagan religion) and directly stated: "He hoʻomana Karistiano nō ka hoʻomana Katolika Roma; ʻaʻole i like ka hoʻomana Katolika Roma me nā hoʻomana pegana o Hawaiʻi nei" (The Roman Catholic religion is a Christian religion; the Roman Catholic religion is not like the pagan religions of Hawai'i).³³ This raises a conundrum: Kamakau simultaneously disavows Hawaiian religion and uses likenesses to it as a defense of Catholicism. The surprising resolution of this conflict that Kamakau offers is that Catholicism (and, he says, Judaism) had actually merged into Hawaiian religion earlier than anyone suspected. Kamakau writes that in ancient days, in the time of the ali'i 'Auanini, Jews had come to Hawai'i and had come to affect Hawaiian religion. Kamakau also argues for an early influence of Judaism in Hawai'i, claiming for example that some prohibitions (which he terms "kapu") in the Hebrew Bible were the same as kapu in Hawai'i. The impact of Catholicism he traced to the coming of Spanish ships to the islands in the time of Kahoukapu, long before Captain Cook's arrival. Kamakau writes that "ua hui pū 'ia ka ho'omana Karistiano me ka ho'omana kama'āina o ka lāhui i ia wa" (the Christian religion merged with the native religion of the Hawaiian people in that time). Kamakau did not endorse the Hawaiian religion, which he called "pagan." Still, by drawing a connection to Judaism and Catholicism, Kamakau in effect suggests that they give Kānaka a long-established connection to sacred power in religions recognized by the Haole. Even while denouncing Kanaka religion, the result is an affirmation of the spiritual power of Kānaka and their land. In short, Hawaiians do not need to depend on Congregationalism or New England for access to the sacred. They can turn to Catholicism, as Kamakau did, redrawing his place on the map of spiritual power, rejecting the efforts of American Congregationalists to suggest Hawaiians were in a position of spiritual dependency.³⁴

Hawai'i's ali'i nui (high chiefs) and monarchs also chose Christian denominations in ways that recast Hawai'i's place in geographies of sacred power. In 1862, Kamehameha IV (known as Alexander Liholiho before his accession to the throne) created the denomination that would become the Church of Hawai'i. More generally known as the Episcopal Church, the religious body was in conference with the Church of England. It enjoyed considerable success in royal circles. Kamehameha IV's embrace of the Episcopal Church was the outcome of a series of personal experiences that demonstrate how his denominational choice was an intervention in the politics of geography. Alexander Liholiho was largely raised at the Royal School. The strict discipline of Amos Starr Cooke and Juliette Montague Cooke, the New England missionary teachers who ran it, may have already soured him to Congregationalism, and perhaps also to Americans. His opinion did not improve when, in 1849 at the age of fifteen, he had the occasion to experience Panama, the United States, England, and France firsthand on an extended trip with his brother, Lot Kapuāiwa, and Gerrit P. Judd, the kingdom's minister of finance. Alexander Liholiho was already familiar, even expert, in Western ways: like other royals, he could speak, read, and write English, dressed in Western attire, lived in and frequented Western-style homes, and had a command of Western protocol. He thus understood precisely what it meant when the Haole honored him or insulted him. On the trip, Alexander Liholiho preferred Britain over the United States. Whereas he was received with respect and ceremony in court by Prince Albert in England, in the United States he faced racist slights—most notably when a conductor tried to remove him from a train car reserved for whites and when a butler insulted him by giving him a bib at dinner. Whereas he found England to be refined, the United States was crude. And whereas he was impressed by the ritualism of the Church of England, he knew well the austere simplicity of Congregationalist worship from his experiences in Hawai'i. But more was at issue than courtesy, ceremony, or even racism. Hawai'i, like Great Britain, was a monarchy. Alexander Liholiho took the throne in 1855. Because the Church of England was a church headed by a monarch that had demonstrated respect for his own royal status, he could judge Britain to be conducive to the Hawaiian Kingdom, whereas he believed that the United States was a threat to it. Granted, Britain also posed dangers to the kingdom: during the reign of his hānai adoptive father Kamehameha III (Kauikeaouli), Britain had occupied Hawai'i for five months in an event remembered as the Paulet Affair.[35] Americans had participated in the successful effort to secure Hawai'i's sovereignty. But during Kamehameha IV's reign (1855–63), it was American power that seemed a threat. The American population and American commercial influence was growing in Hawai'i. To counter U.S. power and the threat of dependence upon it, the king sought to deepen diplomatic and commercial ties with Britain and other European powers. He invited the Church of England to send a full-scale mission to the islands. Weeks after the arrival of the new Anglican bishop, "Kamehameha IV, his cabinet, and his heir, Kamehameha V, all converted to Anglicanism."[36] The foundation and growth of a church in communion with the Church of England must be seen as a religious initiative in this broader effort to strategically stave off dependency on the United States and shore up the sovereignty of the Hawaiian Kingdom. Kamehameha IV's act worked to move Hawai'i from the periphery to the center of a Christian geography of sacred power. It implied that rather than being a peripheral "pagan" land in need of Christian enlightenment, Hawai'i was a Christian kingdom in fellowship with other Christian kingdoms.

Seen in this light, it is not surprising that the Episcopal Church made such inroads among ali'i and royals. Kealani Cook notes that all but two of the monarchs who reigned after Liholiho became Episcopalians. In fact, "only the two with the shortest reigns, Lunalilo and Lili'uokalani, remained within the Congregationalist flock," and the queen actually demonstrated considerable openness to other denominations: she attended services at St. Andrew's Episcopal Cathedral next door to her home, allowed the nuns there to bring her gifts and meals, may have attended a Catholic church, and accepted Mormonism's presence in Hawai'i.[37]

Mormonism offered Kānaka the chance to declare that they held a privileged place in the geography of sacred power. The Church of Jesus Christ of Latter-Day Saints first came to the islands in 1850, when missionaries were dispatched from California to Hawai'i. As Hokulani Aikau demonstrates, they initially intended to focus their efforts on Westerners in the islands. This was in keeping with the young church's already established base among white Americans and Europeans and Mormon beliefs that American Indians and Africans were descendants of tribes who had rejected Jesus Christ in ancient days. Given the anti-Mormonism of Haole in Hawai'i (both Protestants and Catholics), the church enjoyed little success among them. At a point when the Mormons might have given up on the islands, the Mormon missionary George Q. Cannon had what Aikau calls a "fortuitous" vision: Hawaiians and other Polynesians were descendants of a branch of the House of Israel. The Book of Mormon described how the House of Israel had migrated in ancient days to North America. One of their number, Hagoth, built a ship and sailed with his people into "the west sea," never to be seen again. In Cannon's vision, the west sea was the Pacific, and Hagoth's group were the progenitors of the Polynesians. Aikau notes that Cannon's vision "became the bedrock upon which the church built a genealogical connection between Polynesians and the Book of Mormon." In Mormon thought, far from being a despised, marginal, or inferior race, Hawaiians and other Polynesians became nothing short of "a Chosen People" with "a primordial genealogical connection" to the Book of Mormon.[38] The ramifications for a people as genealogically oriented as the Kānaka were enormous. Kanaka Mormons were uniquely positioned in Christianity to answer the challenge that Congregationalism had posed when it situated Kānaka as benighted and ignorant and distant from sources of sacred power and enlightenment—the Holy Land and Western Christian nations. By becoming Mormons, they entered a faith that named them as descendants of a holy people from the Holy Land. This was a privileged position in the world of spiritual power. Armed with this message, the church grew with remarkable speed, claiming more than four thousand members in Hawai'i by 1854. The Church of Jesus Christ of Latter-Day Saints would become an important and prosperous minority church in the islands and a site of connection between Hawaiians and other Polynesians who joined the church. In other words, embracing Mormonism redrew the geography of sacred power in Christianity, centering what Americans called "Polynesians" in one unified chosen people linked by sacred genealogy.

CONGREGATIONALIST KANAKA MISSIONARY OUTREACH

Yet neither the notion that deep genealogical ties connected Kānaka to other Pacific Islanders nor the idea that Kānaka had a role to play in their conversion was unique to Mormons: after all, Iosepa Opunui was a Congregationalist, and his words at the beginning of this chapter make clear that he believed Nuku Hivans to be bone of Hawaiians' bone and flesh of their flesh. Thus his call to support Kanaka missionaries in Nuku Hiva asserted a geography that preceded and exceeded the notion of the unity of Christian believers. For all they differed from Catholic converts, their Episcopalian king, and the new Mormon faithful, Hawaiian Congregationalists in the 1850s were also recasting Hawaiians' place in the global geography of sacred power. Hawaiian Congregationalist churches supported Kanaka missionaries in Micronesia and the Nuku Hiva archipelago from 1853 to the first decade of the twentieth century. Over more than fifty years, "nearly forty Hawaiian families set out as missionaries, in addition to others who went out as domestics, unofficial helpers, or teachers."[39] The missionary enterprise represented both the work of the people who went abroad and also the enthusiastic support from Hawaiian congregations and associations, including the one in Coloma. In doing so, they figured themselves, again, at a central place in the geography of the sacred, as a source of enlightenment for other places.

As Opunui's words demonstrate, a sense of kinship was important to this effort, and sometimes those bonds were quite recent—not the result of the ancient migrations. For example, the reason that the Hawaiian missionaries went to the Nuku Hiva archipelago was that a person bound to Hawai'i by kinship asked them to. In 1853, a man named Matunui (sometimes spelled Matuunui or Makunui) came via a Western ship to Lāhainā on Maui to meet with the missionaries there. He was a chief from the valley of Omoa on Fatu Hiva. A Hawaiian known only as Puu had came to live in Omoa some time before. Puu was a sailor from Olowalu, Maui, who jumped ship from a whaling vessel that harbored there. He settled in Omoa and built kinship ties. Puu became the kāne (male partner or husband) of Matunui's daughter, Vaitatu—and thus became Matunui's hūnōna (at least as described by Samuel Kauwealoha, a Hawaiian missionary), which translates as "son-in-law." The translation is imperfect, as it suggests a Western and legal marital bond rather than a bond arranged between a Hawaiian man, a Fatu Hivan woman, and perhaps her family.

Matunui had learned that Hawai'i was rich in Western goods that Fatu Hiva did not have. He had also been led to believe by Puu that he was highly ranked and related to Kauikeaouli (Kamehameha III) and the kuhina nui, Ka'ahumanu. Matunui therefore decided that if he could get American missionaries with Puu's help, he might be able to get American goods—especially the guns he wanted to fight Tiiteihipe's people in the uplands, with which he was involved in a long-running conflict. Matunui traveled with Puu to Hawai'i and requested missionaries. Soon, a missionary party of eight Kānaka sailed southward toward Matunui's home islands—the Rev. James Kekela and his wife Naomi Kekela, the Rev. Samuel Kauwealoha and his wife Louisa Kauwealoha, and two married lay couples.[40]

This extension of Hawaiian Christian missions to the Nuku Hivan archipelago continued a pattern that had begun with Tahitian missionaries to Hawai'i in the 1820s. Hawaiians and other Pacific Islanders used migration and kinship connections to turn Christianity into a means to rebuild ancient ties and create new bonds and new opportunities within a geography of sacred power. Note that kinship through Vaitatu, a Fatu Hivan woman, was the basis of the kinship bond between

Puu and Matunui, and those two men's initiative lay at the origin of the first Hawaiian missionary endeavor in Fatu Hiva. The story of Vaitatu, Matunui, and Puu thus centers indigenous Pacific initiative in the history of missions. This was an appealing story for American and even European missionary journals, a number of which remarked upon it: as an example of indigenous people reaching out for Christianity, it reaffirmed their belief that those people were dependent upon them for enlightenment. One French-language journal in Switzerland, for example, declared that when Matunui had come to Hawaiʻi, it was an example of how "these savage peoples" ("ces peuples sauvages") yearned for white missionaries—though when Matunui learned he could not have them, he settled for "an indigenous one" rather than have "none at all."[41] Yet Pacific Islanders were not merely a second choice. Pacific Islanders used bonds of Pacific kinship between them as a means to interact with (and control interactions with) the outside world. Matunui's actual goal may have been Western goods rather than the Christian god, but his vehicle was the mission, and the link was a Pacific kinship.

That sense of kinship had very particular limits, however. First, it was stronger in the case of the Nuku Hivan archipelago than as regarded Micronesia. Nuku Hiva and its immediate neighbors were points of origin of the people who had settled Hawaiʻi centuries before. Kānaka and Haole both frequently remarked upon the similarity between the languages, cultures, and physical appearances of people from Hawaiʻi and Nuku Hivans. Comparisons like these were rarer in relation to the Micronesians, whom Haole emphasized were darker and even less civilized than Nuku Hivans. Not coincidentally, Kanaka church people and members of the Ahahui Euanelio Hawaiʻi showed a particular enthusiasm for the Nuku Hiva mission, funding it in its entirety, whereas the Micronesian mission received support from Americans.[42]

The second important factor that limited this sense of kinship, and gave it a colonialist cast, was the tendency of Kanaka missionaries to identify other islanders as being less enlightened versions of themselves. Kanaka missionaries wrote about the people to whom they missioned in distinctly colonialist and disparaging terms—terms that suggest that in some ways their efforts to other islanders followed the structure of the colonialism of the Americans who had preached to them and their parents. Kealani Cook argues that, just as the Americans portrayed themselves as coming from a land of enlightenment to a land of darkness, the Hawaiian missionaries wrote of traveling from a 'āina na'auao to a 'āina na'aupō. Much as the American missionaries had pejoratively emphasized the smallness of the Hawaiian landmass, Kanaka missionaries similarly derided the small size of islands like Hiva Oa and Manaia. Much as Americans had derided Kānaka as ignorant, stupid, wild, and savage, Kānaka heaped similar slurs (na'aupō, and also hūpō meaning "ignorant," and hihiu meaning "wild") upon the islanders among whom they worked in their reports back to Hawai'i. Kealani Cook delineates a causal relationship between three elements: Congregationalist Kanaka anxieties about whether their nation was truly enlightened, their missionary impulse, and the demeaning and even dehumanizing terms that they leveled at other Pacific Islanders. Hawaiian Congregationalists often fretted over whether their homeland was truly saved: was Hawai'i firm and fixed in its enlightenment, or might it slide back into na'aupō? This anxiety "not only encouraged Hawaiians to undertake mission work, but it also encouraged them to do so in a manner that rhetorically and conceptually distanced them from the Islanders who they traveled thousands of miles to work among."[43] Thus, although Opunui invoked a belief in Hawaiian kinship to the people of Nuku Hiva in his 1860 *Hoku Loa* letter of support for the missionaries working there, he also declared that a gulf lay between the enlightened Hawaiians and the na'aupō people of Nuku Hiva. Indeed, to the Kanaka missionaries who derided the people they worked among, the kinship of Kānaka to other Pacific Islanders could also contain a threat of Hawaiian moral and religious backsliding.

NEWSPAPER PUBLISHING, MO'OLELO, AND WAHI PANA

Increasingly in the 1860s, Kanaka writers gave Native Christians who feared a return to "paganism" another potential sign to worry about: the frequent publishing of Hawaiian stories and songs about akua and wahi pana in the Hawaiian-language newspapers. The initiative to create written versions of the oral literature of Hawai'i can be considered another means to reconsider the place of Hawai'i in the geography of sacred power. By recounting the acts of the akua in very specific sites on the islands, this literature emphasized the relationship between the akua and the 'āina (land), answering the missionary desacralization of the Hawaiian landscape.

The 1860s, 1870s, and 1880s saw an intense effort by Hawaiians to research and record Hawaiian history and what we would now call Hawaiian traditions, especially practices, moʻolelo (stories), and mele. This work was anticipated by the labors of Davida Malo and other Kanaka students at Lāhainā in the 1830s, who researched and generated the historical accounts that one of their teachers, Sheldon Dibble, assembled into *Ka Mooolelo Hawaii*, the first book of Hawaiian history ever published.[44] By the 1860s, many Kānaka (and some Haole) worked to record in writing stories, songs, and practices that had fallen into disuse. Many different Kānaka responded by authoring descriptions and narrations of these texts and practices. To get a sense of the breadth of this initiative and the number of writers involved, one need only look at the first chapter of this book, which draws on stories, songs, chants, and practices recorded by Kanaka writers from this period. Only a few of the writers from the 1860s, 1870s, and 1880s are famous, most notably Kamakau, Kepelino, and John Papa ʻĪʻī. Others are well known, but not thought of as writers, such as the high chief (and later king) Davida Kalākaua, whose works include the first published version of the Kumulipo, a Hawaiian origin narrative of profound importance to Hawaiians. But the numbers of writers who are widely remembered is dwarfed by the dozens of little-known Kanaka writers who penned stories, songs, and accounts of Hawaiʻi's past.

These pieces were printed in Hawaiian-language newspapers and collected into archives, and in some cases they were published in books in the twentieth century. As Noenoe Silva has emphasized, this work was pioneered by *Ka Hoku o ka Pakipika* (The Star of the Pacific), the first newspaper under Kanaka editorial control. She writes, "The battle over whether or not, or how severely, Kanaka Maoli were to be subjugated was largely a discursive one, fought with paper and ink through weekly newspapers." Since their arrival in the 1820s, missionaries had controlled printing in the Hawaiian language. In 1861, a group of Kānaka Maoli put forth the first Hawaiian newspaper that was free of missionary or other foreign control: *Ka Hoku o ka Pakipika*. Those Kānaka demonstrated their pride in their "ways of life and their traditions" and soon began publishing moʻolelo and mele that asserted Kanaka traditions.[45]

Other newspapers followed suit—including some newspapers that had accused *Ka Hoku o ka Pakipika* of promoting pagan, ignorant, and benighted beliefs by publishing moʻolelo and mele. Whites in Hawaiʻi also played a role in this preservationist effort, much as Sheldon Dibble had by editing his students' work into *Ka Mooolelo Hawaii* in the 1830s. One of the largest works was a monumental collection of materials (stories, histories, songs, chants, accounts of practices in fishing and tattooing and kapa [bark cloth] making, and all manner of activities that were deemed "tradition") that was researched and written by Kānaka and assembled by a Swede named Abraham Fornander. Fornander jumped a whaleship in 1844, became a loyal subject to the king and minister to the government, and married Alanakapu Kauapinao (Pinao), a female aliʻi from the island of Molokaʻi, the daughter of a medical kahuna, and the descendant of Maui's illustrious Piʻilani line and of aliʻi who were closely allied with Kamehameha.[46] Her moʻokūʻauhau (genealogy) and her family's learning would have been an asset to Fornander in building his career and in his studies of Hawaiʻi. The collection of materials Kānaka had researched and written and that Fornander assembled into an archive was later translated and published in 1919 and 1920 as the *Fornander Collection of Hawaiian Antiquities*. Like most archives, this one speaks in multiple registers: Fornander's dominant voice emphasizes particularly his interest in the ancient migrations of the peoples he called Polynesians—the subject of the book he published using the archive, the *Account of the History of the Polynesian Race*. Written and published with the sponsorship of the kingdom, it expressed a theme that was becoming increasingly explicit in the 1870s: the effort to place Hawaiʻi not as an outlier or isolated aberration, but rather as part of a larger geography and history of nations that together were called Polynesia. The *Fornander Collection* thus reflects this concern, but it also speaks in the register of the many Kānaka who researched and wrote the materials included in it—Kānaka who worked to preserve stories, songs, and practices from the Hawaiian past in writing.[47]

As Silva has demonstrated, this writerly effort was just one part of a broader cultural movement beginning in the 1860s that involved both popular cultural action by Kānaka and official initiatives of the kingdom. That movement included the revindication of hula and related practices (song, chant, assembly of costume, and instruments) and the learned traditions of the Hale Nauā, a royally supported secret society dedicated to the preservation and advancement of knowledge through both "traditional" means (such as genealogy) and "modern" means (such as Western science).[48] Efforts such as these had stirred in the 1860s with the support of Kamehameha V (Lot Kapuāiwa, who reigned 1863–72), and flourished during the reign of Kalākaua (1874–91). Under Kalākaua, a number of cultural initiatives affirmed the nation through what art historian Stacy L. Kamehiro has termed "the arts of kingship."[49] In a movement that had a classicist dimension, Kānaka preserved, promulgated, and performed hula, songs, stories, and genealogies that situated Hawai'i in narrations stretching back to the ancient migrations of people and gods and beyond—to the very origins of the cosmos. They responded to the belittlement of colonialist discourses that would describe Hawaiians as a people without history who lived in small and isolated places. Hula, story, chant, and genealogy preserved the stories about the movement of gods and people between places in Pacific geography: Hāmoa (Sāmoa), Kahiki (Tahiti), Polapola (Borabora), and so on. For Kānaka, the supposed discovery of the Pacific Islands was in fact a remembering—a reassembling of spaces the names of which had been preserved in sacred stories, but where Hawaiians had not visited for generations. Thus the effort after 1860 to tell and record these stories and songs in writing (which created the documents upon which the first chapter of this book depends) places Hawai'i back into an extensive and organically connected Pacific geography. The production of the many writers and other cultural actors involved in this work was so vast and varied that they cannot be described in the course of one chapter of one book.

We can, however, dwell on one motif that appears vividly in much of this printed work: relating sites in the Hawaiian Islands to the gods and other spiritually powerful beings. This theme reveals the profoundly geographic dimension in the cultural politics of recording Hawaiian story and song in print. Publishing stories and songs about the actions of the gods and other beings in Hawaiian places was a doubly potent means of countering colonialism. First, those texts present Hawai'i as a place imbued with mana through the past actions and present residence of spiritually powerful beings at particular wahi pana. Second, by publishing them, Kānaka seized a technology (the printing press) that had been used to malign Hawai'i and put that technology to sovereign Hawaiian purposes. The published stories were many, but two may serve as examples. On September 26, 1861, *Ka Hoku o ka Pakipika* published "Mooolelo no Kawelo," the story of Kawelo, an ancient oral tradition about an ali'i, his struggle to rise to leadership of Kaua'i, his adventures and loves on O'ahu and Kaua'i, and his struggles at sea with Uhumākaʻikaʻi, a supernatural fish. The story of Kawelo, like all of Hawaiʻi's many moʻolelo, happens in specific sites that were still identifiable at the time they were recorded in print (and still today). The author of this newspaper story, S. K. Kawailiula, recounts how Kawelo was born at Hanamāʻulu on Kauaʻi, lived with his grandparents at Wailua, traveled to Waikīkī on Oʻahu, and swam in the ʻĀpuakēhau stream there (now paved over by Kaʻiulani Avenue in Honolulu), until one day when he netted the enormous magical fish Uhumākaʻikaʻi, who asked to be released. When Kawelo would not let him go, the fish pulled his canoe far out to sea, past Waiʻanae, to Kauaʻi, where they went past Manawaikeao, and so forth. In other words, the story marked the places and the places marked the story, so to print and preserve the story was to remind Kānaka Maoli of the very special nature of their places.[50] Similarly, the epic tale of the female god Hiʻiakaikapoliopele, which was first published in installments in *Ka Hoku o ka Pakipika*, is the story of the central deities of the hula, but it is also the story of places: of Kailua, Puna, and Puʻupāhoehoe on Hawaiʻi, of a place called Hāʻena on Kauaʻi and another place called Hāʻena on Hawaiʻi, and of dozens of other sites where the events of the tale unfold. The most famous of these, of course, is Halemaʻumaʻu Crater on Mount Kīlauea on Hawaiʻi Island, the present residence of Pele.[51] To Kānaka in the nineteenth century, reading the story of Hiʻiakaikapoliopele meant reading a narrative of a familiar geography populated by gods.

Despite some Kanaka Christians' objections, many Kānaka seem to have found that promoting these stories in print was not inconsistent with Christianity. In part, some Kānaka (like many Westerners) may have considered the stories and songs about the gods to be safely relegated to a realm of antiquarian folklore. But there was more to it than that. In Kanaka practice, the effort to preserve the stories of Hawaiʻi that sacralized the land through the narration of the actions of the gods and other beings was consistent with Christian labors that marked Hawaiʻi as an enlightened place in the geography of sacred power. This was as true in California as it was in Hawaiʻi. The Kanaka community in Coloma supported the effort to evangelize the Christian message in Nuku Hiva, but they also (as we saw in chapter 5) supported *Ka Hoku o ka Pakipika*, the newspaper that popularized the practice of publishing moʻolelo and mele. Similarly, in Hawaiʻi, many of the most famous of the authors of these supposedly "pagan" stories—men such as Kamakau or Kānepuʻu or Kepelino—who referred to or assembled stories in which the gods were active on Hawaiian soil were dedicated Christians.

An excellent example of the ways that publishing moʻolelo and traditions about the gods overlapped with the geography of sacred power of Kanaka Christianity is apparent in an article titled "Na Wahi Pana o Kaliuwaa" (The Storied Sites of Kaliuwaʻa) that appeared on the pages of *Ka Hoku o ka Pakipika* in November 1861. The piece fell into the tradition of telling histories by describing events in a particularly potent place. Kaliuwaʻa is a gorge and waterfall in the Koʻolau range of Oʻahu that is referred to in English today as Sacred Falls. As the article reports:

> O Kaliuwaa, oia no ke awawa kaulana loa o na awawa a pau ma ka apana o Koolau. O ke kumu o kona kaulana ana, o ka noho ana o Kamapuaa ilaila, a ua nui ka poe i hele ilaila e makaikai, mai ka wa kahiko mai a hiki i keia manawa, o na ʻlii, na kanaka, a me na malihini mai na aina e mai; a he mau tausani o lakou.[52]

> Kaliuwaʻa is the most famous of all the valleys of the Koʻolau district. The reason that it became famous was the fact that Kamapuaʻa lived there, and many people have gone there to visit it, from olden times up to now—aliʻi, people, and visitors or newcomers from foreign lands, thousands of them.

Kamapuaʻa is the pig demigod whose stories Kānaka savor for Kamapuaʻa's insurgent trickster qualities, earthiness, and cleverness.[53] His mother is Hina, female god of the moon, and his father is Kahikiʻula. In the newspaper article, the author describes a number of sites in Kaliuwaʻa, tying each to the story of Kamapuaʻa. He concludes in a way that deserves close scrutiny for what it reveals about the way geographically specific stories and traditions of spiritual power offered resources to Kānaka facing colonialism—even Christian Kānaka. Through the whole story, the author describes the sites of the Kaliuwaʻa Gorge. Then, the author writes a seeming non sequitur:

> Pela no o ka Baibala, eia no ka buke kaulana loa o na buke a pau ma ka honua nei. Ua kaulana ia buke no na wanana, a me na moolelo no Iesu. Ka wailele kiekie mai ka lani mai, oia ke kiowai olu o ke ola mau loa, kahi e maemae ai na uhane, a loaa ka maha,

a me ka malu i ka inaina wela o kona makua. Malaila e inu wai ai, aole make wai hou aku.⁵⁴

Thus the Bible is the most famous of all the books in the world. This book is famous for the prophecies and words of Jesus. The lofty waterfall from heaven is the cool pool of water of eternal life, the place to cleanse souls and find rest and shelter from the hot wrath of the Lord. One who drinks water there will never thirst again.

Clearly this is a Christian completion to a newspaper story about places steeped in Kanaka Maoli culture, history, and tradition. Clearly it asserts the sacred power of the Bible. But at the same time, the mana of Kaliuwaʻa is undeniable in the story. In fact, the Bible benefits from being linked and implicitly compared to Kaliuwaʻa—the subject of the bulk of the article. According to the author, both Kaliuwaʻa and the Bible have water that refreshes and cleanses and gives life. The author of the article even demonstrated his or her commitment to this place in how he or she signed: M. K. Palikoolauloa. The surname means Koʻolauloa Cliff, Koʻolauloa being the district in which Kaliuwaʻa is located.

In this article and in countless other moʻolelo and mele that newspapers printed, that kumu hula (hula teachers) told and to which Kānaka danced, that Kamehameha V and Kalākaua and other aliʻi and mōʻī sponsored, Kānaka Maoli countered the slanders of colonialist geography's pejorative comparisons and classifications. They asserted the mana of the sites of their archipelago against a colonialist geography that situated sacred power far away, in Western nations and biblical lands. The Hawaiian Islands were not just another "pagan" backwater, as the comparative and classificatory systems of colonialist geography textbooks contended. Hawaiʻi was naʻauao, not naʻaupō.

This very assertion, however, did reveal that the colonialist geography that the missionaries and their textbooks promulgated had effects in Hawaiʻi. That is most evident in the pejorative language that some Kanaka missionaries used to describe other Pacific Islanders as ignorant or primitive. Thinking about other people and other nations, and how Kānaka were like or unlike them, was one way that Kānaka reflected on their place in a world in which Western colonialism was a reality. Because of the deep connections that were rediscovered between Hawaiians and other Pacific Islanders, many Hawaiians in the nineteenth century reflected on their place in the world by reflecting on them. Sometimes they affirmed enduring bonds of kinship to Pacific people who were "bone of our bone." Sometimes they insulted supposedly benighted Pacific "pagans" who showed how far Kānaka had climbed and how far they could fall. Other Pacific Islanders, then, were both kin and Other to Kānaka, a means to reflect on their place in the world. Over the course of the nineteenth century, as the United States become an increasing (and increasingly threatening) presence in Hawai'i's world, another group also came to figure prominently in how Kānaka compared themselves to others. That group was American Indians.

7

"WE WILL BE COMPARABLE TO THE INDIAN PEOPLES"

RECOGNIZING LIKENESS BETWEEN KĀNAKA AND AMERICAN INDIANS, 1832–1895

In December 1895, two years after a coup by American businessmen had overthrown Queen Liliʻuokalani and installed a new government under their own control, a Hawaiian nationalist newspaper titled *Ka Makaainana* (The Citizen or The Commoner, published 1887 to 1899) asked its readers a question: "E Hoohuiia Anei Kakou?" (Should We Be Annexed?). The question was urgent: the American usurpers had created the new government with the goal of annexation by the United States and were working to realize that intention. The editors of *Ka Makaainana* suggested that the best way to understand what would happen to Kānaka if the United States annexed their nation would be to look at the situation of American Indian people. If the United States annexed Hawaiʻi, "e hoohalikeia aku ana kakou me na poe Ilikini o Amerika Huipuia" (we will be comparable to the Indian peoples of the United States).[1] Given the desperate circumstances of the Native people of the United States, the newspaper proffered a resounding "no" to the question of annexation.

Embedded in this editorial are a genealogy and history of Native Hawaiian thought about the globe in which, since the 1830s, American Indian people became a central site around which Kānaka reflected on colonialism and their own situation. These themes emerge particularly from a focused reading of nineteenth-century Hawaiian-language newspapers, although there are considerable resonances with earlier discussions of textbooks. What Kānaka read about Native American people in the nineteenth century and what they wrote about them reveal a series of overlapping shifts over the course of the century in the representation of Indians. In a first phase, American Christian missionaries taught

Kānaka that "the Indian" was a model of all things Kānaka must not be. From the time that missionaries introduced writing in the Hawaiian language in the 1820s, the textbooks and newspapers that they produced portrayed Indians to Kānaka as a negative model. Depicting Indians as ignorant, benighted, and savage, these missionary-produced documents made Indians an object lesson to Kānaka as to why they must embrace the missionary message of Christianity and civilization. By the 1850s, however, there began to be a shift in Hawaiian-language newspapers. Direct social contact between Kānaka and American Indians due to Kanaka work in the fur trade, the gold rush, and other laboring domains meant that new knowledge from Kanaka sources supplemented the missionaries' messages, and depictions became more nuanced and sympathetic. In the 1860s, an independent press under indigenous Hawaiian control emerged for the first time. As Kānaka resisted American colonialism in the press, Indians were portrayed increasingly sympathetically. They remained a negative model for Kānaka who were engaged in the defense of their national sovereignty, but in a new way: they became the sign of the colonized, what Kānaka must not become. Ultimately, of course, white Americans overthrew the Hawaiian Kingdom in 1893 in order to seek annexation to the United States. In the process of the illegal overthrow and annexation, a third shift occurred. That shift moved Indians from being what Kānaka must not become to something Kānaka had now become like. Kānaka had not become Indians, but they increasingly saw a *likeness* between their situation and that of American Indian people. In the strand of nineteenth-century Kanaka thought that is an intellectual ancestor of contemporary Native Hawaiian politics and studies, American Indians became an important site for reflecting upon what it meant to be Native Hawaiian.

In Hawaiian writing on American Indians, we see a "politics of recognition" much larger, bolder, and more indigenous-centered than ones that are legible to the settler state. Currently, in response to federal and Office of Hawaiian Affairs pressure, Kānaka are repeatedly being forced to respond to questions about recognition narrowly construed: Shall Kānaka accept a federally circumscribed governing entity over them that would be the object of federal recognition? Shall Kānaka accept a tribal-like government?[2] In contrast, in the late nineteenth and early twentieth centuries, Kānaka recognized their likeness to American Indians in a more profound sense. The intellectual work of Native Hawaiians' creative acts of identification as being *like* American Indian people demonstrates that today's notions of the indigenous as a global category is part of a process and a conversation that Kānaka have been engaged in for well over a century. Most scholarly writing on the twentieth-century history of the indigenous as a category rightly emphasizes the work of George Manuel and others in building the World Congress of Indigenous People (WCIP) in the 1970s, and also the now more famous 2007 United Nations Declaration on the Rights of Indigenous Peoples (UNDRIP). As Chadwick Allen writes, "the WCIP's international vision of indigeneity was a genuinely new step" in indigenous politics.[3] Yet if the WCIP and the UNDRIP are the sources of today's category of "indigenous," why were people in local circumstances prepared to accept this global term? The fact that Kānaka in the nineteenth century and early twentieth century decided that there was a *likeness* between themselves and American Indian people suggests that it would be useful to consider our current notion of the indigenous as part of a story far older than the WCIP. Current notions of the indigenous may have emerged out of many separate such stories of connection, as people came to think about themselves as sharing characteristics with other, distant people we now call "indigenous"—even before that word became common. Recent scholars have for excellent reasons emphasized the ways that Native Hawaiians and other Pacific Islanders have looked at one another and emphasized the genealogies and bonds that join them together in what Epeli Hauʻofa calls a "sea of islands."[4] This chapter suggests that it would also be fruitful to investigate the way different indigenous peoples across the globe came to see the likenesses that lay between them, and the global effects of many different instances of recognition of likeness.[5]

1820S FORWARD: MISSIONARY LESSONS AND THE WHITE MAN'S INDIAN

To understand the significance of changing images of American Indian people in the Hawaiian-language press, it is essential to note that although a number of kinds of newspapers were published in Hawaiian, all of the newspapers published from 1834 (when the first Hawaiian-language newspaper appeared) to the early 1860s were under the editorial control of white missionaries or white settlers closely tied to them. These newspapers reflected European and (especially) American missionary perspectives and served missionaries' agendas in Hawai'i. It would therefore be a mistake to imagine that just because a newspaper was published in the Hawaiian language, it expressed the points of view of Kānaka. It would also be a mistake to imagine that it did not, as Kānaka wrote for these papers even from the earliest days, operating within the constrained space of Haole editorial control. These newspapers' perspectives were not uniform. In their editorial stances one can trace many tensions: between the dominant Calvinist Protestants and minority Catholics, and between Christians who opposed commercial development and those who favored it, just to name two areas of disagreement.

The trope of the Indian emerges vividly in the newspapers published by Calvinist missionaries and the dominant American faction that traced its origins (and often its literal parentage) to missionaries. For decades after the beginning of publishing in Hawaiian, these newspapers represented Indians to Kānaka in ways that constituted a racial and colonial education. In the 1830s, 1840s, 1850s, and forward, these publications taught Kānaka that Indians were degraded, dangerous, and benighted "pagans" in need of the civilization and salvation that would come from colonialism and conversion to Christianity. In effect, the missionary newspapers were introducing Kānaka to what Robert F. Berkhofer has called "the white man's Indian." As Berkhofer puts it, "Native Americans were and are real, but the *Indian* was a White invention."[6] This discursive creation, "the Indian," was translated into Hawaiian just as surely as its name was transliterated as "ka Inikini" or "ka Ilikini."

This translation into Hawaiian of this idea of "the Indian" was initially indirect, because in the 1830s, 1840s, and 1850s, newspapers generally mentioned American Indians only in the context of items that the editors deemed more interesting or important. This was the case for the first mentions of American Indian people in the first Hawaiian-language newspaper, *Ka Lama Hawaii* (The Hawaiian Luminary, published 1834 and 1841). In 1834, the missionary Lorrin Andrews began publishing *Ka Lama Hawaii* at the Lāhaināluna school. It was, in effect, a school newspaper that functioned as "a text-book and forum for composition."[7] In keeping with its pedagogical goals, it presented its student readers with a series of illustrated articles describing animals that did not exist in Hawai'i—and repeatedly noting that Indians used these strange beasts for food. These were translations from John Lee Comstock's 1829 *Natural History of Quadrupeds*.[8] In keeping with Western "natural history" that placed nonwhites in the same natural categories as nonhumans, the Hawaiian text translated a white American notion of what Indians were like along with information on four-legged mammals. After describing the massive and heavily antlered deer that is the North American elk, *Ka Lama Hawaii* wrote: "Aole he mea waiwai ka Eleka, ma kana hana ana, aka, he mea waiwai no kona kino ke loaa. Oia kekahi mea i ai nui ia e na Inikini, a me na kanaka hele ma kahi loihi iloko o ka ululaau" (The Elk is not valuable for what it does, but for what is gotten from its body. It is something that is much eaten by the Indians and the people that go to places far inside the forest).[9] In doing so, the story introduced two important ways of describing Indians: their strange eating habits, and that they were hunters who dwelled in the forest. The newspaper reinforced the same theme, with nearly the same wording, in the description of the American bison: "ua ai nui ia, e na Inikini a me ka poe hele ma ka ululaau o Amerika akau" (it is much eaten by the Indians and the people that go into the forest of North America).[10] The image of Indians is one familiar in North American representations: these are forest-dwelling hunters who eat wild game. At least in the case of the bison, Indians are described as eating a meat that is "ono," or delicious. The same cannot be said of the anteater: "Hohono no kona io i ka ai ia, ai no nae ka poe Inikini" (Its flesh is quite acrid when eaten; nonetheless, the Indian people eat it).[11]

What did it mean to describe Indians to Kānaka as forest dwellers and eaters of wild animals, whether delicious ones like the bison or acrid-tasting ones like the anteater? This spoke to whites' self-conceptions, but also those of Kānaka. To whites, the forest (*selva* in Latin) was literally the etymological origin of the word "savage"; living in the forest and living off game placed Indians outside of the realm of the civilized (etymologically: farm- or town-dwelling) eaters of cultivated crops and farm-raised animals.[12] But Kānaka, too, were conscious of themselves as a predominantly agricultural people, for whom crops (especially taro and sweet potato) were staple foods. Again, the act of translating an American text that described Indians as hunters who consumed game also meant translating an American idea: that Indians were nonfarmers. Naming Indians as nonfarmers meant representing them (falsely) to Kānaka as people who wandered the wilderness in search of game. These descriptions of wandering ignored the agricultural and horticultural practices that fed much of Native North America and misrepresented the elaborate strategies of cyclical migration (not mere random wanderings) that hunting and foraging societies practiced. Then again, the point of these portrayals was not descriptive accuracy. Rather, they served to lend legitimacy to the dispossession of Indian lands and, by contrast, to glorify the agricultural basis of a land-hungry American society. These portrayals demeaned Indians and elevated white Americans.

Yet demeaning Indians in this manner was also a lesson to Kānaka on their place in what whites considered to be a civilizational hierarchy. The depiction of Indians was just one part of the missionaries' efforts to instruct Kānaka in the tiered hierarchy of civilizations and races that placed Kānaka at an intermediary stage between supposedly naʻaupō (ignorant and benighted) Indians and naʻauao (enlightened) Americans and Europeans. This insidious hierarchy encouraged Kānaka to throw in their lot with Americans and Europeans, lest they be relegated to the realm of the naʻaupō Indians. Because Kānaka practiced extensive and intensive agriculture, agriculture was thus an ideal way to portray their supposed superiority over Indians. This was made quite clear in 1844 in *Ka Nonanona* (The Ant, published 1841–45), a newspaper edited by missionary and Minister of Public Instruction Richard Armstrong. Although (unlike earlier newspapers in Hawaiian) Armstrong's paper actually published news of Hawaiʻi in order to disseminate information about events in the kingdom's capital, this American editor nonetheless used his newspaper to pursue his religious goals, advocate for the Westernization of Hawaiʻi, and represent American interests.[13] In an article on agriculture, the author declared farming to be the very source of Hawaiʻi's well-being: "Oia no ke kumu e waiwai ai ko keia pae aina. Nolaila mai ka ai" (It is the source that enriches the people of this archipelago. From it comes the food). He drew a sharp contrast with allegedly nonfarming societies where game and wild plants constituted the diet. In such places, he wrote, "Uuku kahi ai, hiki pinepine mai ka wi, a he poe hoomolowa lakou, a me ka palaualelo" (Food is scarce, famine often comes, and they are a slothful and lazy people). As examples of such people, the author named two populations: "na Inikini ma Amerika, a me ko Kahiki poe" (the Indians in America, and the people of Kahiki).[14] In this context, "Kahiki" describes the people of Tahiti. The description of Indians and Tahitians as non-farmers was part of a broader project of inculcating Kānaka into Western beliefs about hierarchies of civilized and savage, enlightened and ignorant. Note that there was flattery here, as Kānaka were agriculturalists, and thus allegedly superior to American Indians and Tahitians. But note also that the knife cut both ways, because the hierarchy was to be accepted as a whole piece. For Kānaka to accept their position of alleged superiority over Indians and Tahitians also would have meant accepting that they were inferior to Americans and Europeans. White missionary and merchant newspapers showed Kānaka what Indians lacked in order to tell Kānaka to what they should aspire: true civilization as the missionaries defined it. The distinction that white-controlled newspapers made between those peoples who had agriculture and those who did not was part of a broader array of statements of what Indians were lacking and was entirely consistent with what Berkhofer has identified as "the white man's Indian." According to this discourse and according to missionary-controlled newspapers, Indians lacked Christianity: like other benighted people around the world, Indians were "lahuikanaka i ike ole ia Iehova" (peoples who do not know Jehovah).[15] Indians lacked roads: without them, well-meaning whites could carry neither Jehovah nor the goods of white civilization to Indian Country, and thus "ua mau ka hupo a me ka poino a hiki i keia la" (ignorance and misery continue up to today).[16] Indians lacked a system of individual property in lands: without this spur to individual profit, Indians failed to use land properly, and land that could have supported a million people fed only a thousand (a claim that legitimated the dispossession of Indian lands in the United States).[17] Indians lacked homes: Victorian domesticity was beyond their comprehension, because "he poe hele io ia nei, e hahai i na bupalo, a me ke dia, a me kamano" (this is a wandering people that pursues buffalo, and deer, and salmon).[18]

The list of what white-controlled newspapers told Kānaka that Indians lacked can serve as a catalog of what the white editors wanted Kānaka to value: farms, Christianity, trade, private property, and a gauzily sentimental but rigidly restrictive ideology of domesticity. Note that this is true despite the varying allegiances of whites. Even those most loyal to the kingdom envisaged its future as a "modern" and "civilized" nation, and to them modernity and civilization were unimaginable outside of Western values and behaviors. In the thinking of the nineteenth-century white American middle class from which the missionaries and merchants sprang, these things created and ennobled privately held wealth, constrained sexuality, and enforced gender norms.

These same newspaper accounts used Indians to demonstrate to Kānaka that to achieve these things, they themselves needed missionaries.

In 1835, the missionary editors of *Ke Kumu Hawaii* (The Hawaiian Teacher, published 1834–39) established the pattern of using Indians to demonstrate to Kānaka their need for missionaries.[19] In a series of articles titled "He Ui Misionary" (A Missionary Catechism), the missionary editors asked and answered a sequence of questions that they believed established the need for their dominance in Hawaiʻi. "Pehea ka noho ana o kanaka o Hawaii nei i ka wa i pae mai ai ka poe misionari?" (What was the condition of the people of Hawaiʻi at the time that the missionaries landed?). To their own question, the missionaries responded that Hawaiʻi had been in a violent state of disarray. The kapu (sacralizing restriction) had been broken, the old gods had been torn down, the old religion had been abandoned, and in the resultant disorder, thievery, warfare, deceit, and all manner of wickedness reigned. At that point, rather than narrating the activities of the missionaries in Hawaiʻi or tracing the rapid changes in Hawaiʻi (as the majority embraced literacy and Christianity at least to some degree and as the economy rapidly shifted), the catechism turns to the Americas, again using Indians as a crucial tool in its pedagogy. "E noho ana anei ka poe ike ole ia Iehova ma Amerika Huipuia?" (Are there people living in the United States who do not know Jehovah?). Yes, there were. The catechism explained that "ko Europa poe" (Europeans) had called those people Indians. Most Indians, it claimed, had long refused to heed the missionaries and "noho molowa no" (lived slothful lives), but in recent years many had begun to listen to the missionaries, and as a result had repented, abandoned their sinful ways, and embraced schooling, hard work, and "good" (i.e., Western-style) houses.[20] In the context of the preceding description of immediate premissionary Hawaiʻi as a land of disorder and disgrace, the message to Kānaka could not be more clear: just as American Indians needed to heed missionaries to achieve civilization, so too did Kānaka. Note that this injunction was to listen to missionaries specifically, rather than whites in general. As was true in the United States, missionaries in Hawaiʻi were often in conflict with less religious whites (such as rum sellers or other merchants), whom missionaries feared would lead Native people along a sinful path. Thus missionary newspapers in Hawaiʻi warned Hawaiian readers that just as Indians needed to associate only with religious white people, so too must Kānaka be sure to mingle only with godly Haole.[21]

The 1835 catechism's emphasis on Indians' (and Kānaka's) need for missionaries was the norm in the missionary and merchant press for decades to come, but the statement that Indians were progressing was not. Rather, most newspaper accounts depicted Indians as still immersed in sin and ignorance. The American missionaries to Hawaiʻi were part of a network that included missionaries in North America, and the Hawaiian-language newspapers often included accounts of evangelization efforts there. This was the case for the Oregon Country, site of the famous mission of Narcissa and Marcus Whitman among the Cayuse at Waiilatpu. Although an 1842 article about Oregon in *Ka Nonanona* does not mention Waiilatpu or the Whitmans, it was probably about the Cayuse that it reported that the Indians were "paakiki" (unyielding). The author admitted that some sent their children to school and some had become "Inikini haipule" (devout Indians). But, the article intoned, "aole nae he oiaio ka lakou pule, no ka mea, aole o lakou haalele i ka hewa, a malama i ka ke Akua" (theirs is not genuine prayer [or worship], because they do not abandon sin and heed the words of God).[22] In this and other articles, the missionary and merchant press reinforced that Indians were making little progress toward civilization and Christianity and still needed missionaries.

By the middle of the century, many Americans' concerns about Indians' souls were overshadowed by fears of Indians' weapons. The depictions of Indians that whites presented to Kānaka shifted accordingly, but these portrayals continued to be shaped by Americans' preoccupations. By the 1850s, American territorial domination was shifting increasingly westward, and military domination overwhelmed religious conversion as a concern. Accounts in the missionary-sponsored Hawaiian-language press shifted accordingly. In article after article, Hawaiian readers read that Indians were at war with whites in Oregon, California, Florida, Utah, Mexico, and other places, and that Indians were dangerous and brutal.[23] In 1857, *Ka Hae Hawaii* (The Hawaiian Banner or The Hawaiian Flag, a missionary-sponsored newspaper published 1856–61) informed its readers that when Indians in Durango, Mexico, raided the town of Mapimí, they carried off women, children, and animals.[24] In 1860, the same newspaper reported that two thousand Indians had attacked one hundred whites in the Rocky Mountains, slaughtering them all.[25] *Ka Hae Hawaii* was (like *Ka Nonanona*, which is mentioned above) a newspaper controlled by missionary Richard Armstrong, the minister of public instruction. Although its very title might suggest an indigenous nationalist perspective, *Ka Hae Hawaii* resembled Armstrong's other papers in the way it was used as a tool to promote his vision of transforming Native Hawaiian people. Armstrong's daughter recalled that her father "understood the native character": "He saw that Kings, chiefs, and people were mere children, governed by impulse, untrained to thinking."[26] He believed it was his role as minister of the gospel, minister of public instruction, and director of the board of education to transform these "children" into modern and civilized subjects along American lines. Thus it is not surprising that the stories about Indian warfare that *Ka Hae Hawaii* published for its Hawaiian readers—stories featuring Indians raiding white towns and massacring white settlers—were familiar in writing on Indian war in U.S. newspapers.

Equally familiar from American papers was the notion that American Indian warfare was a bloody but futile prelude to the disappearance of Indian people and might even hasten their demise. In 1856, *Ka Hae Hawaii* reported that though wars raged between Indians and whites in California, Oregon, and Florida, the ultimate outcome of the wars was certain: "ka hoopau ana o ka lahui Inikini" (the elimination of the Indian people).[27] Four years later, the same newspaper told Kānaka that the defeat and "elimination" of Indians would be the natural result of the supremacy of the United States in all things. *Ka Hae Hawaii* reported that "ke kipi hou nei na Ilikini ma Amerika Huipuia" (the Indians are again rebelling in the United States). But victory was impossible for the Indians: "Heaha la ka lakou loaa e hoaano aku nei ia Amerika? Heaha la ko lakou wahi ikaika?" (What do they have that allows them to challenge America? What is their little strength? or possibly: What is their stronghold?). These questions were rhetorical; they required no response. *Ka Hae Hawaii* informed Kānaka that American Indians had lost America to the United States, a superior nation—"he Aupuni naauao, a he Aupuni ikaika no hoi" (an enlightened Country and a powerful Country, too).[28]

SYMPATHY FOR THE UNFORTUNATES

Yet even in this early period of the 1830s and 1840s, when the press participated in teaching Kānaka what was wrong with Indians, notes of sympathetic identification appeared in print as Kānaka looked at Indians and discerned a likeness between them and Hawaiians. In 1839, for example, *Ke Kumu Hawaii* published a piece that hinted at how the demographic collapse of American Indian nations struck chords of recognition among Kānaka, who were suffering greatly from virgin soil epidemics of foreign diseases. *Ke Kumu Hawaii* reported that a terrible smallpox epidemic had devastated "ka poe inikini a ka Haole i kapa aku ai Inikini wawae eleele" (the Indian people that the Haole called Blackfoot Indians). Note that the phrasing "that the Haole called" either expresses, or at least leaves room for, the Kanaka reader to imagine perspectives outside of Haole points of view. In fact, this phrase may suggest that this author of this piece was Hawaiian, given that *Ke Kumu Hawaii* included writings by Kanaka students and teachers. The report was followed by interjections of grief that suggest that those non-Haole perspectives encompassed sympathy and an identification of likeness between Kanaka and Indian: "Aloha ino lakou, Emi loa lakou" (What a pity for them. They are much reduced in number).[29] Kānaka were painfully aware that their numbers were also falling precipitously, with many succumbing to smallpox and other foreign diseases. Indeed, the same newspaper would publish an article only weeks later that attempted to quantify the terrible truth that Kānaka were already aware of: "ka Emi Ana o na Kanaka" (the Reduction in Numbers of the Kānaka).[30] Even in the early days of the missionary press, room for sympathetic identification by Kānaka for American Indians could be found on the newspaper page.

Such expressions of likeness became more important in the 1850s and 1860s, a period of transition in the way the Hawaiian-language press depicted Indian people. Increasingly, perspectives that suggested Hawaiian sympathy for and even identification with American Indian people began to appear in press accounts. They did not fully replace demeaning depictions of Indians, and often they were rife with ambiguity, but nonetheless these articles constituted an important shift in the messages Kānaka received about Indians—and about their conflicts with whites.

This shift was especially apparent in coverage of warfare. The brutality of the Civil War prompted a reconsideration of what white men were capable of in warfare, which in turn suggested that Indian warfare might not be, after all, so exceptional in its violence. This shift was quite apparent in accounts of the Dakota War of 1862 in Minnesota, in which Dakota in the Minnesota River valley attempted unsuccessfully to remove white settlers from ceded lands. In the fall of 1862, brief accounts in *Ka Nupepa Kuokoa* had hewed closely to the lines of depicting Indian warfare as Indian brutality: battles amounted to "hana ino" (mistreatment or abuse) of government soldiers, in which the latter risked being taken captive. There was no discussion of the causes of the war, only its allegedly brutal prosecution by Dakota warriors. But as the Civil War progressed, accounts of the now-completed Dakota War shifted. Indian violence began to be judged alongside violence by whites against nonwhite people. In May 1864, *Ka Nupepa Kuokoa* carried an account of the Fort Pillow Massacre, in which white Confederate troops slaughtered black Union troops who were trying to surrender—a war crime that was widely denounced in the northern United States. The author says that in comparison to the merciless brutality of the Confederate army, the attacks two years before by the Dakota in Minnesota looked minor. The devil himself, "e noho ana ma ka lua hohonu o ka make a me ka ino" (living in the deep hole of death and wickedness), would approve of the Southern atrocities.[31] The next year, in 1865, the same newspaper went still further in an article on the treatment of Union prisoners at Confederate prison camps at Andersonville, Richmond, Columbia, and elsewhere: "Ua oi aku ka lokoino o keia hana a na'Lii Kipi mamua o ka poe aikanaka o ka Mokupuni Fiki (Feejee Island) a me na Ilikini o Amerika Akau" (What the Rebel officers do is more merciless than what the cannibals of Fiji Island and the Indians of North America do).[32] It is difficult to know who penned these words. They could be translations of articles from the Northern press in the United States, or original pieces by a white American writer in Hawai'i. Most white Americans in Hawai'i were from the Northeast and loyal to the Union. The pieces could also have been written by Kānaka, however. Whoever the authors or translators might have been, they were presenting Kānaka with startlingly different notions of Indian warfare than had been the norm in the Hawaiian-language press over the decades. It is crucial to note here that two things were being reconsidered: whites as well as Indians.

Such perspectives were evident in letters to the editor from Kānaka from the 1850s forward. Although newspapers were edited and mediated forums, they did sometimes allow Hawaiian writers to publish views that challenged missionaries and their role among Indians and Kānaka. In 1862, a girl from Nuku Hiva traveled to Hawai'i with Paulo Kapohaku, a Hawaiian missionary who had adopted her as his hānai daughter while he was living there. According to a letter to *Ka Hae Hawaii* from J. H. Kānepu'u, though the girl was impressed by Honolulu and Kānaka, she was not well received in Hawai'i. She was teased for her clothes and appearance, and even mocked as pupule, or crazy. Kānepu'u—the schoolteacher, writer, and vocal opponent of American annexation of Hawai'i whom we have already encountered—lamented that Hawaiians cruelly teased her. Kānepu'u warned Kānaka to treat Nuku Hivans better than whites treated Indians. American Indians could tell the Haole missionaries, "'E hoi oukou e ao aku i na Haole o oukou,' no ka ike no o na Inikini, ua hana ino na Haole hewa ia lakou, a pela 'ku." ("Return and teach your own Haoles," because the Indians know that evil Haoles have abused them, and so on).³³ Kānepu'u challenged the hierarchy that the missionaries promoted. Perhaps whites were not such good missionaries, and Indians not so needy of them, as earlier reports had told Kānaka.

Face to Face: Native–Native Contact between Hawai'i and North America

Like stories about the Civil War, articles about connections between Indians and Kānaka in Hawai'i and in North America marked an important shift toward seeing likeness in the larger story of Hawaiian–Indian relations. Over the preceding years, as increasing numbers of Kānaka encountered actual Native American people, face-to-face social relations and stories about them in the press undermined the figurative Indian that the missionary press in Hawai'i had portrayed. These encounters happened both in North America and in Hawai'i and traced back to the travels of Ka'iana and Ka Wahine to Yuquot and the arrival of American Indian mariners aboard vessels. The fur trade, the gold rush, whaling, and seafaring labor brought Hawaiian men (and a few Hawaiian women) to North America, where many of them lived, worked with, and made families with American Indian people.³⁴ Via reporting in the Hawaiian-language press, direct encounters for Kānaka in North America became vicarious encounters for Hawaiian readers in Hawai'i, and genetic kinship between certain Kānaka and certain Indians fed the emergence of a narrative of kinship between Kānaka and Indians more generally.

Because of the significant presence of New England American Indian men as crew on American whaling and shipping vessels, it is likely that encounters with Indian people trace back to some of the earliest encounters of Kānaka with outsiders in the eighteenth century. Hawaiian readers first could begin reading about such encounters in the late 1830s, when newspapers carried reports of American Indians landing in Hawaiian ports as workers in the sea otter hunting trade.[35] Assuming such workers and sailors were afforded shore leave, the arrival of ships with American Indian crew would likely have resulted in relatively brief encounters between Indians and Kānaka. One can only speculate on the nature of any such encounters, as Kānaka met Indians across boundaries of language. Putting into port at Lāhainā or Honolulu also created opportunities to "jump ship"—to escape the oppressive labor conditions of the ships, into which crewmembers had sometimes been forced, rather than signing on voluntarily. As the Australian historian of the Pacific Greg Dening has argued, the ship-jumpers and beachcombers that lived on the fringes of Pacific societies were among the most important ways that islanders first gained perspectives on other places and peoples.[36]

By about 1850, American Indian beachcombers begin to appear in the written record in Hawaiʻi. Speculation by Kānaka about them is emblematic of the ways that Kānaka in these years were expressing interest in American Indians, but as of yet were still learning about them. In 1856, Kānepuʻu wrote to *Ka Hae Hawaii* to report that a stranger had been living for seven years in the mountains above Niu (east of Honolulu on Oʻahu), and that the man may have been an "Inikini no Amerika mai" (Indian from America). Residents of the area evidently found his presence disquieting. Kānepuʻu referred to him as "kekahi kanaka hihiu" (a wild man). The konohiki (district official) of the area had captured the unknown man to try to force him "e noho pu me na kanaka, a e hana ma na hana pono, e launa me ka oluolu" (to live together with people, and act properly, and associate pleasantly). The captive escaped his chains and returned to the upland. But for all the talk of his wildness, the man seems to have caused no trouble for Kānaka. Kānepuʻu himself admitted that the man lived peacefully in the mountains, cultivated kalo (taro), and if anyone came up to bother him, he simply fled "a he mama maoli no kona me he popoki la" (just as quickly as a cat).[37] The fact that this man was cultivating kalo and was able to survive up in the mountains suggests that he was not as isolated as Kānepuʻu evidently thought: kalo was not a crop that was known in North America or Mexico, so Kānaka presumably had taught the man to grow it and perhaps were associating with him. Nine years later, B. L. Koko wrote to *Ka Nupepa Kuokoa* to report a similar story about a person in the Niu area (perhaps the same man) whom he identified as a "kanaka Ilikini" (an Indian man).[38] This mysterious man was certainly not socially integrated into Kanaka society, but his presence in Hawai'i and in the imaginations of Kānaka speaks to the way that Indians were coming to occupy a special place in Kānaka's thinking by the middle of the nineteenth century.

By the 1850s, as detailed earlier, Hawaiian men and American Indian women were making families in North America. Some of their children moved to Hawai'i, and soon Kānaka were encountering American Indian people and part-Hawaiian, part-Indian people in the islands. People of mixed heritage made plain to Kānaka—a people very oriented to genealogy—that they now literally had a kinship to Indian people. In 1866, Olepau returned from years of labor in California, bringing his Indian "kokoolua" (companion), their two daughters, and their son.[39] Most likely, Olepau had been one of the thousands of young Kānaka who had flocked to California in the wake of the discovery of gold in 1849. His sons were part of the history of the gold rush, as were two boys of mixed heritage whose names bore the mark of the rush for gold: Dala (meaning Dollar) and Imikula (meaning Seek-gold). Dala Kauanamano, a part-Indian child who won the speech contest at the Sunday school competition in central Hāmākua on Hawai'i Island in 1868, was probably the child of a similar union.[40] A. E. Mahuka and a Wintu woman (whose name I have not been able to locate) gave similar gold rush–derived names to their two sons, William Imikula and John Elikula (meaning Dig-gold). They, too, moved to Hawai'i, studied at the Oahu Charity School, and lived out their lives in Honolulu. Other people of mixed Kanaka–California Indian descent lived parts of their lives in Hawai'i and parts in California, such as John Paaniani, son of Robert Paaniani and a Maidu woman. Though certainly fewer in number than Hawaiian people of part-Haole or part-Chinese ancestry, Hawaiian people of part-Indian descent were becoming a presence in Hawai'i by the early 1850s.[41]

Kinship between Kanaka individuals and Indian individuals was talked about frequently in articles reporting on the lives of Kānaka living in North America, especially those who lived in California, Oregon, and Washington. After 1860, newspapers frequently reported on the relations between Hawaiian men overseas and Indian people. Common topics included the marriages of Hawaiian men and Indian women, the deaths of Hawaiian men, and the birth of children of mixed Hawaiian and Indian heritage. Such stories frequently included the "one hanau" ("birth sands," meaning birthplace) of the Hawaiian man in question.[42] In this manner, these individuals' relatives and former neighbors in Hawaiʻi were able to learn of their activities. But just as importantly, the many other readers of these newspapers were given a sense of the connection that Kānaka and American Indians had built in North America.

This did not mean that these connections were without conflict or even violence. In the 1860s, the Hawaiian-language newspapers reported on several such conflicts in California: Hawaiian men allegedly killed by Indians, a Hawaiian man arrested for stealing from an Indian, and so forth.[43] Similarly, newspaper stories make clear that relations in Hawaiʻi between Kānaka and Indians were not free of conflict. In 1876, the newspaper *Ka Lahui Hawaii* reported that an Indian who worked at a sugar plantation near Lāhainā had killed a Hawaiian coworker in a fight.[44] In these accounts, one senses the enduring power of the images of Indian brutishness that the missionary and merchant press had promulgated among Kānaka for decades. Placed in the larger context of stories of Kānaka and American Indians who worked together, lived together, and made families together, these stories of conflict give nuance to a broader sense that Indians were people to whom Kānaka felt they were connected. As the story in *Ka Lahui Hawaii* made clear, that connection was not without difficulties, but it was nonetheless becoming established.

1860s Forward: What Kānaka Must Not Become

Beginning in the 1860s, a new generation of nationalist newspapers began to point to ways that Hawaiians were being impoverished and dispossessed much as Indian people had been and to urge Kānaka to resist that trend. The pages of *Ka Hoku o ka Pakipika* and of other Kanaka-controlled papers that followed it—*Ke Au Okoa* (The New Era), *Ka Makaainana* (The Citizen or The Commoner), *Ka Oiaio* (The Truth), *Ke Aloha Aina* (The Patriot), and others—allow us to shift the gaze from what whites said about Kānaka to what Kānaka said about themselves and others. Using these newspapers, groundbreaking studies have dismantled the myth of Kanaka passivity in the face of American colonialism and revealed the resistant politics of Kanaka cultural and intellectual endeavor in the second half of the nineteenth century. Kanaka editorial control was not sufficient to ensure that Indians would be portrayed differently than they had been in newspapers under white missionary control. For example, in 1862, *Ka Hoku o ka Pakipika* published an account of savage Indian warfare and scalping that clearly showed how much its editors were shaped by the American newspapers upon which they depended for sources and by the years of seeing Indian people through American representations such as those in the missionary-controlled press.[45]

But a real shift was underway, and Hawaiian voices in the press were already pointing out that Kānaka were being dispossessed and must resist dispossession lest they be reduced to the state of American Indians. It would be a mistake to assume that this shift occurred suddenly in the 1860s or was apparent only in Kanaka-controlled newspapers. First, there was a fine line between depictions of Indians that served as a negative model for Kānaka (whether portrayed as the white man's Indian or as objects of pity) and looking at Indians as indicators of the danger that colonialism posed for Kānaka, and articles sometimes blurred that line. Moreover, even in the missionary-controlled press, Kānaka used articles and letters to express the fear that colonialism could reduce them to the status of Indians. Indians were generally portrayed sympathetically, but functioned as a negative referent for Kānaka. To be Indian meant to be impoverished, landless, and colonized. In 1861, J. B. Nakea wrote to *Ka Hae Hawaii* about a group of Kānaka whom he had encountered in the vicinity of Waipi'o-uka (upper Waipi'o) on O'ahu. He came upon them as they ended a long search in the mountains for an abandoned 'awa (kava) field they had heard about—probably because they were desperately poor and needed something to sell. They were so impoverished that Nakea could compare them only to Indians: "Ua like ko lakou ano, a me ko lakou helehelena i ka nana aku, me he poe Ilikini la, aohe wahi lole wawae, aohe papale o lakou a pau. He mau wahi palaka wale no. A ua kunahihi lakou i ka make i ke anu" (Their nature and their appearance was like that of the Indians: none of them had pants or hats. Just shirts. And they were numbed to death by the cold).[46] Nakea's portrayal of these poor people wandering through the cold mountains in search of an abandoned field is painfully evocative in the aftermath of the increasing landlessness of the time. Following an 1848 land privatization that is known in Hawaiian history as the Māhele ("the Division"), some makaʻāinana (commoners) were able to live for a time on small plots they had been assigned. But over time, as Lilikalā Kameʻeleihiwa has demonstrated, the privatization of land removed the bulk of the Hawaiian population from the small farms on which they had made their living.[47] Increasing numbers were reduced to a poverty so desperate that Nakea could describe it only as "like that of the Indians." Although Nakea's comparison, like the colonial pedagogy of the missionary and merchant press, emphasized what Indians lacked, there was an important difference. The missionary press dedicated most of its energies to portraying Indians as what Kānaka should eschew: undomesticated hunters, brutish warriors, and unapologetic non-Christians. Nakea evoked a new fear that Kānaka were coming to be like Indians: impoverished and dispossessed. Perhaps Indians were not something distant and strange, but something to which Kānaka were related, something that Kānaka could imagine being mapped onto their own bodies and perhaps their own nation.

It appears that in the late nineteenth century, some Kānaka began to imagine their own struggles against colonialism as akin to those of Indians. This comparison was already politically contentious in 1868, when it was mobilized around the issue of the election of whites to the Hawaiian legislature. Hawaiʻi's monarchs were Native Hawaiian, but a perceived need for Western expertise (encouraged by Haole in Hawaiʻi), debt to Western creditors, and the leading place of foreigners in Hawaiʻi's economy combined to make Haole a major presence in government posts, including the legislature. In elections in late 1867, whites (who already held a number of seats) increased in numbers in the legislature. This outcome deeply worried many Kānaka, including a man who wrote to the nationalist *Ke Au Okoa* under the name "Hawaii Ponoi" (Hawaiʻi's own, a patriotic catchphrase that became the title of the Hawaiian national anthem). Hawaii Ponoi warned Kānaka that their sovereignty was at stake, and they must elect only Kānaka in the future.[48] The author had chosen an appropriate newspaper for his piece. One of the new generation of nationalist newspapers, *Ke Au Okoa* was edited by John Makini Kapena. Kapena was an active figure in governmental affairs, especially during the reign of his relative King Davida Kalākaua (who himself was instrumental to the emergence of the nationalist *Ka Hoku o ka Pakipika* newspaper).[49] Hawaii Ponoi's article set off a series of dismissive editorial replies in the more establishment-oriented *Ka Hae Hawaii* and

Ka Nupepa Kuokoa. Most interesting among these was an editorial in *Ka Nupepa Kuokoa* that scornfully portrayed the opposition party as saying: "A e kaili ku ia ana ka kakou Kuokoa, e lawe wale ia ana ko kakou mau loi kalo, a o kakou o na makaainana e kipakuia ana ma na kuahiwi e like me na Ilikini" (Our Kuokoa is being grabbed up, and our kalo [taro] patches are being taken, and we are the children of the land that are being driven away into the mountains like the Indians).[50] There was a thinly veiled kaona (double meaning) here: whites "grabbed up" the kūʻokoʻa, a reference to *Ka Nupepa Kuokoa*, a newspaper that served whites' interests, and also to Hawaiʻi's independence, as "independence" is what kūʻokoʻa means. The larger ramifications of being "like the Indians" are clear: it meant the loss of lands and of sovereignty. The very fact that the editorial took the time to mockingly make this comparison between Kānaka and Indians suggests that the comparison had currency among Kānaka.

Indeed, the danger of being "like the Indians" caused concern even to those who celebrated the changes that were remaking much of rural Hawaiʻi into an enormous American sugar plantation in the 1870s. In 1877, A. W. Wekeweke of North Kohala on Hawaiʻi Island wrote to *Ka Lahui Hawaii* (The Hawaiian Nation, a church-oriented newspaper published 1875–77) to inform the paper of the rapid progress of sugar production in the area, a progress that in some ways he celebrated. White-owned plantations were renting vast amounts of land for sugar production, with over two-thirds of the area given over to cane production. Wekeweke noted that such lands had probably passed forever into white hands. With plantations covering all of the lands all the way up to the side of Mauna Kea, most Kānaka would have nowhere to turn: "e like aku ana me na Ilikini o Amerika, ka auwana iloko o ka ululaau" (our wanderings in the forest would be like those of the Indians of America).[51] With all the lands taken by Haole, Kānaka would have to go live on the slopes of Mauna Kea. This statement echoes the depictions of Indians as wandering hunters that dominated the colonialist teachings of the missionary and merchant press, but there was a very different politics at play when a Hawaiian linked those wanderings to the story of Kānaka's own dispossession in their own nation. Rather than existing as a savage Other, Indians were becoming an unfortunate object of identification—a fate Hawaiians could already see but must resist.

Because dispossession was the undesirable shared experience behind this identification of Kānaka with Indians, naming Kānaka as being like Indians was both politically potent and inherently unstable. In some ways, this unstable identification resembled the way Hawaiian Congregationalists both identified with other Pacific Islanders (as "bone of our bone") and believed themselves to be superior to them. American Indians had functioned as a negative referent for Kānaka through the nineteenth century—from the 1820s when missionaries held Indians up as a model of how not to live, to later in the century when aloha 'āina (patriots) declared that Kānaka were like Indians when they were dispossessed. Kānaka could, therefore, identify with Indians, but the connotations of this identification were frequently negative. It could spur Kānaka to resist colonization, but it could also encourage them to declare themselves to be different from Indians, who were the very sign of the colonized.

1890S FORWARD: ANTIANNEXATION POLITICS AND THE POLITICS OF LIKENESS

Identifying with American Indians became even more compelling and even more fraught in 1893, when American planter and commercial interests overthrew the sovereign Hawaiian Kingdom and instituted the Republic of Hawai'i under their own control in hopes of rapid annexation to the United States. The likeness between Kānaka and American Indians was made especially clear on the pages of *Ka Makaainana*. This pro–Hawaiian sovereignty and antiannexation newspaper announced its politics in its very name. "Ka maka'āinana" translates into English as "the commoner," "the citizen," or "the people," but can be more literally translated as "the one who is on the land." The title is rich with nationalism and especially with attachment to the 'āina, the land. The masthead of the paper thus gave particular power to articles about Indian lands, such as a piece reporting that the Nicaraguan government had decided to "alapoho ae i kela okana-aina kuokoa o na Ilikini a ua hoopauia aku ka heaia ana o ia inoa ma nei mua aku" (swallow up the [Miskito] Indians' independent land district and put a halt to it being called by that name from this time forward).[52] For Kānaka, who had seen their constitutional monarchy "swallowed up" and forcibly renamed a republic less than two years earlier, the seizure and renaming of Miskito lands helped to frame their own condition as colonial. *Ka*

Makaainana pointed out that "Ke manao la na poe Ilikini o Iukatana e hakaka aku ia Mekiko no ka hoihoi hou ia mai o ko lakou mau aina" (The Indian peoples of the Yucatan are planning to fight Mexico for the return of their lands).[53]

In the aftermath of the overthrow of the Hawaiian Kingdom and the threat of annexation by the United States, comparisons to other colonized people became more powerful and geographically expansive. The articles above referred to Ilikini (Indians) in Nicaragua and Mexico, whereas most earlier references to Ilikini in the Hawaiian-language press spoke about the United States. This geographical expansion was an important shift, given that *Ka Makaainana* was using its pages to draw similarities between Kānaka and American Indians around the issue of dispossession and colonialism. The newspaper made this even clearer in an 1896 article in which it reported (perhaps figuratively, as other sources do not refer to the event) that Americans had stolen the bones of Kamehameha I. The newspaper used colonialism and dispossession to draw even more expansive connections between Kānaka and other peoples. *Ka Makaainana* reported that in both Latin America and Africa, invading Europeans had not been content to plunder the land. They also plundered the most sacred things, including the bones of the deceased. Now, the newspaper declared, the minister of finance of "keia Aupuni Repubalika oehaa" (this crooked Republican Government) was figuratively plundering the bones of Kamehemeha. This was the sort of geographically more expansive comparison that now expressed the situation of the Kānaka: a comparison to Indians of Latin America and Africans.[54]

Two months earlier, in December 1895, *Ka Makaainana* had posed the question in the editorial that opens this chapter: "E Hoohuiia Anei Kakou?" (Should We Be Annexed?). The clear answer was "no," Kānaka must resist annexation. If the United States annexed Hawai'i, "e hoohalikeia aku ana kakou me na poe Ilikini o Amerika Huipuia" (we will be comparable to the Indian peoples of the United States): dispossessed economically and politically in their own native land. The only possible difference: Kānaka might get to vote on the issue of statehood.[55] This offered the editors little consolation.

Hawai'i was annexed as a territory by the United States in 1898 in contravention of both international law and American constitutional law. In 1959, when powerful forces in Hawai'i and Washington came to favor statehood, the question of statehood was (as the editors of *Ka*

Makaainana had said it might be) put to a vote. But the voters for that referendum were not just Kānaka. Rather, all of the residents of the territory, only a minority of whom were Kānaka, were permitted to vote. Nor did the ballot give voters the chance to cast their vote for independence. The only option was to approve or deny whether Hawai'i should "immediately be admitted into the Union as a State."[56] In this way, the loss of effective sovereign control by Kānaka over their homeland that resulted from annexation was even more devastating than the editors of *Ka Makaainana* had predicted in 1895.

Because they had lost their effective sovereignty, because their lands had been taken, and because they had been impoverished, Kānaka had been made "comparable to the Indian peoples," but it took a *creative act* for Kānaka to see that likeness and to point it out in writing. Kānaka did not resemble Indians in many of the ways they had been described in the missionary-controlled Hawaiian-language press: they ate neither elk nor bison nor anteater, did not roam the forest primeval, and did not lack a farming tradition or homes or a sense of property. Then again, most American Indian people did not fit those essentialist and racist descriptions either. For all the differences between American Indians and Kānaka, the most important force that made them like one another was the problem of American colonialism—or, in the broader vision of *Ka Makaainana*'s articles in the 1890s, colonialism more broadly. It took an act of intellectual creation for Kānaka to understand themselves to be like Indians. This was a positive affirmation of identification, but it was fraught because it was based on a shared experience of dispossession by American colonialism. This complex and unstable act of identification emerged over a century's time as Kānaka read about and encountered American Indian people. These acts of identifying likeness serve as important reminders to us that we are engaged in a very old conversation when we talk about the notion of global indigeneity, which is powerful in political movements and academia worldwide today. This idea was implicit in late nineteenth-century Hawaiian critiques of colonialism that skillfully appropriated and deployed American texts and ideas for sovereign Hawaiian purposes.

EPILOGUE

GENEALOGIES OF THE PRESENT IN OCCUPIED HAWAI'I

On June 12, 1923, Eli Keolanui graduated from Hilo Boarding School, a school American missionaries had opened nearly a century before. He made the first of the student presentations at the graduation ceremony that day, on a topic one suspects was assigned to him: "The Immigration Question." Keolanui's opening words are provocative:

> In proper discussion of the immigration question it is often said that all who came to this continent since its discovery should be considered equally as immigrants and that only the aboriginal inhabitants can properly be called natives. The Indians only are considered natives of this continent while all who came to the so-called new world should be considered as immigrants.[1]

Together these two sentences capture a tension between the colonialist education system in the U.S.-occupied Territory of Hawai'i and the response to it by Kanaka Maoli students such as Keolanui. As we have seen through the history of geography education, the education system was a site of contention: it was the product of Kanaka initiative, but it also bore the imprint of the American missionaries, American business interests, and American territorial officials who had each tried to shape the schools to serve their own purposes, with considerable success since the 1870s and especially after the overthrow of the kingdom in 1893 and annexation to the United States in 1898. In the early twentieth century, the resulting education system made it difficult for Kanaka students to name American colonialism as colonialism in their classrooms, but it opened up opportunities to do so through indirect comparison. On the one hand, in the early territorial era success in the colonialist education system required writing from the American perspective that schools taught as the norm. This made naming colonialism in Hawai'i difficult. Keolanui writes "all those who came *to this continent,*" adopting the North American perspective that the colonial educational system has required of him. Moreover, the immigration question to which he refers in his presentation—the mass migration of working-class people from southern and eastern Europe that arose after 1880—was a phenomenon of the continental United States, not Hawai'i. To the extent that he could not explicitly speak as a Kanaka, the North American voice that his school expected of him made it hard for his essay to address these issues of colonialism.

But on the other hand, taking on a North American stance opened new possibilities for Keolanui to pose an indigenous critique of colonialism in North America and, implicitly, in Hawai'i. Keolanui reminded his listeners, "The Indians only are considered natives of this continent while all who came to the so-called new world should be considered as immigrants." Keolanui is using different words to lay out a central proposition of scholarship in settler colonial studies: only indigenous people deserve the name Native, and all others who live in settler colonial societies from South Africa to Israel to the United States to Hawai'i are properly understood as settlers.[2] Keolanui further emphasizes his indigenous stance by referring to the Americas as "the so-called new world." Keolanui's political intervention is much the same as that made by settler colonial studies in recent years. Although he retains the word "immigrant" (whereas recent literature rightly insists that immigrants are settlers) he deploys the word to emphasize that Europeans are fundamentally foreign to North America. He writes that although "the class of people that lived" at the time of the Revolution "have been called colonists, nevertheless they also were immigrants." The "immigrants ... later overspread the other colonies" and much of the continent. Indeed, Keolanui writes, "the pioneers were practically all English[,] Irish, Dutch and Germans."[3] The emphasis on the foreign emphasizes what today we would call a settler/indigenous binary.

Reading Keolanui's polished prose and taking note of the extensive research in his high school graduation address, one is struck by a suspicion that is familiar to anyone used to grading essays: did he really compose these paragraphs? A quick check of text databases confirms the hunch. Keolanui created almost the entire speech from unattributed pieces of three articles that were published in the early 1920s. What is more, all three had recently been reprinted together in one book, the likely source for Keolanui's essay. "History of Immigration" by Prescott Farnsworth Hall, "Immigration: A Review" by Senator Henry Cabot Lodge, and Philip Davis's "What America Means to the Immigrant" had all appeared in Davis's 1920 collection, *Immigration and Americanization: Selected Readings*.[4]

But this is no common case of plagiarism. It is a colonized subject's appropriation of the discursive tools of the colonial power, and the nature of the sources makes Keolanui's insistently indigenous critique of colonialism nothing short of astonishing. Keolanui uses the sentences of three white American writers nearly word for word but edits them deftly to advance a critique that at times stands in nearly direct opposition to those authors' intent. His use of passages from Hall's essay is a prime example. As the secretary of the Immigration Restriction League, Hall was a leader of one of the United States' largest eugenicist organizations. Hall wrote his essay as part of his effort to defend what he called the "Nordic race" in America from what he considered to be the danger of replacement by inferior "race stocks"—"Alpine, Mediterranean, Semitic, and Asiatic" people. Hall asserts that white Americans with ancestors from western and northern Europe are the real and true Americans, and anyone who questions that is wrong. To advance this point, Hall admits that "in *popular* discussions of the immigration question it is often said" that all non-Indians (including whites like himself) are immigrants but declares that this is merely a layperson's misconception. Keolanui takes Hall's sentence and transforms its meaning by changing one word: he writes that "in *proper* discussions" of immigration it is said that all non-Indians are immigrants (emphasis added). The difference is crucial. Hall contends that European "colonists" built the United States and they are therefore the truest of natives. Keolanui specifically rejects the term "colonists," because over a century of settler colonial rhetoric had prepared American readers to believe Hall's argument. Keolanui writes that "immigrant" is the better term, because it emphasizes settlers' enduring foreignness to the land: while the settlers "have been called colonists, nevertheless they also were immigrants." By insisting that Indians are the only true Americans, Keolanui remakes Hall's argument for racially restrictive immigration laws into a subtle indictment of white

Americans' claims to Native status. And he does this, let us remember, in a graduation ceremony at a boarding school whose founders were dedicated to remaking Kānaka in the image of white Americans.[5]

How was it possible for a young person—who was raised in the decades after Hawai'i's annexation to the United States in 1898, trained in an educational system dominated by colonialist administrators, and placed in a position where he was supposed to demonstrate his mastery of the discourse of the colonizer in a speech to his teachers and school administrators—to take a racist white American source and so adroitly turn it into an indigenous critique of colonialism that remaps North America as fundamentally Indian? This outcome is not to be taken for granted: resistance is not a given in any context, and the adoption of an indigenous stance has not been a given in Kanaka political discourse.[6]

Keolanui was capable of this remarkable act of resistance because he was an heir to and exemplar of a genealogy and history of Native Hawaiian thought about the globe. It has been the purpose of this book to trace that genealogy and narrate that history. As moʻolelo and mele (stories and songs) demonstrate, Kānaka were thinking about global geography and generating their own understandings of it long before Cook arrived on the shores of Hawai'i. After the Westerners' incursion, that process of exploration continued as Kānaka like Kaʻiana and Ka Wahine explored the world directly by going abroad, as labor overseas landed Kānaka in the United States, and as missionary work sent Kānaka to Nuku Hiva and other Pacific Islands. The process of exploring global geography and shaping the place of Hawai'i in it also continued at home in the islands. There, Kānaka shaped the encounter with Christianity into a means of using text to explore the terrestrial globe as well as the world of religion and to reconnect with other Pacific Islanders. All of these actions were the inheritance of Kānaka of Keolanui's generation. But of all the nineteenth-century Hawaiian geographical initiatives, Keolanui's graduation address most clearly bore the marks of Kanaka efforts in education and in writing about American Indians. Kanaka students and teachers navigated and shaped geographical education in Hawai'i in order to pursue their own goals, and Kanaka writers took the image of the Indian that missionary texts gave them and remade it into a means to think about their relationship to colonialism and other indigenous people—a feat much like Keolanui's appropriation of settler texts. Keolanui was clearly a powerful thinker and writer, but the piece he wrote also derives great power from the intellectual kūpuna (ancestors) from which he descended. From them, he inherited the power to use American materials to forward a thesis that would have startled most white Americans in 1923—and all in the heavily circumscribed context of a boarding school graduation ceremony.

He was remarkable but not unique. At the Kamehameha School in Honolulu, a boarding school specifically established for Native Hawaiian students, young Kānaka who were given a very restrictive and colonialist education used the smallest openings to make room to place Hawaiʻi and Hawaiians in a broader geography than their school's curriculum envisioned. School administrators worked to Americanize their students. The curriculum of the school—both the formal curriculum in the classroom and the informal cultural curriculum taught through means ranging from military-style uniforms to rules of comportment—was focused on this goal of Americanization. And yet there were always openings though which students could consider the world and their place in it differently.

Perhaps the most powerful of these openings was the idea of folklore. In the 1930s, school president Frank Midkiff and teacher Donald Mitchell demonstrated an attachment to Hawaiian language and folklore and sincerely worked for their preservation. When school administrators and teachers turned Hawaiian culture into something called "folklore"— for example, organizing a series of "Hawaiian Culture Lectures" for the students in 1935—it was an action rich with geographic meaning. It preserved Hawaiian culture, but also restricted it within a folkloric space of charming stories and quaint practices, a space that might safeguard normative colonial geographies from disruption. But Kanaka students practiced folklore in expansive ways that asserted the place of Hawaiʻi in a broader Pacific geography.[7] Students largely pursued these activities in gendered groupings, reflecting the gender segregation of the Kamehameha School at the time and also a pattern familiar in Kanaka life. In the 1930s, the boys of the Hui Oiwi (Native Club, a meaningful name given that the entire school was intended to educate Native Hawaiian students) practiced "old-time wrestling" in malo (loincloths), learned and performed mele, and told "folk tales." They hosted "a regular luau," and built a Hawaiian hale, which the school newspaper described as "a grass hut." They interested themselves in other Pacific Islander peoples, watching a movie about Sāmoan rock sliding and learning Sāmoan knife dancing.[8] Meanwhile, girls took the lead in writing about Hawai'i from Kanaka perspectives—a leadership role in which we can discern a stage in the moʻokūʻauhau (genealogy) of Hawaiian female intellectuals that stretches up the present-day prominence of women in Hawaiian studies. Young wāhine (women) like Ululani Weight, Vivian Crockett, Amy Crockett, Mary Kauimeheiwa, and Haunani Cooper used the school newspaper and the school's annual literary supplement to publish moʻolelo and poetry and articles on Hawaiian history that incorporated short passages in Hawaiian. Some young kāne (men) like David White and Eugene Burke joined the effort. Burke even published a moʻolelo in Hawaiian—a language that had once been forbidden at the school but that was, under Midkiff, briefly taught again.[9] This was more than the nostalgic and folkloric cultural preservation that school administrators seemed to have envisioned. It is revealing that students' decisions to write Hawaiian literature coincided with their interest in learning about Sāmoan people, language, and culture. Like Keolanui's implicit reflection on indigeneity in his graduation address at Hilo Boarding School, the Kamehameha School students' activities and writings in the 1930s provided a means for making Kanaka knowledge and practices relevant in the contemporary age and rethinking Hawaiians' place in the world. Paradoxically, students made folklore a way to challenge the notion of Hawai'i as a small and isolated place that was stuck in the past and assert that it was an important part of a still-living world region in the Pacific.

Also like Keolanui, the Kamehameha School students who engaged in these activities were heirs to kūpuna before them. The initiative of the many nineteenth-century Kanaka writers (famous ones like ʻĪʻī, Malo, Kamakau, and Kepelino, and less famous ones like those who wrote the many newspaper articles that this book looks to as sources) and teachers (in schools, in hālau hula [hula schools], and in informal contexts of home and community) had nurtured and preserved Hawaiian knowledge that could be used to assert the embeddedness of Hawai'i in the extensive geography of ʻĀinamoana and the world beyond. The remarkable young Kānaka of the early twentieth century were heirs to those who had gone before them.

To speak of Keolanui and the young people of the Kamehameha Schools in the 1920s and 1930s as heirs is to invoke moʻokūʻauhau. Moʻokūʻauhau is a practice at the very core of Hawaiian life. It is a methodology at the center of Hawaiian studies. In this case, moʻokūʻauhau

Members of the Hui Oiwi at Kamehameha School with a hale pili (grass house) they built in 1933. *Left to right:* Carl Kahalewai, William Paalani, Walter Burke, and Folinga Faufata. Courtesy of and copyright Kamehameha Schools Archives, Honolulu.

can remind us that in addition to being intellectual, cultural, and political descendants of kūpuna Hawaiʻi (Hawaiian ancestors), the young people of the early twentieth century also became ancestors in their own right. A moʻokūʻauhau (whether genetic, intellectual, cultural, or political) can be thought of as a lei. Each blossom (pua, a word that also means offspring) follows another blossom, from whose seed it has grown. But each blossom, in turn, also makes possible a new blooming, a new pua. The young people of the first half of the twentieth century, the youth of the territorial era, were descendants, but they are also ancestors.

Who are their pua and their mamo—their offspring and descendants? As we look into the twentieth century from the perspective of its earliest years, we see the blossomings of this inheritance. At midcentury, we find tireless documenters of tradition and language like Mary Kawena Pukui, indefatigable historians and teachers of the Hawaiian martial art lua like Charles Kenn, brilliant kumu hula (hula teachers) like Edith Kanakaʻole, and countless others who persevered in the face of a consolidated U.S. occupation. In the 1970s, we see an uluwehi (a lush flourishing) of pua in the young people whose energies fueled a Hawaiian Renaissance that was both cultural and political, revitalizing the language, invigorating the defense of Kahoʻolawe and Kalama Valley and Waiāhole and wahi pana (storied places), reclaiming portions of the occupied landscape for loʻi kalo (taro patches), infusing new energy into Hawaiian music, founding new hālau hula, and awakening an activist movement that made clear that the American occupation would not go unchallenged. In the 1980s and beyond, we see the seeds of those pua take root and blossom again in the proliferation of political action around the defense of the land and of the food that grows from it, the building of institutions of education from preschool to graduate school that preserve Hawaiian language and transmit and elaborate Hawaiian knowledge, and the movement of Kanaka intellectuals into the center of academic life in Hawaiʻi, North America, and the global indigenous world.

Perhaps most pertinent of all for this book, beginning in the 1970s and stretching to the current day, we see the reblossoming of the Hawaiian long-distance canoe voyaging tradition. The waʻa kaulua (double-hulled canoes) of the Polynesian Voyaging Society are the modern incarnations of the canoes in which the ancestors of today's Kānaka first traveled to Hawaiʻi and then voyaged back and forth through ʻĀinamoana. Ever since the first voyage of *Hōkūleʻa* to Tahiti in 1976, the oceangoing canoe has been a powerful symbol of the reassertion of Hawaiʻi's place in the Pacific, demonstrating that Hawaiʻi's most fundamental connection is to ʻĀinamoana, not America. As I complete this book, my children regularly go to the Polynesian Voyaging Society website to follow the progress of *Hōkūleʻa* and *Hikianalia,* which have begun a new endeavor beyond the Pacific: a worldwide voyage in Hawaiian oceangoing canoes.

All of these are part of a moʻokūʻauhau of exploration and of efforts by Kānaka to define their place in the world. They explored the world through the generations stretching from Kāʻeokūlani's first exploration of James Cook's vessel in 1778 up to the travels of Kaʻiana, Ka Wahine, and those who came after them. They considered the shape of the world in their encounters with new ideas and systems of knowledge at schools at home and abroad, from their first encounters with Tahitian missionaries to Eli Keolanui's graduation at Hilo Boarding School in 1923. They shaped their place in the world through labors stretching from California goldfields to New England harbor towns to Nuku Hiva missionary churches to ships across the globe. Through all of these decades and in all of these ways, Kānaka have used the resources available to them— even such unlikely resources as missionary churches, colonialist textbooks, and boarding schools—to explore the world, construct their own understandings of it, and shape their place in it. They continue to do so today in classrooms and aboard canoes, in government chambers and in church pews, in talk-story with other Pacific Islanders and in conversations with American Indians, in libraries and in loʻi kalo, in the islands and in the diaspora, in hālau hula and on the streets. And given the surprising history of the Kanaka engagement with global geography over more than two centuries, Kānaka are undoubtedly reshaping their place in the globe in ways and spaces we do not yet even suspect.

ACKNOWLEDGMENTS

This book has been nurtured by the love and friendship and brilliant ideas of friends and family and colleagues across years of labor and oceans of space. The few words of thanks I can offer here could never capture all the generous support I received, but I hope they convey a sense of how grateful I am to everyone who helped me along the way. I owe a profound debt to the generosity of scholars who helped me find my way in Hawaiian studies. Noenoe Silva, your scholarship and your example were what first gave me hope that I might make some contribution to Hawaiian studies. Your encouragement and support have meant the world to me, and I will be forever grateful. Hokulani Aikau, you helped me to think through the issues of being both 'ōiwi and malihini to Hawai'i and offered kind words, a warm welcome, and much-needed practical assistance in Honolulu. Mahalo to you and your wonderful 'ohana. Thanks also to Lisa Kahaleole Hall: our long-distance writing sessions kept me working, your reading of key passages made this book much better, and your friendship is one of the most precious things to come from this project. Noelani Arista, mahalo for your support and interest and helping me to refine a key translation and to think through central issues. Kealani Cook, your timely suggestions were as important to me as your stunning dissertation. Leilani Basham, I am grateful for your comradeship on the Native American and Indigenous Studies Association Council and for the model of your incisive scholarship on nineteenth-century mele. Mahalo to Candace Fujikane, Ty Kāwika Tengan, J. Kēhaulani Kauanui, Dean Saranillio, Kali Fermantez, and Nālani Wilson-Hokuwhitu for your kind encouragement, and to Alohalani Brown for your warm support and acute linguistic insights. Christine Manganaro, I am grateful for our conversations that helped me think through early phases of this project. Thanks also to emerging scholars whose kindness to me has been as inspiring as the promise of their work: Maya Saffery, Leon Noʻeau Peralto, and Hiʻilei Hobart, I am eager to read more of your work. Mahalo to 'Ekela Kanī'aupi'o Crozier, my valued first kumu of Hawaiian language whom I never met in person, and to Lalepa Koga, whose intensive teaching prepared me for close readings of nineteenth-century texts.

Mark Rifkin, I cannot thank you enough for your invaluable support and your good humor and incredible capacity for work. They were crucial to finishing this book while serving with you on NAISA Council. William Bauer was an absolutely essential reader of this manuscript, a guide to a deeper understanding of Native California; this book is far better because of your guidance. Louise Pubols, you helped me to understand the relationship of California to Hawai'i, and your friendship helped me to endure the rigors of research. Thanks to Eric Sheppard, Heiga Leitner, Ryan Holifield, Kristin Sziarto, and Moira Macdonald, who helped me in the world of geography. Mahalo to Circe Sturm, Erika Bsumek, and Luís Cárcamo-Huechante and the participants in the seminar "Territorial Roots and Diasporic Routes" at the University of Texas at Austin; Joseph Hall at Bates College; David Freund, Wendy Cheng, Ana Elizabeth Rosas, and other organizers and participants in the "Migrant Metropolis" conference at the University of Maryland; Mia Bay, Ann Fabian, Nayan Shah, and other participants at the "Borders and Belonging" seminar at Rutgers; Maria Montoya, Linda Gordon, and other participants at the "Gender and Internal Colonialism" workshop at New York University; and Jordan Sand, Katherine Benton-Cohen, Paul Kramer, Jun Uchida, Takashi Fujitani, Eiichiro Azuma, and other participants in the "Pacific Empires" working group at Georgetown. Thanks also to Nancy Shoemaker and Susan Lebo, whose work taught me about the worlds of whaling. Peggy Pascoe's early encouragement was essential to the origins of this project. She is no longer with us, but my gratitude to her remains.

I am blessed with remarkable colleagues at the University of Minnesota. Thanks to Ann Waltner, Mary Jo Maynes, and Jean O'Brien for your thoughtful insights and support since the early days of this project. Many, many thanks to friends and comrades Karen Ho, Jigna Desai, Kevin Murphy, Katherine Hayes, Sarah Chambers, Michael Lower, Lianna Farber, Erika Lee, Malinda Lindquist, Susanna Blumenthal, Katharine Gerbner, J. B. Shank, Ruth Mazo Karras, Susannah Smith, Brenda Child, Lisa Norling, the participants in the Geography Department Coffee Hour and in the Anthropology Department Socio-

Cultural Workshop, and the Center for Study of the Premodern World. Mahalo to Clint Carroll, Angelica Lawson, and all the participants in the American Indian and Indigenous Studies Workshop who worked through chapter drafts with me. To the amazing students and former students I have been privileged to talk about my work with, including Bernadette Pérez, Jameson Sweet, Jessica Arnett, Kent Weber, and Juliana Hu-Pegues: thank you.

This work would not have been possible without generous support and research leave time funded by the National Endowment for the Humanities and a number of entities within the University of Minnesota: the Imagine Fund, the McMillan Travel Grant, the Institute for Advanced Study, and the Grants-in-Aid of Research, Artistry, and Scholarship Program.

For skillful help in research, heartfelt thanks to Barbara Dunn at the Hawaiian Historical Society, Stacy Naipo at the Midkiff Learning Center at the Kamehameha Schools, Joan Hori and Dore Minatodani at the Hawaiian Collection at the University of Hawaiʻi–Mānoa Hamilton Library, Marilyn Reppun and John Barker at the Hawaiʻi Mission Children's Society Archives, Margarite Ragnow and Margaret Borg at the James Ford Bell Library, and the staffs of the Hawaiʻi State Archives, the Bishop Museum Library and Archives, the Lyman House Museum and Archives, the National Archives in San Francisco, and the Meriam Library Special Collections at California State University, Chico.

The love and help of my family sustains me. Thank you to my parents, Joe Chang and Anne Craig, my brothers and sisters Aileen, Sarah, Joe, and John, their spouses, my nieces and nephews, Uncle Tommy, Aunty Vivian, Uncle Toshi, Aunty Ronnie, Uncle Henry, Aunty Amy, Uncle Francis, Aunty Jenny, and Aunty Barbara. Your encouragement means so much to me. Mahalo to cousins Cathy Chang, Harry Palmer, Cecelia Chang, Jason and Kathy Chang, Pele Bauch, and Williamson Chang. Bill, mahalo piha for your encouragement and commitment. And thanks to friends who are like family: Deedee Terzian, Mark Felton, Suzanne Herko, Simone, and Nicholas. You made work trips feel like homecomings.

No words in English or Hawaiian or any other language could express my gratitude to Tracey Deutsch, Gabriel Kaʻimipono Chang-Deutsch, and Nathaniel Kahawainui Chang-Deutsch. Tracey, this book is about knowing the world from one's own perspective, and you constantly challenge me to see the world anew. Kaʻimi and Kahawainui, this book is about exploring the world and making your place in it. I love to watch you explore the world, and I look forward to watching as you find and make places of peace, justice, and joy.

NOTES

INTRODUCTION

1. In matters relating to Hawai'i, Kanaka (plural: Kānaka) is a synonym for Native Hawaiian. Meaning "person," the term is sometimes modified by such adjectives as "maoli" (real or original), "'ōiwi" (indigenous), or "kama'āina" (child of the land). The term was commonly used in the nineteenth century, and in contemporary Hawaiian studies generally signals the centering of indigenous perspectives on Hawai'i. I will use all of these terms in this book as synonyms for "Native Hawaiian" and "Hawaiian" to refer to the indigenous people of Hawai'i and their descendants, both in Hawai'i and in diaspora.

2. Stuart B. Schwartz, ed., *Implicit Understandings: Observing, Reporting, and Reflecting on the Encounters between Europeans and Other Peoples in the Early Modern Era* (New York: Cambridge University Press, 1994).

3. David Tavares and Mark Brosseau, "The Representation of Mongolia in Contemporary Travel Writing: Imaginative Geographies of a Travellers' Frontier," *Social and Cultural Geography* 7 (2006): 299–317; Joan M. Schwartz, "The Geography Lesson: Photographs and the Construction of Imaginative Geographies," *Journal of Historical Geography* 22 (1996): 16–45; Anne Godlewska, "Map, Text, and Image: The Mentality of Enlightened Conquerors. A New Look at the *Description d'Egypte*," *Transactions of the Institute of British Geographers*, n.s. 20 (1995): 5–28; Emma Teng, *Taiwan's Imagined Geography: Chinese Colonial Travel Writing and Pictures, 1683–1895* (Cambridge, Mass.: Harvard University Press, 2004).

4. A noteworthy exception exists in the literature on the Japanese exploration of the world in order to appropriate ideas and technologies useful for Japanese purposes. Akira Iriye, *Japan and the Wider World: From the Mid-Nineteenth Century to the Present* (London: Longman, 1997); Ronald P. Toby, *State and Diplomacy in Early Modern Japan: Asia in the Development of the Tokugawa Bakufu* (Princeton, N.J.: Princeton University Press, 1994).

5. Gregory T. Cushman, *Guano and the Opening of the Pacific World: A Global Ecological History*, Studies in Environment and History (Cambridge: Cambridge University Press, 2014); David Igler, *The Great Ocean: Pacific Worlds from Captain Cook to the Gold Rush* (Oxford: Oxford University Press, 2013); Matt K. Matsuda, *Pacific Worlds: A History of Seas, Peoples, and Cultures*

(Cambridge: Cambridge University Press, 2012); Nicholas Thomas, *Islanders: The Pacific in the Age of Empire* (New Haven, Conn.: Yale University Press, 2012); Matt K. Matsuda, *Empire of Love: Histories of France and the Pacific* (New York: Oxford University Press, 2005).

6. Noenoe K. Silva, *Aloha Betrayed: Native Hawaiian Resistance to American Colonialism* (Durham, N.C.: Duke University Press, 2004), Introduction.

7. Lisa Kahaleole Hall, "Strategies of Erasure: U.S. Colonialism and Native Hawaiian Feminism," *American Quarterly* 60 (2008): 274.

8. On loss of political power and of a kingdom with effective political sovereignty, see Michael Dougherty, *To Steal a Kingdom* (Waimanalo, Hawaiʻi: Island Style Press, 2000) and Rich Budnick, *Stolen Kingdom: An American Conspiracy* (Honolulu: Aloha Press, 1992). The emphasis on loss of culture has been a particularly powerful theme in trade books by writers who researched and wrote their works without learning to read Hawaiian-language sources and are thus poorly prepared to understand cultural persistence, change, and innovation among Kānaka. Despite this liability, these books have often achieved popular success and influence, such as Julia Siler, *Lost Kingdom: Hawaii's Last Queen, the Sugar Kings, and America's First Imperial Venture* (New York: Grove Press, 2013); James L. Haley, *Captive Paradise: A History of Hawaii* (New York: St. Martin's, 2015); Susanna Moore, *Paradise of the Pacific: Approaching Hawaii* (New York: Farrar, Straus and Giroux, 2015).

9. Lilikalā Kameʻeleihiwa, *Native Land and Foreign Desires: Pehea Lā e Pono Ai? How Shall We Live in Harmony?* (Honolulu: Bishop Museum Press, 1992); Jocelyn Linnekin, *Sacred Queens and Women of Consequence: Rank, Gender, and Colonialism in the Hawaiian Islands* (Ann Arbor: University of Michigan Press, 1990); Sally Engle Merry, *Colonizing Hawaiʻi: The Cultural Power of Law* (Princeton, N.J.: Princeton University Press, 2000); Jonathan Kay Kamakawiwoʻole Osorio, *Dismembering Lāhui: A History of the Hawaiian Nation to 1887* (Honolulu: University of Hawaiʻi Press, 2002); Silva, *Aloha Betrayed*; John Charlot, *Classical Hawaiian Education: Generations of Hawaiian Culture* (Lāʻie: Pacific Institute, Brigham Young University, 2005); J. Kēhaulani Kauanui, *Hawaiian Blood: Colonialism and the Politics of Sovereignty and Indigeneity* (Durham, N.C.: Duke University Press, 2008); Noelani Arista, "Captive Women in Paradise, 1796–1826: The Kapu on Prostitution in Hawaiian Historical Legal Context," *Native American Culture and Research Journal* 35 (2011): 39–55; Hokulani Aikau, *Chosen People, Promised Land: Mormonism and Race in Hawaiʻi* (Minneapolis: University of Minnesota Press, 2012); Kamanamaikalani Beamer, *No Mākou Ka Mana: Liberating the Nation* (Honolulu: Kamehameha Publishing, 2014).

10. Katrina-Ann R. Kapāʻanaokalāokeola Nākoa Oliveira, *Ancestral Places: Understanding Kanaka Geographies* (Corvallis: Oregon State University Press, 2014); Carlos Andrade, *Haʻena: Through the Eyes of the Ancestors* (Honolulu: University of Hawaiʻi Press, 2009); B. Kamanamaikalani Beamer and T. Kaeo Duarte,

"I palapala no ia aina—Documenting the Hawaiian Kingdom, A Colonial Venture?," *Journal of Historical Geography* 35 (2009): 68; Kalei Nuʻuhiwa, "Papakū Makawalu: An Analytical Methodology and Pedagoy for Understanding the Hawaiian Universe," paper presented at the Native American and Indigenous Studies Annual Meeting, Austin, Texas, May 30, 2014; Renee Pualani Louis, "Can You Hear Us Now? Voices from the Margin: Using Indigenous Methodologies in Geographic Research," *Geographical Research* 45 (2007): 130–39.

11. Kali Fermantez, "Re-Placing Hawaiians in Dis Place We Call Home," *Hūlili* 8 (2012): 1; Cole Harris, *Making Native Space: Colonialism, Resistance and Reserves in British Columbia* (Vancouver: University of British Columbia Press, 2002); Sarah De Leeuw, "Intimate Colonialisms: The Material and Experienced Places of British Columbia's Residential Schools," *Canadian Geographer/Le Géographe canadien* 51 (2007): 339–59; Evelyn J. Peters, "Focus: Making Native Space; a Review Symposium," *Canadian Geographer/Le Géographe canadien* 47 (2003): 75–87; Lucy Gale, "Defining a 'Maori Geography': Issues Explained, and One Personal Opinion on What It Could Consist Of," *New Zealand Journal of Geography* 101 (1996): 14–20.

12. Carlos Andrade, "A Hawaiian Geography or a Geography of Hawaiʻi?," in *I Ulu I Ka ʻĀina: Land*, ed. Jonathan K. Osorio (Honolulu: University of Hawaiʻi Press, 2014).

13. Kealani Cook, "Kalākaua's Polynesian Confederacy: Teaching World History in Hawaiʻi and Hawaiʻi in World History," *World History Connected* 8 (October 2011); Kealani R. Cook, "Kahiki: Native Hawaiian Relationships with Other Pacific Islanders, 1850–1915" (PhD diss., University of Michigan, 2011), 188–99, 335–95; Richard A. Greer, "The Royal Tourist: Kalākaua's Letters Home from Tokio to London," *Hawaiian Journal of History* 5 (1971): 75–109; William N. Armstrong, *Around the World with a King* (Rutland, Vt.: Charles E. Tuttle, 1977); Alexander Liholiho, *The Journal of Prince Alexander Liholiho: The Voyages Made to the United States, England, and France in 1849–50* (Honolulu: University of Hawaiʻi Press, 1967); Stacy L. Kamehiro, *The Arts of Kingship: Hawaiian Art and National Culture of the Kalākaua Era* (Honolulu: University of Hawaiʻi Press, 2009), 24–25; Stacy Kamehiro, "Hawaiʻi at the World Fairs, 1867–1893," *World History Connected* 8 (October 2011). The story of Kānaka who traveled to London will also be part of forthcoming work by historian Coll Thrush.

14. [Samuel Manaiākalani Kamakau], "No ke Kaapuni Makaikai i na Wahi Kaulana a me na Kupua, a me naʻLii Kahiko mai Hawaii a Niihau," *Ka Nupepa Kuokoa*, June 15, 1865, 1. Kamakau identifies that ancestral place as two large islands "kapa ia e na haole o Nu Zelani" (called New Zealand by the Haole).

15. Hokulani Aikau, "Indigeneity in the Diaspora: The Case of Native Hawaiians at Iosepa, Utah," *American Quarterly* 62 (2010): 477–500. Mahalo nui to Hokulani Aikau for helpful conversations about this topic.

1. LOOKING OUT FROM HAWAI'I'S SHORE

1. "The Coming of Pele," in *The Echo of Our Song: Chants and Poems of the Hawaiians*, ed. and trans. Mary K. Pukui and Alfons L. Korn (Honolulu: University of Hawai'i Press, 1973), 53.

2. Mary Kawena Pukui and Samuel H. Elbert, *Hawaiian Dictionary: Hawaiian–English, English–Hawaiian*, rev. and enl. ed. (Honolulu: University of Hawai'i Press, 1986), s.vv. "honua," "ākea"; Yves Lemaître, *Lexique du Tahitien Contemporain: Tahitien–Français, Français–Tahitien* (Paris: Éditions de l'IRD, 1995), s.vv. "fenua," "atea."

3. Andrade, *Ha'ena*; Oliveira, *Ancestral Places*; Beamer and Duarte, "I palapala no ia aina," 68; Nu'uhiwa, "Papakū Makawalu."

4. On the rejection of the term in contemporary Aotearoa / New Zealand because of the ways it carves up the peoples of the Pacific into Western-imposed categories of Micronesian, Melanesian, and Polynesian, see Alice Te Punga Somerville, *Once Were Pacific: Māori Connections to Oceania* (Minneapolis: University of Minnesota Press, 2012), 112.

5. Gavan Daws, *Shoal of Time: A History of the Hawaiian Islands* (Honolulu: University of Hawai'i Press, 1989), xii–xiii; Gary Y. Okihiro, *Island World: A History of Hawai'i and the United States* (Berkeley: University of California Press, 2009), 45–47; Kenneth D. Collerson and Marshall I. Weisler, "Stone Adze Compositions and the Extent of Ancient Polynesian Voyaging and Trade," *Science* 317 (2007): 1907–11; H. A. Bridgman, "Could Climate Change Have Had an Influence on the Polynesian Migrations?," *Palaeogeography, Palaeoclimatology, Palaeoecology* 41 (1983): 193–206. For an estimate that places the first human settlement of Hawai'i well after 1000 CE, see Janet M. Wilmshurst et al., "High-Precision Radiocarbon Dating Shows Recent and Rapid Initial Human Colonization of East Polynesia," *Proceedings of the National Academy of Sciences of the United States of America* 108 (2011): 1817.

6. James Cook, *The Journals of Captain James Cook on His Voyages of Discovery* (Cambridge: Published for the Hakluyt Society at the University Press, 1955–67), 3, pt. 2:265.

7. Greg Dening, *Islands and Beaches: Discourse on a Silent Land, Marquesas 1774–1880* (Honolulu: University Press of Hawai'i, 1980), 48.

8. Linda K. Menton et al., *History of Hawaii—Student Edition*, 2nd ed. (Honolulu: University of Hawai'i Press, 1999), 34. Rona Tamiko Halualani notes that this emphasis on isolation sets Hawaiians apart chronologically as well as spatially—relegating them to a primitive "spatialized time" that corresponds strongly with what I describe here as the implication that Kānaka were unaware of the outside world. Rona Tamiko Halualani, *In the Name of Hawaiians: Native Identities and Cultural Politics* (Minneapolis: University of Minnesota Press, 2002), 15. Work on "virgin soil" epidemics has similarly given an emphasis on isolation that carries cultural, as well as biological, resonances: Alfred W. Crosby, *Germs, Seeds*

and *Animals: Studies in Ecological History* (Armonk, N.Y.: M. E. Sharpe, 1994), 121–23. In fact, the emphases on ecological and cultural isolation often reinforce one another, as in Patrick Vinton Kirch, *Feathered Gods and Fishhooks* (Honolulu: University of Hawai'i Press, 1998), 22, 23, 308.

9. The classic discussion of the distinction between belief and knowledge can be found in Edmund L. Gettier, "Is Justified True Belief Knowledge?," *Analysis* 23 (1963): 121–23. See also Eroc Schwitzgebel, "Belief," in *The Stanford Encyclopedia of Philosophy*, ed. Edward N. Zalta (Spring 2014 Edition), http://plato.stanford.edu/archives/spr2014/entries/belief/; Yudhijit Bhattacharjee, "New NSF Survey Tries to Separate Knowledge and Belief," *Science* 333, no. 6041 (July 22, 2011): 394.

10. Pukui and Elbert, *Hawaiian Dictionary*, s.vv. "'ike," "mana'o." On 'ike as sight, knowledge, and insight, see Oliveira, *Ancestral Places*, 95.

11. Silva, *Aloha Betrayed*, 18–21.

12. Hawai'iloa's name is sometimes given as Ke Kowa i Hawai'i or (by Kepelino) as Hawaii-nui. Kepelino, "Ka Lahui Hawaii, Kona Ano," in *Kepelino's Traditions of Hawaii*, ed. Martha Warren Beckwith (Honolulu: Bernice Pauahi Bishop Museum, 1932), 76–77; Davida Malo, *Ka Mo'olelo Hawaii: Hawaiian Antiquities*, ed. Malcolm Nāea Chun (Honolulu: Folk Press, Kapiolani Community College, 1987), 5; Martha Beckwith, *Hawaiian Mythology* (Honolulu: University of Hawaii Press, 1970), 363–65; Abraham Fornander, *Account of the Origins of the Polynesian Race, Its Origin and Migrations* (1880; repr. Rutland, Vt.: Tuttle, 1969), 1:23–24.

13. Note that different narratives that recount these oceanic migrations often disagree with one another. Like other people around the world, different accounts favored by different Kānaka proposed different understandings of the world and history—or, in academic terms, there were various schools of geographical and historical interpretation among Hawaiians. These disagreements, moreover, were pointed out by the earliest Kanaka authors who took up the pen in the 1830s, just a few years after American missionaries introduced written language to the islands—writers such as Davida Malo. Davida Malo, *Hawaiian Antiquities: Moolelo Hawaii*, trans. Nathaniel Emerson (Honolulu: Bishop Museum Press, 1987), 1–3. Malo attributes divergent accounts of the past partly to error introduced by imperfections of memory in an oral culture. The statement suggests that he was influenced in this belief by the missionaries and other Westerners who emphasized the superiority of text over orality.

14. Silva, *Aloha Betrayed*.

15. Pakui, "Ka Mele a Pakui," in *Fornander Collection of Hawaiian Antiquities*, ed. Abraham Fornander (Honolulu: Bishop Museum Press, 1919–20), 4:12, n. 2; see also Kepelino and S[amuel] M[anaiākalani] Kamakau, "Legend of Hawaiiloa," in Fornander, *Fornander Collection*, 6:267.

16. Pakui, "Ka Mele a Pakui," 4:17. The author of the description of Pāku'i was most likely Samuel Manaiākalani Kamakau, Kepelino, or S. N. Haleole,

whose work constitutes the bulk of what was later published as the *Fornander Collection* (see 4:1). Here I use the translation in Fornander, which was most likely prepared by Thomas George Thrum, because while it is inexact, it contains specificities that are not present in the Hawaiian original—specific identification as a historian, for example. This likely occurs because the Kānaka who assembled this material and Fornander were familiar with Pākuʻi's reputation, as he had lived only a couple decades before this collection was assembled. Note that work by Kapāʻanaokalāokeola Oliveira that has just appeared as of this writing also emphasizes the importance of "Ka Mele a Pakui" to understanding genealogical relationships between the islands, the gods, and the Kānaka. Oliveira, *Ancestral Places*, 6–15.

17. S[amuel] M[anaiākalani] Kamakau, "Na olelo pane a S. M. Kamakau," *Ka Nupepa Kuokoa*, October 31, 1864, 4.

18. The translation in these paragraphs is based closely on Fornander's, although I update spelling of some of the personal and place names. My translation of the verses regarding Niʻihau and Kaʻula differs from Fornander's. Fornander offers "Niʻihau is the last droppings." Other plausible translations would render Niʻihau as the sprout, the afterbirth, or the rootlet. For Lehua, Fornander offers that Lehua was "the closing one." Pukui and Elbert, *Hawaiian Dictionary*, s.vv. "ēwe," "ēweewe," "panina."

19. Pakui, "Ka Mele a Pakui," 4:14–15.

20. Kaleikuahulu, "A Wakea Creation Chant," in Fornander, *Fornander Collection*, 6:360; biographical information in n. 1. Kauhane, "Ka Hoomana Kahiko. Helu 4: Ka Moolelo no Ku, Kona Ano, a me Kana Hana," *Ka Nupepa Kuokoa*, January 26, 1865, 1.

21. Malo, *Ka Moʻolelo Hawaiʻi*, 10. For more on Kahiki as references to different areas of the heavens, see Samuel Manaiākalani Kamakau, *Works of the People of Old: Na Hana a Ka Poʻe Kahiko* (Honolulu: Bishop Museum Press, 1976), 5, 10; Valerio Valeri, *Kingship and Sacrifice: Ritual and Society in Ancient Hawaii* (Chicago: University of Chicago Press, 1985), 9.

22. Joseph K. Kahele, Jr., "He Moolelo no ka Ohelo," in Fornander, *Fornander Collection*, 5:577. "Nā Kūkulu o Kahiki" refers to part of a wall holding up the heavens. Note that the same word, "kukulu," was used to refer to the four directions, for example in the passage above in which Kamakau explains the terms for the four directions. Pukui and Elbert, *Hawaiian Dictionary*, 178.

23. Anonymous, "Ka Hoonoho Ana o ka Mookuauhau o ka Aina Ana ma Hawaii Nei mai a Wakea mai," in Fornander, *Fornander Collection*, 4:21.

24. Anonymous, "Ka Moolelo o ka Aina Ana ma Keia Mau Mokupuni a me ka Laha Ana o Keia Lahuikanaka," in Fornander, *Fornander Collection*, 4:8.

25. Anonymous, "Ka Moolelo o Moikeha," in Fornander, *Fornander Collection*, 4:157.

26. Pakui, "Ka Mele a Pakui," 4:13, and Fornander, "Traditional and Genealogical Notes," in Fornander, *Fornander Collection*, 6:321. See also "Wahanui

at http://www2.hawaii.edu/~dennisk/voyaging_chiefs/wahanui.html for Kanehunamoku.

27. The first translation is from Pukui. The second, by this author, is suggested by Pukui's capitalization of the word "Kahiki," which suggests a reference to the specific place Kahiki rather than foreign lands in general. Mary Kawena Pukui, 'Ōlelo Noʻeau : Hawaiian Proverbs and Poetical Sayings (Honolulu: Bishop Museum Press, 1983), 263.

28. Pukui, 'Ōlelo Noʻeau, 274. In this quotation, I have changed "ia" to "iā" in both usages to reflect contemporary spelling.

29. Nathaniel B. Emerson, Unwritten Literature of Hawaii: The Sacred Songs of the Hula (Washington, D.C.: Government Printing Office, 1909), 202–3. Here I reproduce Emerson's transcription of the song, but not his translation. The translation is my own, but it is influenced by Emerson's work.

30. Note, moreover, that prior to Western influence, the sacralizing prohibitions of Hawaiian kapu required men and women to eat separately; thus, whatever the genders involved in this poem, we cannot assume that the beloved is not the same sex as the voice in the poem.

31. Lorrin Andrews's dictionary declares, "Hawaiians express a strong internal glow of love for a person by maeele, equivalent to the external feeling of a limb when the flow of blood has for a time been stopped or retarded and the limb, in common language, is said to be asleep." Among the usages he includes one finds, "Ua maeele kona puuwai i ke aloha." This could be rendered as "His or her heart was benumbed by love." Andrews, A Dictionary of the Hawaiian Language (Honolulu: Henry M. Whitney, 1865), s.v. "maeele."

32. Kauwenaole, "No ka Niu," in Fornander, Fornander Collection, 5:591–97. "Niu," at Hawaiian Ethnobotany Online Database, http://data.bishopmuseum.org/ethnobotanydb.

33. Kahele, "He Moolelo no ka Ohelo," 577–83; "'Ōhelo," at Hawaiian Ethnobotany Online Database, http://data.bishopmuseum.org/ethnobotanydb.

34. [Samuel Manaiākalani Kamakau], "Ka Mookuauahu o Nanaulu no Kakuhihewa ke Alii o Oahu," in "Ka Moolelo o Hawaii Nei, Helu 14," Ka Nupepa Kuokoa, September 30, 1865, 1; Anonymous, "Moolelo no Kualii," in Fornander, Fornander Collection, 4:392, n. 1; "'Ulu," at Hawaiian Ethnobotany Online Database, http://data.bishopmuseum.org/ethnobotany.

35. W. S. Lokai, "Moolelo no ka Ulu," in Fornander, Fornander Collection, 6:679.

36. John Mana, "Moolelo no ka Awa," in Fornander, Fornander Collection, 5:607.

37. Anonymous, "Kaao no Kana a Me Niheu," in Fornander, Fornander Collection, 4:446–44.

38. This treatment has been persistent. In 1916, W. D. Westervelt wrote that to Kānaka, far-off lands were "located in the mysterious outside world Kahiki." Westervelt even treated identifiable places like Nuʻuhiwa (Nuku

Hiva), Polapola (Bora Bora), and Upolu as part of that world. W. D. Westervelt, *Hawaiian Legends of Ghosts and Ghost Gods* (Boston: Ellis Press, 1916), 260–61. In 1940, Martha Beckwith referred to Kaʻulu (in the stories, the place where the gods Kāne and Kanaloa drank ʻawa with the spirits, the birthplace of Pele, and the homeland of the human traveler Olopana) as a "mythical land." Beckwith, *Hawaiian Mythology*, 78. In 1989, Robert D. Craig declared Kāne-huna-moku to be a "mythical island." Robert D. Craig, *Dictionary of Polynesian Mythology* (Westport, Conn.: Greenwood Press, 1989), 4, 103. Valerio Valeri does much better than the other scholars, treating Kahiki as a place "where heroes go . . . to approach the gods and acquire power." Valeri, *Kingship and Sacrifice*, 8.

39. For early publications of this story, see B. K. Hauola, "He Wahi Kaao a me ka Mele Pu. Helu 2," *Ka Hae Hawaii*, August 15, 1860, 1; [M. J. Kapihenui], "He Mooolelo no Hiiakaikapoliopele, Helu 1," *Ka Hoku o ka Pakipika*, December 26, 1861, 1. On Kapihenui and the Pele and Hiʻiaka epic, see Silva, *Aloha Betrayed*, 76.

40. Kapihenui, "He Mooolelo no Hiiakaikapoliopele, Helu 7," *Ka Hoku o ka Pakipika*, February 13, 1862, 4.

41. M. K. Palikoolauloa, "Na Wahi Pana o Kaliuwaa," *Ka Hoku o ka Pakipika*, November 14, 1861, 4; Anonymous, "Kaao no Kamapuaa," in Fornander, *Fornander Collection*, 5:321. The essential guide to and scholarship on the Kamapuaʻa tradition is Lilikalā K. Kameʻeleihiwa, *A Legendary Tradition of Kamapuaʻa, the Hawaiian Pig God: He Moolelo Kaʻao o Kamapuaʻa* (Honolulu: Bishop Museum Press, 1996).

42. Another version of the story maintains that Pele was killed by a male named Namakaeha: John Moo, "Ka Iwi o Pele," in Fornander, *Fornander Collection*, 5:506.

43. Pukui and Elbert, *Hawaiian Dictionary*, s.vv. "akau," "hema," "hikina," "komohana."

44. As Venuti writes, "By producing the illusion of transparency, a fluent translation masquerades as a true semantic equivalence when it in fact inscribes the foreign text with a partial interpretation, partial to English-language values, reducing if not simply excluding the very differences that translation is called on to convey." Lawrence Venuti, *The Translator's Invisibility: A History of Translation* (New York: Routledge, 2008), 16. For critiques of translation practice in Hawaiian studies, see Larry Kimura, "Native Hawaiian Culture," in *Report on the Culture, Needs, and Concerns of Native Hawaiians* (Washington, D.C.: Native Hawaiians Study Commission, 1983), 1:182; Silva, *Aloha Betrayed*, 12; Bryan Kamaoli Kuwada, "To Translate or Not to Translate: Revising the Translating of Hawaiian Language Texts," *Biography* 32 (2009): 54–65.

45. Note that although Descartes was important in the history of European thought on perspective, the system of Cartesian graphing also expressed a dedication to categorical abstraction. Daniel Garber writes that "Cartesian bodies are just the objects of geometry made real, purely geometrical objects that exist

outside of the minds that conceive them." Daniel Garber, "Descartes' Physics," in *The Cambridge Companion to Descartes*, ed. John Cottingham (Cambridge: Cambridge University Press, 1992), 294; Ron Broglio, "Connecting Renaissance Linear Perspective and Cartesian Geometry and Optics," http://lmc .gatech.edu/~broglio/1102/desc_paint.html.

46. Using his own phonetic system 'Ōpūkahaʻia renders ka hikinalā as "k3-he-ke-n3 l3", komohana lā as "ko-mo-ho-n3-l3," o kai as "o-ki, and o uka as "o-8 k3." Henry Obookiah [ʻŌpūkahaʻia], *A Short Elementary Grammar of the Owhyhe Language* (Honolulu: Manoa Press, 1993), no page.

47. S[amuel] M[anaiākalani] Kamakau, "Ka Moolelo Hawaii, Helu 4," *Ke Au Okoa*, November 4, 1869, 1. In this passage, Kamakau is refining a passage Davida Malo had written decades before: "Ina huli ke kanaka ma kahi a ka la e komo aku ai kona alo, e maopopo auanei, o ka aoao ma kona lima hema he Kukulu hema ia, o ka aoao ma kona lima akau, he Kukulu akau ia." Malo, *Ka Moʻolelo Hawaii*, 8. Nathaniel E. Emerson translates this as "If a man faces towards the sunset his left hand will point toward the south, *kukulu hema*, his right to the north, *kukulu akau*." Malo, *Hawaiian Antiquities*, 9.

2. PADDLING OUT TO SEE

1. Silva, *Aloha Betrayed*, 18–20.

2. S[amuel] M[anaiākalani] Kamakau, "Ka Moolelo o Kamehameha I, Helu 11," *Ka Nupepa Kuokoa*, January 19, 1867, 1; Kāʻeokūlani's ancestry: [Samuel Manaiākalani Kamakau], "Ka Moolelo o Hawaii Nei, Helu 14," *Ka Nupepa Kuokoa*, September 30, 1865, 1; "Mookuauhau Alii," in Edith Kawelohea McKinzie, *Hawaiian Genealogies: Extracted from Hawaiian Language Newspapers*, ed. Ishmael W. Stagner II (Honolulu: University of Hawaiʻi Press for Institute for Polynesian Studies, 1983–86), 2:31.

3. Kamakau, "Ka Moolelo o Kamehameha I, Helu 11."

4. Pekka Hämäläinen, *The Comanche Empire* (New Haven, Conn.: Yale University Press, 2008), 7.

5. With the lack of humility characteristic of his narrative and centuries of European relations with Kānaka and other indigenous people since, he claims that the Englishmen were delighted to find that they could understand the Kānaka, as they spoke Tahitian. This is surely a great exaggeration born of overconfidence, given the substantial differences between Hawaiian and Tahitian and the fact that his crew's linguistic skills were limited to the halting Tahitian of a few officers. See Gananath Obeyesekere, *The Apotheosis of Captain Cook: European Mythmaking in the Pacific* (Princeton, N.J.: Princeton University Press, 1997), 66–67.

6. James Cook, *The Voyages of Captain James Cook* (London: William Smith, 1842), 2:234.

7. Although now decades old, the most influential standard narratives of

the events of 1778 continue to be Ralph S. Kuykendall, *The Hawaiian Kingdom: 1778–1854, Foundation and Transformation* (Honolulu: University of Hawai'i Press, 1938–67), 1:12–28; Daws, *Shoal of Time*, 1–8.

8. Marshall Sahlins, *Historical Metaphors and Mythical Realities: Structure in the Early History of the Sandwich Islands Kingdom* (Ann Arbor: University of Michigan Press, 1981); Marshall Sahlins, *How* (Chicago: University of Chicago Press, 1996); Obeyesekere, *Apotheosis of Captain Cook*.

9. [Samuel Manaiākalani Kamakau], "Ka Moolelo Hawaii, Helu 8," *Ka Nupepa Kuokoa*, August 5, 1865, 1. See also Kameʻeleihiwa, *Native Land and Foreign Desires*, 42; Pukui and Elbert, *Hawaiian Dictionary*, 15.

10. S[amuel] M[anaiākalani] Kamakau, "Ka Moolelo o Kamehameha I, Helu 12," *Ka Nupepa Kuokoa*, January 26, 1867, 1; S[amuel] M[anaiākalani] Kamakau, "Ka Moolelo o Kamehameha I, Helu 13," *Ka Nupepa Kuokoa*, February 2, 1867, 1.

11. On the initial hypotheses regarding coconuts and bull hides, see Kamakau, "Ka Moolelo o Kamehameha I, Helu 11" as reprinted in Kamakau, *Ke Kumu Aupuni* (Honolulu: Native Books, 1996), 46. On Niuolahiki, see Beckwith, *Hawaiian Mythology*, 478–79 and 486–87. On Kūʻīlioloa, see [S. W. Kahiolo], "He Moolelo no Kamapuaa, Helu 4," *Ka Hae Hawaii*, July 17, 1861, 64; S[amuel] M[anaiākalani] Kamakau, "Ka Moolelo o Kamehameha, Helu 10," *Ka Nupepa Kuokoa*, January 12, 1867, 1; Anonymous, "Kaua Akolu a Kamapuaa me Kuilioloa," in Fornander, *Fornander Collection*, 5:333.

12. Kuykendall, *The Hawaiian Kingdom*, 1:21–22.

13. Kameʻeleihiwa, *Native Land and Foreign Desires*, 44–45.

14. Kameʻeleihiwa, *Native Land and Foreign Desires*, 46–47.

15. S. N. Haleole, "Moolelo no Ko Hawaii Oihana Kahuha I Kapaia ka Oihana Hoomanamana," in Fornander, *Fornander Collection*, 6:57–65; S. N. Haleole, "Moolelo no Ko Hawaii Oihana Kahuna i ka Wa Mamua, I Kapa Ia he Hoomanamana," in Fornander, *Fornander Collection*, 6:67–159.

16. Beth Hill and Cathy Converse, *The Remarkable World of Frances Barkley* (Surrey, B.C.: TouchWood Editions, 2003), 40.

17. S[amuel] M[anaiākalani] Kamakau, "Ka Moolelo of Kamehameha I, Helu 22," *Ka Nupepa Kuokoa*, April 6, 1867, 1.

18. David Igler, "Diseased Goods: Global Exchanges in the Eastern Pacific Basin, 1770–1850," *American Historical Review* 109, no. 3 (June 2004): 697–98.

19. As David A. Chappell and Caroline Ralston both argue, treating all of these sexual contacts as prostitution reduces their agency to narrowly conceived economic dimensions. It was only in the 1850s that "one could finally speak of Euroamerican-style prostitution, with all of its pejorative connotations." David A. Chappell, "Shipboard Relations between Pacific Island Women and Euroamerican Men 1767–1887," *Journal of Pacific History* 27 (1992): 132.

20. Hill and Converse, *Remarkable World*, 40.

21. Capt. [John T.] Walbran, "The Cruise of the Imperial Eagle," *Victoria Daily Colonist*, March 2, 1901, 10; Hill and Converse, *Remarkable World*, 17.

22. Emma Christopher, Cassandra Pybus, and Marcus Buford Rediker, eds., *Many Middle Passages: Forced Migration and the Making of the Modern World* (Berkeley: University of California Press, 2007); H. E. Maude, *Slavers in Paradise: The Peruvian Labour Trade in Polynesia, 1862–1864* (Canberra: Australian National University Press, 1981); Peter Corris, "Blackbirding in New Guinea Waters, 1883–84: An Episode in the Queensland Labour Trade," *Journal of Pacific History* 3 (1968): 85–103. For the use of the term as it applies to Hawai'i, see J. A. Bennett, "Immigration, 'Blackbirding', Labour Recruiting? The Hawaiian Experience, 1877–1887," *Journal of Pacific History* 11 (1976): 3–27.

23. Chappell, "Shipboard Relations," 143.

24. Hill and Converse, *Remarkable World*, 54.

25. John Meares, *Voyages Made in the Years 1788 and 1789, from China to the North West Coast of America. To Which Are Prefixed, an Introductory Narrative of a Voyage Performed in 1786, from Bengal, in the Ship Nootka; Observations on the Probable Existence of a North West Passage; and Some Account of the Trade Between the North West Coast of America and China; and the Latter Country and Great Britain* (London: Logographic Press, 1790), 28.

26. Meares, *Voyages Made in the Years 1788 and 1789*, 10.

27. Meares, *Voyages Made in the Years 1788 and 1789*, xxxix.

28. Chappell, "Shipboard Relations," 143–45.

29. [William Douglas], "Voyage of the Iphegenia," in Meares, *Voyages Made in the Years 1788 and 1789*, 293.

30. Karl Marx, *Capital*, vol. 1, *A Critique of Political Economy*, ed. Friedrich Engels (1906; repr. Mineola, N.Y.: Dover, 2011), 185–89. On the functioning of production and distribution within ahupua'a and the system of communal production more broadly, see Kame'eleihiwa, *Native Land and Foreign Desires*; Carlos Andrade, "Ha'ena, Ahupua'a: Towards a Hawaiian Geography" (PhD diss., University of Hawai'i–Mānoa, 2001); Marshall Sahlins and Patrick Vinton Kirch, *Anahulu: The Anthropology of History in the Kingdom of Hawaii*, vol. 1, *Historical Ethnography* (Chicago: University of Chicago Press, 1994), 17–20.

31. S[amuel] M[anaiākalani] Kamakau, "Ka Moolelo o Kamehameha I, Helu 24," *Ka Nupepa Kuokoa*, April 20, 1867, 1.

32. Cook, *The Journals of Captain James Cook*, 3, pt. 2:1228–30.

33. David G. Miller, "Ka'iana, the Once Famous 'Prince of Kaua'i,'" *Hawaiian Journal of History* 22 (1988): 3.

34. Meares, *Voyages Made in the Years 1788 and 1789*, xxxix.

35. Ibid., 7. The idea of the noble savage is most closely associated with Jean-Jacques Rousseau, but Meares would likely have been more familiar with the myth's many iterations in English drama, verse, and prose. On the noble savage, see Terry Jay Ellingson, *The Myth of the Nobel Savage* (Berkeley: University of California Press, 2001).

36. My thoughts on the power of claiming the right to judge have been

influenced by Seth Edward David Epstein, "Tolerance, Governance, and Surveillance in the Jim Crow South: Asheville, North Carolina, 1876–1946" (PhD diss., University of Minnesota, 2013).

37. Meares, *Voyages Made in the Years 1788 and 1789*, 5–6.

38. Ibid., xxxix, 5–7.

39. Kamakau, "Ka Moolelo o Kamehameha I, Helu 24" (April 20), 1. Mahalo piha to Noelani Arista for crucial help in refining this translation.

40. In the original, this passage is printed in normal prose style. I added line breaks to call attention to the verse-like form of this helu. Thanks to Noelani Arista for identifying a portion of this passage as a helu; personal communication, November 14, 2014. On helu as genre, see Charlot, *Classical Hawaiian Education*, 239–41.

41. Daniel Kahāʻulelio, *Ka ʻOihana Lawaiʻa: Hawaiian Fishing Traditions*, ed. M. Puakea Nogelmeier (Honolulu: Bishop Museum Press, 2006); Mary Kawena Pukui and Noelani Arista, "No ka Mahiʻai ʻAna, Māhele 1: ʻŌlelo Mua no ka ʻOhina HEN / Agricultural Lore, Part 1: Introduction to the HEN Collection," *Ka Hoʻolina / The Legacy* 1 (March 2002): 4–6; Haleole, "Moolelo no Ko Hawaii Oihana Kahuna i ka Wa Mamua, I Kapa Ia he Hoomanamana," 119–25.

42. Meares, *Voyages Made in the Years 1788 and 1789*, 8, 28, 224.

43. Here I differ with Robert J. Morris, who suggests that these relationships were nonhierarchical. Robert J. Morris (Kapaʼihiahilina), "Configuring the Bo(u)nds of Marriage: The Implications of Hawaiian Culture & Values for the Debate about Homogamy," *Yale Journal of Law & the Humanities* 8 (2013): 105–57. For foundational work on "homosexual" and "gay" as terms connoting a stable position in a sexual binary that emerged in Western thought in the late nineteenth century, see Eve Kosofsky Sedgwick, *The Epistemology of the Closet* (Berkeley: University of California Press, 2008), 1–5, 67–90.

44. On Kaʻiana's age, see Miller, "Kaʻiana," 2.

45. Meares, *Voyages Made in the Years 1788 and 1789*, 224; Kekahi Mau Haumana o Ke Kulanui [Some Students of the High School], *Ka Mooolelo Hawaii* (Lahainaluna: Mea Pai Palapala no ke Kulanui, 1838); reprinted as *Ka Mooolelo Hawaii: The History of Hawaii*, Hawaiian Language Reprint Series 3 (Honolulu: Hawaiian Historical Society, 2005), 19; Kamakau, "Ka Moolelo o Kamehameha I, Helu 22," 1.

46. Meares, *Voyages Made in the Years 1788 and 1789*, 6.

47. Departure from Hawaiʻi: Meares, *Voyages Made in the Years 1788 and 1789*, xxxix, 10, 207, 225; Miller, "Kaʻiana," 8.

48. Meares, *Voyages Made in the Years 1788 and 1789*, 28–29.

49. Caroline Ralston, "Hawaii 1778–1854: Some Aspects of Makaʻainana Response to Rapid Cultural Change," *Journal of Pacific History* 19 (1984): 50.

50. George Mortimer, *Observations and Remarks Made During a Voyage to the Islands of Teneriffe, Amsterdam, etc.* (London: George Mortimer, 1791), 52; Nathaniel Portlock, *A Voyage Round the World, but More Particularly to the*

North-West Coast of America (London: John Stockdale, 1789), 360; Meares, *Voyages Made in the Years 1788 and 1789*, 7; Leah Pualahaole Caldeira, *Royal Hawaiian Featherwork: Nā Hulu Aliʻi* (Honolulu: University of Hawaiʻi Press, 2015).

51. Portlock, *A Voyage Round the World*, 360.

52. Ibid., 361–62. In the English of the time, "Tatar" (as a variant of "Tartar") referred to any number of groups stretching from the Caspian Sea to northeast Asia; the word was "vaguely applied" to people that Europeans believed were descendants of "the mingled host of Mongols, Tartars, Turks, etc." that expanded rapidly under the Mongol leader known as Genghis Khan: *The Oxford English Dictionary*, s.v. "Tartar." For Manchu as "Tatar," see "Asia Polyglotta," *The Quarterly Oriental Magazine*, December 1824, 155.

53. Anonymous, "Ka Moolelo o Moikeha," 4:157; Kekahi Mau Haumana o Ke Kulanui, *Ka Mooolelo Hawaii*, 19; Kamakau, "Ka Moolelo o Kamehameha I, Helu 22," 1.

54. Captain James Colnett as quoted in Hill and Converse, *Remarkable World*, 42.

55. On the diverse and cosmopolitan population of Macao, see C. X. George Wei, ed., *Routledge Studies in the Modern History of Asia*, vol. 87, *Macao: The Formation of a Global City* (Milton Park: Routledge, 2014); Zhidong Hao, *Macau History and Society* (Hong Kong: Hong Kong University Press, 2011).

56. Portlock, *A Voyage Round the World*, 359; Meares, *Voyages Made in the Years 1788 and 1789*, 6; Michael Greenberg, *British Trade and the Opening of China, 1800–1842* (Cambridge: Cambridge University Press, 2008), 23–24.

57. Portlock, *A Voyage Round the World*, 361.

58. Mortimer, *Observations and Remarks*, 51; Portlock, *A Voyage Round the World*, 361; Meares, *Voyages Made in the Years 1788 and 1789*, 6–7.

59. Portlock, *A Voyage Round the World*, 361.

60. Ibid., 360.

61. I have found no record of Portlock's religious affiliation, but his father is buried at St. Paul's Anglican Church in Norfolk, England, and his ancestors were baptized in the Anglican Church: see "Nathaniel Portlock's Origins—Sources?," at http://www.atdesignuk.com/clients/swmaritime.org.uk/1199.

62. Lindsay Ride, May Ride, and Bernard Mellor, *An East India Company Cemetery: Protestant Burials in Macao* (Hong Kong: Hong Kong University Press, 1995), 37–38, 89.

63. "Mosque in Macau," *Al Nahda* 1 (1981), 49.

64. On the churches and temples of Macao, see *Património Arquitectónico / Aomen jian zhu wen wu / Macau Cultural Heritage* ([Macao]: Instituto Cultural de Macau, [1990?]); Chen Weiheng, *Aomen miao yu tu lu / Templos de Macau / Macao Temples* ([Macao]: Min zheng zong shu wen hua kang ti bu, 2002).

65. Rogério Miguel Puga, *The British Presence in Macau: 1635–1793* (Hong Kong: Hong Kong University Press, 2013), 79–83.

66. Opium trade unhindered: ibid., 102.

67. Meares, *Voyages Made in the Years 1788 and 1789*, 10.
68. Ibid., 15.
69. Ibid., 17.
70. Ibid., 22.
71. Ibid., 25.
72. The "man and boy" from Maui were at this point traveling on the *Iphegenia*, the *Imperial Eagle*'s companion ship.
73. Meares, *Voyages Made in the Years 1788 and 1789*, 28.
74. This date seems incorrect, as it is marked as a Sunday, but February 7, 1788, is a Sunday in neither the Julian calendar nor the Gregorian calendar.
75. Meares, *Voyages Made in the Years 1788 and 1789*, 31–32.
76. Ibid., 33–35, 37. "Samboangan," in Walter Hamilton, *The East Indian Gazetteer* (London: Parbury, Allen, 1828), 2:499.
77. Meares, *Voyages Made in the Years 1788 and 1789*, 35.
78. Ibid.; John N. Schumacher, "The Early Filipino Clergy," *Philippine Studies* 51, no. 1 (March 2003): 7–62; Rudy Guevarra, "Filipinos in Nueva España: Filipino-Mexican Relations, Mestizaje, and Identity in Colonial and Contemporary Mexico," *Journal of Asian American Studies* 14, no. 3 (2011): 389–416.
79. Meares, *Voyages Made in the Years 1788 and 1789*, 36. On Maguindanao society and the way it related to empires and nations, see Shinzō Hiyase, *Mindanao Ethnohistory beyond Nations: Maguindanao, Sangir, and Bagobo Societies in East Maritime Southeast Asia* (Quezon City: Ateneo de Manila University Press, 2007).
80. Palauans sent a canoe frantically after the *Iphegenia* as it sailed off. It was only later that Douglas realized why: the chief of that island had in 1784 entrusted his son, whom the English called Lee Boo, to an Englishman to bring to England. He traveled abroad—much like Ka'iana—and was celebrated in London. But like so many islanders who went overseas, like Tawnee and Ka Wahine, Lee Boo perished overseas. Douglas writes that he later realized that Lee Boo's father probably sent the party in pursuit after him in hopes of regaining his son or at least learning about his fate. Meares, *Voyages Made in the Years 1788 and 1789*, 293–96. On "Johnstone's Island" as Tobi/Hahotobei, see Earnest S. Dodge, *Pacific Exploration from Captain Cook to the Challenger, 1776–1877* (New York: Little, Brown, 1971), 68. Daniel J. Peacock, *Lee Book of Belau: A Prince in London* (Honolulu: University of Hawai'i Press, 2007).
81. Cook, *The Journals of Captain James Cook*, 3, pt. 1:296, 298; Barry M. Gough, *The Northwest Coast: British Navigation, Trade, and Discoveries to 1812*, Pacific Maritime Studies Series 9 (Vancouver: University of British Columbia Press, 1992), 34.
82. Heather Harbord, *Nootka Sound and the Surrounding Waters of Maquinna* (Surrey, B.C.: Heritage House, 1996), 9–10.
83. Meares, *Voyages Made in the Years 1788 and 1789*, 184–86, 204–5; Michael

Harkin, "Whales, Chiefs, and Giants: An Exploration into Nuu-Chah-Nulth Political Thought," *Ethnology* 37 (Autumn 1998): 318; Hill and Converse, *Remarkable World*, 38; Robin Fisher, *Contact and Conflict: Indian-European Relations in British Columbia, 1774–1890* (Vancouver: University of British Columbia Press, 1977).

84. Meares, *Voyages Made in the Years 1788 and 1789*, 185, 191, 194–95.
85. Ibid., 210; Coll Thrush, "Vancouver the Cannibal: Cuisine, Encounter, and the Dilemma of Difference on the Northwest Coast, 1774–1808," *Ethnohistory* 58 (2011): 1–35.
86. Meares, *Voyages Made in the Years 1788 and 1789*, 224, 275.
87. Kamakau, "Ka Moolelo o Kamehameha I, Helu 24" (April 20), 1.
88. Kamakau, "Ka Moolelo o Kamehameha I, Helu 24," *Ka Nupepa Kuokoa*, April 27, 1867, 1.
89. Miller, "Ka'iana," 9–10.
90. Kamakau, "Ka Moolelo o Kamehameha I, Helu 24" (April 27), 1.
91. Meares, *Voyages Made in the Years 1788 and 1789*, 4.
92. Beamer, *No Mākou Ka Mana*, 69.
93. George Vancouver, *A Voyage of Discovery to the North Pacific Ocean, and Round the World: In Which the Coast of North-West America Has Been Carefully Examined and Accurately Surveyed* (London: G. G. and J. Robinson, 1798), 2:129; William T. Brigham, *Hawaiian Feather Work*, vol. 1 of *Memoirs of the Bernice Pauahi Bishop Museum* (Honolulu: Bishop Museum Press, 1899), 50.
94. [Fornander], "Things Similar in India, etc., and Polynesia," in *Fornander Collection*, 6:347; Anonymous, "Mele no Kauikealui," in *Fornander Collection*, 6:485; George Hu'eu Kanahele, *Kū Kanaka: Stand Tall; a Search for Hawaiian Values* (Honolulu: University of Hawai'i Press, 1993), 48.
95. Brandy Nālani McDougall and Georganne Nordstrom, "Ma ka Hana ka Ike (In the Work Is the Knowledge): Kaona as Rhetorical Action," *College Composition and Communication* 63 (2011): 100–101; Kame'eleihiwa, *A Legendary Tradition of Kamapua'a, the Hawaiian Pig God*, viii–ix.
96. Vancouver, *A Voyage of Discovery*, 165–66.
97. Ibid., 130, 140, 141.
98. David Kalakaua, *The Legends and Myths of Hawaii*, ed. Rollin Mallory Daggett (1888; repr. Honolulu: Mutual Publishing, 1990), 384.

3. A NEW RELIGION FROM KAHIKI

1. Ephraim Spaulding Diaries, MS Am 1390, Houghton Library Special Collections, Harvard University, Cambridge, Massachusetts, no page; [William Channing Woodbridge and unnamed others], *He Hoikehonua, He Mea Ia e Hoakaka'i i ke Ano o ka Honua Nei, a me Na Mea Maluna Iho* (Oahu: Na na Misioneri i Pai, 1832). Author's translation of title. Woodbridge is considered

by many bibliographers to be the author of the book, because the bulk of the text is a translation of his work in English. However, some passages in *He Hoikehonua* (notably those on the Hawaiian Islands and the Pacific) are entirely new or substantially different from those in Woodbridge's English-language book. David W. Forbes notes that Samuel Whitney and William Richards prepared the book using the English-language geographies of "Woodbridge, Worcester, Morse, and Malte-Brun, and also Morse's *Gazetteer*, the *Encyclopedia Americana*, the *Edinburgh Encyclopedia*, Lempiere's *Classical Dictionary*, the *History of the United States, History of England, Naval Chronicle, American Almanac, History of Boston*, and the *Missionary Herald*." See David W. Forbes, *Hawaiian National Bibliography* (Honolulu: University of Hawai'i Press, 1999), 2:30. I consider the book to be the product of Woodbridge and other Western authors, American missionaries, and Native Hawaiians working with them. On Spaulding's background, see Hawaiian Mission Children's Society, *Portraits of American Protestant Missionaries to Hawaii* (Honolulu: Hawaii Gazette, 1901), 43.

2. Works that emphasize the role of American missionaries in this manner include Kuykendall, *The Hawaiian Kingdom*, 1:100–116; Daws, *Shoal of Time*, 61–65, 97–105; Patricia Grimshaw, *Paths of Duty: American Missionary Wives in Nineteenth-Century Hawaii* (Honolulu: University of Hawai'i Press, 1989); LaRue W. Piercy, *Hawaii's Missionary Saga: Sacrifice and Godliness in Paradise* (Honolulu: Mutual Publishing, 1992); Mary Zwiep, *Pilgrim Path: The First Company of Women Missionaries to Hawaii* (Madison: University of Wisconsin Press, 1991). Jennifer Thigpen also places American missionaries at the heart of the story but explores the relations of Hawaiian women with missionary women. Jennifer Thigpen, *Island Queens and Mission Wives: How Gender and Empire Remade Hawai'i's Pacific World* (Chapel Hill: University of North Carolina Press, 2014). For Malo as translator with the missionaries, see Malcolm Nāea Chun, *Nā Kukui Pio'ole: The Inextinguishable Torches* (Honolulu: First Peoples' Press, 1993), 2; Charlot, *Classical Hawaiian Education*, 616.

3. Prime examples of this approach include Lawrence E. Fuchs, *Hawaii Pono, Hawaii the Excellent: An Ethnic and Political History* (Honolulu: Bess Press, 1961), 8–12; Francine Du Plessix Gray, *Hawaii: The Sugar-Coated Fortress* (New York: Random House, 1972), 39.

4. Walter D. Mignolo, "Epistemic Disobedience, Independent Thought and Decolonial Freedom," *Theory, Culture, and Society* 26 (2009): 159–81.

5. Kame'eleihiwa, *Native Land and Foreign Desires*, 142–45; Silva, *Aloha Betrayed*. See also Osorio, *Dismembering Lāhui*.

6. Edith Wolfe, "Introduction," in Edwin W. Dwight, *Memoirs of Henry Obookiah, a Native of Owhyhee* (Honolulu: Women's Board of Missions for the Pacific Islands, 2012), xiii.

7. Daws, *Shoal of Time*, 61, 381.

8. John Demos, *The Heathen School: A Story of Hope and Betrayal in the Age of the Early Republic* (New York: Knopf, 2014), 5, 20.

9. [Edwin W. Dwight with corrections from S. W. Papaula], *Ka Moolelo o Heneri Opukahaia* (New York: Ko Amerika Ahahui Teraka [American Tract Society], 1867), 20–23; Dwight, *Memoirs of Henry Obookiah*, 1–5, 11–12, 14, 16.

10. Silva, *Aloha Betrayed*, 45–86.

11. S. W. Papaula data at "Hawaii, Deaths and Burials, 1862–1919," database, *FamilySearch*, https://familysearch.org, February 10, 1883, citing Oahu, Hawai'i, reference, FHL microfilm 1,027,534; G. B. Kalaau, "Na Oihana a me ke Kumukanawai, o ka Ahahui Mahiai, no Kona Hema, Mokupuni o Hawaii, 'Ko Hawaii Pae Aina,'" *Ka Hae Hawaii*, April 23, 1856, 31; "I-a Pae," *Ka Nupepa Kuokoa*, February 2, 1865, 3; S. W. Papaula, "No ka Ahahui Euanelio ma Kona Hema, Hawaii," *Ka Nupepa Kuokoa*, September 20, 1862, 4; "Papa Inoa," *Ke Alaula*, September 1, 1867, 23; [Dwight and Papaula], *Moolelo o Heneri Opukahaia*, 9–12; S. W. Papaula, "Moolelo no Opukahaia," *Ka Nupepa Kuokoa*, November 4, 1865, 4. Note that this article by Papaula served as the basis for the early pages of the book published two years later under the title *Moolelo o Heneri Opukahaia*. Although often treated as a translation of the *Memoirs*, the new opening and corrections to the text appear to depend on Papaula's newspaper article, "Moolelo no Opukahaia." In examining the revision and correction of the text as indicative of Kanaka Maoli intellectual agency, I am following the work of John Charlot, *Classical Hawaiian Education*, 617–18.

12. Papaula, "Moolelo no Opukahaia."

13. Ibid.

14. [Dwight and Papaula], *Moolelo o Heneri Opukahaia*, 10.

15. Papaula, "Moolelo no Opukahaia"; Wayne H. Brumaghim, "Life and Legacy of Heneri 'Ōpūkaha'ia, Hawai'i's Prodigal Son" (MA thesis, University of Hawai'i–Mānoa, 2011), 27.

16. Brumaghim, "Life and Legacy," 23–24.

17. Pukui and Elbert, *Hawaiian Dictionary*, s.v. "mo'o kahuna."

18. On metaphor in Hawaiian thought and society, see Kame'eleihiwa, *Native Land and Foreign Desires*, 5–8.

19. Edmund S. Morgan, *Visible Saints: The History of a Puritan Idea* (Ithaca, N.Y.: Cornell University Press, 1963), 72.

20. John Garrett, *To Live Among the Stars: Christian Origins in Oceania* (Suva, Fiji: Oceania Printers, 1982), 39. Kanaka converts and lay preachers remained crucial to the missionary process for years to come, and some who converted in New England traveled with Haole back to Hawai'i to work as missionaries. S[amuel] M[anaiākalani] Kamakau, "Ka Moolelo o na Kamehameha, Helu 17," *Ka Nupepa Kuokoa*, June 6, 1868, 1. Here I use "Tahitian" in the now-common sense of the people of the Society Islands archipelago, not just the people of the island of Tahiti, just as "Hawaiian" is used to refer to people from all of the Hawaiian archipelago, not just the island of Hawai'i.

21. Chun, *Nā Kukui Pio 'Ole*, 1; Kame'eleihiwa, *Native Lands and Foreign Desires*, 143–44; Douglas D. Tzan, "Arioi for Christ: An Exploration of Early

Missions by Society Islanders," *Missiology* 37 (2009): 230–31; Dorothy Barrère, "A Tahitian in the History of Hawai'i: The Journal of Kahikona," *Hawaiian Journal of History* 23 (1989): 75–107; Dorothy Barrère and Marshall Sahlins, "Tahitians in the Early History of Hawaiian Christianity: The Journal of Toketa," *Hawaiian Journal of History* 13 (1979): 19–35.

22. Garrett, *To Live Among the Stars*, 39; Tzan, "Arioi for Christ."

23. Garrett, *To Live Among the Stars*, 40; Barrère and Sahlins, "Tahitians in the Early History of Hawaiian Christianity," 20.

24. Barrère and Sahlins, "Tahitians in the Early History of Hawaiian Christianity," 22–23.

25. William Richards to Jeremiah Evarts, Lahaina, January 10, 1826 [letter dated 1816 in error], Records of the American Board of Commissioners for Foreign Missions, Microfilm, Houghton Library, Harvard University, Cambridge, Mass. (henceforth: ABCFM Records), reel 794, frame 631; Wm. Richards to Jeremiah Evarts Esq., Honolulu, March 10, 1826, ABCFM Records, reel 794, frame 624.

26. Jane Kamensky, *Governing the Tongue: The Politics of Speech in Early New England* (New York: Oxford University Press, 1997); Susan Juster, *Disorderly Women: Sexual Politics and Evangelicalism in Revolutionary New England* (Ithaca, N.Y.: Cornell University Press, 1996).

27. On gender among the *arioi* in Tahiti, see Tzan, "Arioi for Christ." Female kāhuna (in the sense of spiritual experts, understood by Westerners as "priestesses") did exist, but to my knowledge there is not any extensive scholarship about them. See Kepelino, *Kepelino's Traditions of Hawaii*, 23. It is possible that many of the references to female kāhuna in the predominantly Christian press in the nineteenth century are efforts to slander or discredit Hawaiian practice. Still, it is apparent that female kāhuna did not disappear with the advent of the missionaries. One kahuna wahine, Kamaipuupaa, was part of the household of Kamehameha V in the 1870s and was said to be influential upon him. Kuykendall, *The Hawaiian Kingdom*, 2:241–42.

28. Female deities are of such importance in the Hawaiian tradition that publishing moʻolelo (stories) concerning the akua wahine Pele and Hiʻiaka was, as Noenoe K. Silva has noted, an important political act of the first newspaper under truly indigenous control. Silva, *Aloha Betrayed*, 70–76. Recent scholarship in Hawaiian studies has emphasized the collection, preservation, and analysis of moʻolelo concerning female gods. See kuʻualoha hoʻomanawanui, *Voices of Fire: Reweaving the Literary Lei of Pele and Hiʻiaka* (Minneapolis: University of Minnesota Press, 2014); Noenoe Silva, "Pele, Hiʻiaka, and Haumea: Representations of Women in Native Hawaiian Literature, 1861 & 1906," *Pacific Studies* 30, nos. 1/2 (2007): 159–81; Lilikalā Kameʻeleihiwa, *Nā Wāhine Kapu / Divine Hawaiian Women* (Honolulu: ʻAi Pōhaku Press, 1999). Given the place of female–female sexuality in these traditions, notably in the Pele and Hiʻiaka

cycle, this is also a central issue in the study of sexuality and literature, as demonstrated by Lisa Kahaleole Chang Hall and J. Kēhaulani Kauanui, "Same-Sex Sexuality in Pacific Literature," *Amerasia Journal* 20, no. 1 (1994): 75–81. See also Noenoe Silva, "Hawai'i," in *Encyclopedia of Homosexuality*, vol. 1, *Lesbian Histories and Cultures* (New York: Garland, 2000), 157–58. Kēhaulani Kauanui is currently researching the institutionalization of gender hierarchy through law, property, and religion.

29. Ross Cordy, *Exalted Sits the King: The Ancient History of Hawai'i Island* (Honolulu: Mutual Pubishing, 2000), 78.

30. Kame'eleihiwa, *Native Land and Foreign Desires*, 67–94.

31. Auna Diary, June 26, 1822, manuscript translation by William Ellis, in ABCFM Records, reel 793, frame 1247, June 2, 1822.

32. Wm. Ellis to Jerh. Evarts, Oahu, March 10, 1823, ABCFM Records, reel 793.

33. "Journal Kept at Lahaina by Wm. Richards," ABCFM Records, reel 794, frame 635, June 16, 1826.

34. Auna Diary 1822, reel 793, frame 1246, June 1, 1822.

35. Rev. Daniel Tyerman and George Bennet to the American Board of Commissioners for Foreign Missions, August [day illegible], 1822, Woahoo (O'ahu), ABCFM Records, reel 793, frame 1237.

36. Auna Diary 1822, reel 793, frame 1248.

37. Sahlins and Kirch, *Anahulu*, 1:89, n. 16.

38. Ellis to Evarts, Oahu, March 10, 1823.

39. Charles Kenn as cited in Chun, *Nā Kukui Pio 'Ole*, 1.

40. Charlot, *Classical Hawaiian Education*, 616.

41. Chun, *Nā Kukui Pio 'Ole*, 1.

42. King Tamoree [Kaumuali'i] to Samuel Worcester, July 28, 1820, Atooi [Kaua'i], ABCFM Records, reel 793, frames 1162–63.

43. Keariiahonui [Keali'iahonui] to Jeremiah Evarts, November 8, 1823, Oahu, ABCFM Records, reel 793, frames 890–93.

44. Jennifer Noelani Goodyear-Ka'ōpua, *The Seeds We Planted: Portraits of a Native Hawaiian Charter School* (Minneapolis: University of Minnesota Press, 2013), 15; Silva, *Aloha Betrayed*, 32–35; Kuykendall, *The Hawaiian Kingdom*, 1:106–7; Benjamin O. Wist, *A Century of Public Education in Hawaii: October 15, 1840–October 15, 1940* (Honolulu: Hawaii Educational Review, 1940), 22–23.

45. J. H. Kanepuu, "He Moolelo no Kanewailani. Ke Keiki a Maoloha. I Unuhi Ia Mailoko Mai o na Moolelo Kahiko o Hawaii Nei. Helu 2," *Ke Au Okoa*, February 27, 1868, 4, cited in Charlot, *Classical Hawaiian Education*, 227.

46. Lorrin Andrews, quoted in Charlot, *Classical Hawaiian Education*, 620–24.

47. Ibid., 626.

48. Phillip H. Round, *Removable Type: Histories of the Book in Indian Country, 1663–1880* (Chapel Hill: University of North Carolina Press, 2010), 16, 56–58.

49. Jennifer Noelani Goodyear-Kaʻōpua, "Kū i ka Māna: Building Community and Nation through Contemporary Hawaiian Schooling" (PhD diss., University of California–Santa Cruz, 2005), 12.

50. June 28, 1826, entry in diary of William Richards, "Journal Kept at Lahaina, Maui by Wm. Richards," ABCFM Records, reel 794, frame 640.

51. Wm. Richards to Brother [Rufus] Anderson, Honolulu, May 9, 1826, ABCFM Records, reel 794, frame 627.

52. Rev. Wm Richards to Mr. Rufus Anderson, Lahaina, March 31, 1827, ABCFM Records, reel 794, frame 647. This edition of the Sermon on the Mount would be an addition to the Gospel of Matthew, of which it was an excerpt, which first appeared in 1828 in an edition of fifteen thousand copies. *Ka Euanelio a Mataio* (Rochester, N.Y.: Mea Pai Palapala a Lumiki [Elisha Loomis], 1828), described in Forbes, *Hawaiian National Bibliography*, 1:472–73.

53. Charlot, *Classical Hawaiian Education*, 623.

54. Here I include the version published in *O Ka Buke Hua Mua e Ao Ai i Ka Palapala* (Honolulu: Mea Pai Palapala a na Misioneri, 1832), 46–47, as I am unable to obtain a copy of the document that Spaulding refers to in his letter. That is most likely the version described by David W. Forbes, "Ka Olelo a Iesu ma ka Mauna," in *Hawaiian National Bibliography*, 2:467, and described by W. D. Westervelt, "The Printing Press of the Mission under Mr. Loomis," in *Thrum's Hawaiian Almanac and Annual for 1909* (Honolulu: Thomas G. Thrum, 1908), 112.

55. Charlot, *Classical Hawaiian Education*, 239–41.

56. On the ways that deteriorating conditions and demands for labor could greatly exacerbate the ravages of diseases, see Paul Kelton, *Epidemics & Enslavement: Biological Catastrophe in the Native Southeast, 1492–1715* (Lincoln: University of Nebraska Press, 2007). On the attraction of Christianity to Kānaka beset by epidemics, see Kameʻeleihiwa, *Native Land and Foreign Desires*, 80–81. Thanks to William Bauer for emphasizing the need to understand epidemics in the context of colonial processes and disruptions such as aliʻi demands for sandalwood for foreign trade. William Bauer, personal communication, March 16, 2015.

4. THE WORLD AND ALL THE THINGS UPON IT

1. J. H. Kanepuu, "Ka Honua Nei a me na Mea a Pau Maluna Iho." The series ran weekly in *Ka Lahui Hawaii* from June 28, 1877, to December 13, 1877.

2. Kanepuu, "He Moolelo no Kanewailani. . . . Helu 2," 4, cited in Charlot, *Classical Hawaiian Education*, 227.

3. "Ke Kaapuni Ana a ke Kahukula Nui," *Ke Au Okoa*, December 11, 1865,

1; "Hoike o na Kula La Aupuni o Kona Oahu Nei," *Ka Nupepa Kuokoa*, July 4, 1869, 1; J. H. Kanepuu, "Kumumanao," folder 320-1-7, "Essays and Arithmetic Papers, Essays, Oahu, n.d., 1854, 1861," Hawai'i State Archives, Honolulu.

4. Goodyear-Kaʻōpua, *The Seeds We Planted*, 16, 17, 22.

5. Mark Rifkin, *The Imperial Construction of U.S. National Space* (Oxford: Oxford University Press, 2009).

6. Epeli Hauʻofa, "Our Sea of Islands," in *We Are the Ocean: Selected Works* (Honolulu: University of Hawai'i Press, 2008), 27–40.

7. Osorio, *Dismembering Lāhui*.

8. Scott Richard Lyons, *X-Marks: Native Signatures of Assent* (Minneapolis: University of Minnesota Press, 2010).

9. J. B. Harley, *The New Nature of Maps: Essays in the History of Cartography* (Baltimore: Johns Hopkins University Press, 2001); Beamer and Duarte, "I palapala no ia aina," 66–67. This argument does not deny that, as scholars such as Kalani Beyer have argued, missionaries, their descendants, and other Americans used the schools as implements of colonialism. Rather, it contends that Kānaka *also* used them in pursuit of their own goals—learning, the embrace of what they perceived to be modern, and especially the defense of national sovereignty. For the former point, see Kalani Beyer, "A Century of Using Secondary Education to Extend an American Hegemony over Hawai'i," *American Educational History Journal* 39 (2012): 515–35.

10. George Theodore Lecker, "Lahainaluna, 1831–1877: A Study of the History of Hawaii's Pioneer Educational Institution and Its Socio-economic Influence at Home and Abroad" (MA thesis, University of Hawai'i, 1938).

11. The rules for the formation of the school can be found in "He Mau Kanawai no ka Hale Kula no na Keiki Alii" [Honolulu, 1840] in the collection of Hawai'i Mission Children's Society Archives, Honolulu.

12. "Sandwich Islands. Plan of a High School for Teachers," 1832, quoted in Charlot, *Classical Hawaiian Education*, 59–60. On the founding of Lāhaināluna, see Lecker, "Lahainaluna, 1831–1877," 26–72.

13. William Richards, *Report of the Minister of Public Instruction, Ministerial Reports; Read before His Majesty to the Hawaiian Legislature, August 1, 1845* (Honolulu: Printed by Order of the Government, 1846), 53.

14. Ibid.

15. *Annual Reports Read before His Majesty, to the Hawaiian Legislature, April 22, 1850, with the King's Speech to the Legislature, April 10, 1850* (Honolulu: Government Press, 1850), 23.

16. Kamanamaikalani Beamer as quoted in Goodyear-Kaʻōpua, *The Seeds We Planted*, 253, n. 43.

17. Mary A. Richards, *The Hawaiian Chief's Children's School 1839–1850: A Record Compiled from the Diary and Letters of Amos Starr Cooke and Juliette Montague Cooke* (Rutland, Vt.: C. E.Tuttle, 1970), 131.

18. *Report of the Minister of Foreign Relations to the Legislature of 1856* (Honolulu: Government Press, 1856), 3.

19. *Annual Reports Read before His Majesty, to the Hawaiian Legislature, April 22, 1850*, 23.

20. *The King's Speech Read before the Hawaiian Legislature, April 6, 1853: with the Reports of Ministers, to the Same Body* (Honolulu: Government Press, 1853), 54.

21. *Report of the Minister of Foreign Relations to the Legislature of 1856*, 1.

22. *The King's Speech Read before the Hawaiian Legislature, April 6, 1853*, 57. This lack of texts may appear surprising at first, given that historians have rightly emphasized the enormous number of pages of text that were published in the Hawaiian language for Kānaka to use in the nineteenth century. Perhaps atlases were scarce because the missionary press found them particularly expensive and laborious to produce, given that they featured costly colored pages.

23. *He Mau Palapala Aina a me na Niele e Pili Ana* (Lahainaluna: [Mea Pai a na Misionari], 1840). The same year, the press produced *He Mau Palapala Aina a me na Niele no ka Hoikehonua* (Lahainaluna: [Mea Pai a na Misionari], 1840), which David W. Forbes notes was "intended for distribution to schools throughout Hawaiʻi. The 1841 General Meeting of the Sandwich Island Mission apparently resolved to produce one thousand copies of each of these books, but Forbes notes that it is uncertain how many of these books were actually produced, and how many of them contained only the text and not the maps themselves. According to Forbes, the two books drew on different English-language sources: the first on Jessy Olney, *A New and Improved School Atlas, to Accompany the Practical System of Modern Geography* (Hartford, 1830 and later editions), and the second on Samuel Reed Hall, *Child's Book of Geography* (Springfield, 1832 and later editions). Forbes, *Hawaiian National Bibliography*, 2:251–53.

24. M. Puakea Nogelmeier, "Hawaiian Geography, Old and New," in *He Mau Palapala Aina a Me Na Niele e Pili Ana: Maps and Questions Regarding Them*, Hawaiian Language Reprint Series 4 (Honolulu: Hawaiian Historical Society, 2011), xiii; Gary L. Fitzpatrick, "Mapmaking at Lahainaluna," in *He Mau Palapala Aina*, xiv. These maps made up only part of the engraving output of Kanaka students and nonstudents at Lāhaināluna. See David W. Forbes, *Engraved at Lahainaluna: A History of Printmaking by Hawaiians at the Lahainaluna Seminary, 1834–1844* (Honolulu: University of Hawaiʻi Press, 2012).

25. Chun, *Nā Kukui Pioʻole*, i, n. 2.

26. Bernardo A. Michael, *Statemaking and Territory in South Asia: Lessons from the Anglo-Gorkha War (1814–1816)* (London: Anthem Press, 2012), 5, as cited in Evan Taparata, "Placing and Displacing Borderlands: A Historiography of Global Borderlands," April 8, 2014, unpublished manuscript in the author's possession.

27. [Woodbridge and others], *He Hoikehonua* (1832). On this edition and the later editions, see Forbes, *Hawaiian National Bibliography*, 2:29–30, 129–30, 447.

28. *He Vahi Hoikehonua: He Mea ia e Hoakakaʻi ke Ano o ka Honua Nei* (Honolulu: Pai Palapala Katolika, 1842).

29. William Channing Woodbridge, *Rudiments of Geography on a New Plan, Designed to Assist the Memory by Comparison and Classification* (Hartford: Oliver D. Cooke, 1830). This is the thirteenth edition of this often-republished textbook, most editions of which bore the title *Universal Geography*. Given that *He Hoikehonua* was first published in 1832, and the strong resemblance between the two texts, it seems reasonable to assume that the translators depended on the 1830 edition. It is possible that they worked from the 1832 edition of Woodbridge's book, but this seems less likely, given the time needed for translating the text.

30. Daniel H. Calhoun, "Eyes for the Jacksonian World: William C. Woodbridge and Emma Willard," *Journal of the Early Republic* 4 (1984): 4, 14.

31. Hauʻofa, "Our Sea of Islands," 30–32.

32. Thomas Nagel, *The View from Nowhere* (Oxford: Oxford University Press, 1986), 26. Geographer Steven Shapin similarly rejects the "view from nowhere" because it makes it difficult to perceive how scientific ideas are generated in particular spaces and travel to other places. Steven Shapin, "Placing the View from Nowhere: Historical and Sociological Problems in the Location of Science," *Transactions of the Institute of British Geographers*, n.s. 23 (1998): 5–12.

33. Karen Piper, *Cartographic Fictions: Maps, Race, and Identity* (New Brunswick: Rutgers University Press, 2002), 12.

34. [William Channing Woodbridge and unnamed others], *He Hoikehonua, He Mea Ia e Hoakakaʻi i ke Ano o ka Honua Nei, a me Na Mea Maluna Iho* (Oahu: Mea Pai a na Misionari, 1836), 20. In this chapter, I depend on the 1836 edition rather than the 1832 edition, as this longer and more fully illustrated edition served as the basis for later editions.

35. Ibid., 21.

36. *He Vahi Hoikehona*, 13.

37. [Woodbridge and others], *He Hoikehonua* (1836), 21.

38. Ibid., 22.

39. Ibid.

40. Ibid.; *He Vahi Hoikehona*, 12.

41. Kealani Cook translates lāhui in this manner in his discussion of the 1845 edition of *He Hoikehonua*. Cook, "Kahiki," 77–78.

42. [Woodbridge and others], *He Hoikehonua* (1836), 31.

43. Ibid., 32.

44. Pukui and Elbert, *Hawaiian Dictionary*, s.vv. "hoa paio," "hoa kaua," and "hoa pāonioni."

45. Piper, *Cartographic Fictions*, 14.
46. Martin Brückner, "Lessons in Geography," *American Quarterly* 51 (1999): 326.
47. Jedidiah Morse, *Geography Made Easy* (Boston: Samuel Hall, 1791), v; Jedidiah Morse, *American Geography* (Elizabethtown: Shepard Kollock, 1789), vii.
48. Martin Brückner, *The Geographic Revolution in Early America: Maps, Literacy, and National Identity* (Chapel Hill: University of North Carolina Press, 2006), 238.
49. [Woodbridge and others], *He Hoikehonua* (1836), 39.
50. The term might easily be confused for the name that was given to a manmade island in Honolulu in the 1970s, ʻĀina Moana State Recreation Area, more commonly referred to as Magic Island. That name is generally described as a phrase invented to describe a place made in the sea by the use of landfill. While it is possible that the park's name reflects a remembrance of the term ʻĀinamoana as it was taught in the 1800s, it seems far more likely that it is a coincidental reminder of a term that seems never to have appeared in print in Hawaiian after the 1870s.
51. Mahalo nui to Alice Te Punga Somerville for this insight. Conversation with Alice Te Punga Somerville, April 29, 2012.
52. Stephanie Seto Levin, "The Overthrow of the Kapu System in Hawaii," *Journal of the Polynesian Society* 77 (1968): 412.
53. On equilibrium and pono, see Kameʻeleihiwa, *Native Lands and Foreign Desires*, 13, 25.
54. Andrew Matzner, *ʻO Au No Keia: Voices from Hawaiʻi's Mahu and Transgender Communities* (Philadelphia: Xlibris, 2001); Carol E. Robertson, "The Mahu of Hawaiʻi," *Feminist Studies* 15 (1989): 313–26; Kathryn Xian, *Ke Kulana He Mahu: Remembering a Sense of Place* (Honolulu: Zang Pictures, 2001); Jade Snow, "Beyond the Binary: Portrait of Gender and Sexual Identities in the Native Hawaiian Community," *Mana: The Hawaiian Magazine*, March 22–29, 2014. For a discussion of Sāmoa, gender, and *faʻafahine* (gender liminality), see Niko Besnier, "Polynesian Gender Liminality through Time and Space," in *Third Sex, Third Gender: Beyond Sexual Dimorphism In Culture and History*, ed. Gilbert H. Herdt (New York: Zone Books, 1996), ch. 6.
55. The King's Speech Read before the Hawaiian Legislature, April 6, 1853, 57.
56. R. Armstrong to Bro. Hitchcock, Honolulu, June 13, 1853, ABCFM Papers, reel 804, frame 1222.
57. "Poepoe Hikina," *He Ninau no ka Palapala Honua* (n.p., n.d.). There is no known example of the atlas to survive with a title page, so its title is unknown. Existing examples are catalogued by libraries under the title of the question section, "He Ninau no ka Palapala Honua."
58. "Hale Kuai Buke," *Ka Nupepa Kuokoa*, October 1, 1861, 1. Whitney was advertising an item he referred to as "Palapala Honua Mua" (First World Map), which was likely this atlas.

59. Mary L. Hall and Harvey Rexford Hitchcock, *Ka Honua Nei: Oia ka Buke Mua o ka Hoike Honua, no na Kamalii o na Kula Maoli o ke Aupuni* (Honolulu: Papa Hoonaauao, 1873); Forbes, *Hawaiian National Bibliography*, 3:548. Note that *Ka Honua Nei* is generally considered to be and cataloged by libraries as Hitchcock's work, although he translated the bulk of the text from Hall's book, adding and editing passages to it, most notably the final section on the geography of the Hawaiian Islands. On the reorganization of educational administration, see Kuykendall, *The Hawaiian Kingdom*, 2:106–7.

60. Mary L. Hall, *Our World, or, First Lessons in Geography, for Children* (Boston: Crosby and Nichols, 1864). Lucretia Crocker coauthored the book: John F. Ohles, *The Biographical Dictionary of American Educators* (Westport, Conn.: Greenwood, 1978), sv. "Crocker, Lucretia." A number of later printings and editions would appear, including 1870, 1875, 1876, 1889, and 1897.

61. "He Papainoa no na Kahu, a me na Kumu, a me na Haumana o ke Kulanui o Hawaii Nei, Ma Lahainaluna i Maui, 1836," *Ke Kumu Hawaii*, February 15, 1837, 76.

62. Harvey Rexford Hitchcock, *An English–Hawaiian Dictionary: With Various Useful Tables; Prepared for the Use of Hawaiian-English Schools* (San Francisco: Bancroft, 1887); Kuykendall, *The Hawaiian Kingdom*, 2:108–9.

63. Hitchcock, "Na Olelo Mua," in Hall and Hitchcock, *Ka Honua Nei*, no page.

64. Hall and Hitchcock, *Ka Honua Nei*, 45, 47, 56, 64, and unpaginated engraving facing page 83.

65. Ibid., 54. The original English text in Hall reads, "The part of the world that belongs to any one nation is called its *country*, and no other people has a right to it." Hall, *Our World*, 59.

66. Hall and Hitchcock, *Ka Honua Nei*, 54.

67. Ibid., 29, 32.

68. Ibid., 34.

69. Ibid., 116–19.

70. "Hoike Honua," *Ka Lahui Hawaii*, July 5, 1877. Kawainui was the younger sibling of J. U. Kawainui, the editor of *Ka Nupepa Kuokoa* at the time that that newspaper was working for annexation in the early to mid-1890s. Thanks to Noenoe Silva for identifying information on Kuea and Kawainui.

71. See for example Miriama Kahuma, "He Kumumanao no ka Awa"; S. H. Petero, "Kumumanao: He Mea Pono Anei I ke Kumukula ke Hele Ai I ka Nana Hula"; Kahulukae, "Na keia Lahui Hawaii no e hooemi nei ia ia iho": all in folder 320-1-5, "Essays and Arithmetic Papers," Oahu, n.d. [1870s], Hawai'i State Archives, Honolulu.

72. John R. K. Clark, *Hawaii Place Names* (Honolulu: University of Hawai'i Press, 2002), 190.

73. J. W. M. Poohea, "Kumumanao: Heaha la na mea i paa ai ka make nui ana o keia lahui Hawaii," September 28, 1872; "In What Ways as to Prevent the

Increasing Death of the Hawaiian Race" (unattributed typescript translation into English), September 28, 1872; "What are the reasons for the high death rate among Hawaiians?" (unattributed typescript translation into English), September 28, 1872: all in folder 320-1-10, "Essays and Arithmetic Papers," Oahu, 1872, Hawai'i State Archives, Honolulu. On Poohea's church activities and professional activities as a lawyer and notary, see "Na Inoa Kaulan o ka Poe e Holo," *Ka Lahui Hawaii,* January 20, 1876, 2; Thomas G. Thrum, *Almanac and Annual for 1885* (Honolulu: Thrum, 1885), 85, 87; Whalley Nicholson, *From Sword to Share: Or a Fortune in Five Years at Hawaii* (London: W. H. Allen, 1881), 287; *Twenty-first Annual Report of the Hawaiian Evangelical Association, June, 1884* (Honolulu: The Hawaiian Gazette, 1884), 3.

74. S. W. Kahoopii, "Kumumanao: Pehea la e loaa ai i ka lahui Hawaii, ka waihona nui o ka naauao iloko o na buke Haole?" and "Subject: In What Way Can the Hawaiian Race Be Educated and Developed Much More Learning out of English Books" (unattributed typescript translation into English), no date [1870s], folder 320-1-3, "Essays and Arithmetic Papers," Hawai'i State Archives, Honolulu.

75. J. W. Kamoku, "Kumumanao: Na Mana o ka Naaupo . . ." and "Subject: Power of Ignorance . . ." (unattributed partial typescript translation into English), no date [1870s], folder 320-1-3, "Essays and Arithmetic Papers," Hawai'i State Archives, Honolulu. The paraphrase of Kauikeaouli's words is deleted from the English-language translation in the file.

76. Kanepuu, "Ka Honua Nei," part 1, June 28, 1877, 1.

77. Kanepuu, "Ka Honua Nei," part 2, July 5, 1877, 1.

78. Kanepuu, "Ka Honua Nei," part 3, July 12, 1877, 1.

79. Kanepuu, "Ka Honua Nei," part 4, July 19, 1877, 1; part 5, July 28, 1877, 1; part 7, August 9, 1877, 1; part 8, August 16, 1877, 1; parts 9–17, August 30–October 25, 1877.

80. Kanepuu, "Ka Honua Nei," part 15, September 20, 1877, 1; part 17, October 25, 1877, 1.

81. Kanepuu, "Ka Honua Nei," part 17, October 25, 1877, 1.

82. Silva, *Aloha Betrayed,* 73–79.

83. John Charlot, "Pele and Hi'iaka: The Hawaiian-Language Newspaper Series," *Anthropos,* 1998, 55–75.

84. "Catalogue" ledger book, AR-7, Hilo Boarding School Records, Education Files, Enrollment Records, Box 4, Folder 29, page 88, Lyman House Museum and Archives, Hilo, Hawai'i.

85. Michelle Morgan, "Americanizing the Teachers: Identity, Citizenship, and the Teaching Corps in Hawai'i, 1900–1941," *Western Historical Quarterly* 45 (2014): 150–52; Hawaii Department of Public Instruction, *Report of the Minister of Public Instruction to the President of the Republic of Hawaii for the Biennial Period Ending December 31* (Honolulu: Hawaiian Gazette, 1900), 109.

86. The board of education in Hawaii ordered the book in spring 1889. They ordered one thousand copies of the combined U.S. and Hawaii edition and two thousand copies of the Hawaii edition alone to start, with the expectation that more would be ordered later. J. Mott Smith to Messrs Barnes & Co., April 1889, Boston, Board of Education Incoming Letters, Hawai'i State Archives, Honolulu.

87. Because the 1889 edition was not available for this research, these citations are based on the 1883 edition. James Monteith, *Elementary Geography* (New York: A. S. Barnes, 1883), 9, 14, 15, 18, 22, 66, 67, 75.

5. HAWAIIAN INDIANS AND BLACK KANAKAS

1. Kanaka Maoli historian Charles W. Kenn researched this family and others descended from the workers that worked on Sutter's Fort through thorough documentary research and extensive correspondance with descendants and other people in California. Charles W. Kenn, "Descendants of Captain Sutter's Kanakas," in *Proceedings of the Second Annual Meeting of the Conference of California Historical Societies* (Stockton: College of the Pacific, 1956), 87–101; Charles W. Kenn, "Sutter's Canackas," *Newsletter of the Conference of the California Historical Society* 2, no. 2 (1955): 3–6.

2. Margaret A. Ramsland, *The Forgotten Californians* (n.p.: Jensen Graphic, 1974). The story receives a longer and more fictionalized treatment in James D. Houston, *Bird of Another Heaven* (New York: Anchor Books, 2008). The Brooklyn Museum owns jewelry and baskets of Mele's making; her baskets are featured in the collections of the Chico Museum and the California State Parks. See https://www.brooklynmuseum.org.

3. In the final years of his life, in the late 1960s and early 1970s, Henry Azbill dedicated his efforts to perpetuating the culture and history of the Mechoopda and the Maidu, though not neglecting his Kanaka Maoli heritage. Writing articles and pamphlets, he preserved the stories of his Maidu ancestors and the gods and spirits. On film, he recounted the Maidu creation story and the history of the forced removal of his people to a reservation in the 1870s so that they would be known to future generations. In conversations with ethnographic researchers (found in the Dorothy Hill Collection in the California State University Special Collections Library), he shared his knowledge to ensure that they would represent his people accurately. See Henry Keʻaʻaʻla Azbill, *The Mechoopda Legends and Myths* ([San Francisco]: Book Club of California, 1973); Henry Azbill, "They Call us Concow," in "Koyo'ngkaui: We Live in the Open Country," DQ University Indian Education Workshop, Summer 1972, Hehaka Sapa College, Davis, California; Henry Azbill, "World Maker," *Indian Historian* 2, no. 1 (Spring 1969): 20–24; Henry Azbill, "How Death Came to the People," *Indian Historian* 2, no. 2 (Summer 1969): 13–14. Videos of Henry

Keaala Azbill explaining the history and oral tradition of his family and the Concow Maidu people can be found at http://eagle.csuchico.edu. The homepage of the Mechoopda Indian Tribe of Chico Rancheria testifies to his importance to the preservation of the culture and history of his community by prominently quoting from his words: "You have to know who you are." http://www.mechoopda-nsn.gov. Late in life, he began "actively to create objects of personal adornment and regalia." Craig D. Bates, "A Maidu Dough Carrier," *Journal of California and Great Basin Anthropology* 5, no. 2 (1983): 241.

4. Aikau, "Indigeneity in the Diaspora," 478.

5. Nancy Shoemaker, *Native American Whalemen and the World: Indigenous Encounters and the Contingency of Race* (Chapel Hill: University of North Carolina Press, 2015).

6. Liholiho, *The Journal of Prince Alexander Liholiho*, 108.

7. Sandra Bonura and Sally Witmer, "Lydia K. Aholo—Her Story: Recovering the Lost Voice," *Hawaiian Journal of History* 47 (2013): 127.

8. *Ministerial Reports; Read before His Majesty to the Hawaiian Legislature, August 1, 1845* (Honolulu: Printed by Order of the Government, 1846), 7–8.

9. Ibid. On the fur trade, see George I. Quimby, "Hawaiians in the Fur Trade of North-West America, 1785–1820," *Journal of Pacific History* 7 (1972): 92–103.

10. Census data from Sannie Kenton Osborn, "Death in the Daily Life of the Ross Colony: Mortuary Behavior in Frontier Russian America," http://www.fortrossstatepark.org/sannieosborn.htm.

11. On Native people at Fort Ross, see Kent G. Lightfoot, *Indians, Missionaries, and Merchants: The Legacy of Colonial Encounters on the California Frontiers* (Berkeley: University of California Press, 2005), 154–80. Robert H. Jackson, "Intermarriage at Fort Ross: Evidence from the San Rafael Mission Baptismal Register," *Journal of California and Great Basin Anthropology* 5, no. 2 (1983): 240–41.

12. Albert Hurtado, *John Sutter: A Life on the North American Frontier* (Norman: University of Oklahoma Press, 2006), 95–100. Thanks to William Bauer for clarifying to me the relationship between the two forts. William Bauer, personal communication, March 16, 2015.

13. Manuiki and other names of California Kānaka appear in various forms in different sources. When spellings are available from Hawaiian-language newspapers, I use them, as they are most likely to reflect the correct Hawaiian versions. A number of these names, including Manuiki's, appear in Pokue [J. F. Pogue], "No na mea i ike maka ai ma Kaleponi," *Ka Nupepa Kuokoa*, September 19, 1868, 4.

14. Cited in Kenn, "Descendants of Captain Sutter's Kanakas," 88.

15. Ibid., 88, 90.

16. Jean Barman and Bruce McIntyre Watson, *Leaving Paradise: Indigenous*

Hawaiians in the Pacific Northwest, 1787–1898 (Honolulu: University of Hawai'i Press, 2006); Kenn, "Descendants of Captain Sutter's Kanakas."

17. 1860 U.S. census, El Dorado County, California, population schedule, Coloma township, p. 353, digital image, *Ancestry.com*, http://www.ancestry.com/, citing National Archives and Records Administration publication M653 (microfilm, 1,438 reels, National Archives and Records Administration, n.d.), reel 58; Kulika Opio [Theodore Gulick], "No Waiulili, Babling Waters, he Wahine Ilikini no Kaliponia," *Ka Nupepa Kuokoa*, July 12, 1862, 1; L. Kamika [Lowell Smith], "Mai a L. Kamika Mai," *Ka Nupepa Kuokoa*, July 7, 1866, 4; [Theodore Gulick], "No ka Mai Puupuu Liilii ma Irish Creek, Kaliponia," *Ka Nupepa Kuokoa*, July 5, 1862, 3. On the location of Irish Creek, see Charles H. Lee, "An Intensive Study of the Water Resources of a Part of Owens Valley, California," United States Geological Survey Water-Supply Paper 294 (Washington, D.C., 1912), 49. On the Kanaka Maoli settlements, see Richard H. Dillon, "Kanaka Colonies in California," *Pacific Historical Review* 24 (February 1955): 17–23.

18. Harry Laurenz Wells, Frank T. Gilbert, and W. L. Chambers, *History of Butte County, California*, vol. 2, *History of Butte County, from Its Earliest Settlement to the Present Time* (San Francisco: Harry L. Wells, 1882), 254. In using the spelling "Concow," I follow the usage of William J. Bauer Jr., a Wailacki and Concow scholar. He chooses the spelling from the seal of the Round Valley Indian Tribes in order that his scholarship "will be understood by the Round Valley Indian Community," the community where many Concow live today. William J. Bauer Jr., *We Were All Like Migrant Workers Here: Work, Community, and Memory on California's Round Valley Reservation, 1850–1941* (Chapel Hill: University of North Carolina Press, 2009), xiii.

19. An excellent description of the seasonal rounds of the Conkow as seen from the point of view of a Hawai'i-born white missionary who visited them can be found in Kulika Opio [Theodore Gulick], "No Lemaine, ka Makuahine o Waiulili," *Ka Nupepa Kuokoa*, September 13, 1862, 3. See also Roland B. Dixon, "The Huntington California Expedition: The Northern Maidu," *Bulletin of the American Museum of Natural History* 17 (1905): 181–84. For a discussion on bodyguards, see Alonzo Delano, *Life on the Plains and among the Diggings* (Auburn, N.Y.: Miller, Orton and Mulligan, 1854), 298.

20. On population collapse, see Albert L. Hurtado, *Indian Survival on the California Frontier* (New Haven, Conn.: Yale University Press, 1988), 1. On the forced march, see Bauer, *We Were All Like Migrant Workers Here*, 54.

21. Kulika Opio [Gulick], "No Waiulili, Babling Waters, he Wahine Ilikini no Kaliponia," 1. The Kanaka Bar referred to in the text is at a point on the Middle Fork of the Feather River just north of Kanaka Peak. Another Kanaka Bar is currently under the northern point of Lake Oroville, which was created by the damming of the Feather River in 1968. David L. Durham, *Durham's Place*

Names of California's North Sacramento Valley (Clovis, Calif.: Word Dancer Press, 2001), 152; and Wells, Gilbert, and Chambers, *History of Butte County*, 2:254. On hānai, see Karen L. Ito, *Lady Friends: Hawaiian Ways and the Ties that Define* (Ithaca, N.Y.: Cornell University Press, 1999), 27.

22. Ruth Mazo Karras, *Unmarriages: Women, Men, and Sexual Unions in the Middle Ages* (Philadelphia: University of Pennsylvania Press, 2012), introduction.

23. Pokue [J. F. Pogue], "Na Kanaka Hawaii ma Vernona," *Ka Nupepa Kuokoa*, September 19, 1869, 4.

24. Francis A. Riddell, "Maidu and Konkow," in *California*, ed. Robert F. Heizer, vol. 8 of *Handbook of North American Indians*, ed. William C. Sturtevant (Washington, D.C.: Smithsonian Institution Press, 1978), 380.

25. Kulika Opio [Gulick], "No Waiulili, Babling Waters, he Wahine Ilikini no Kaliponia," 1. Young Concow women were "marriageable" shortly after their first menses. Riddell, "Maidu and Konkow," 381. For the location of Irish Creek, see United States Geological Service, "Garden Valley Quadrangle—El Dorado, Co.," 1949, photorevised 1973, map, Borchert Map Library, University of Minnesota, Minneapolis.

26. Riddell, "Maidu and Konkow," 371, 373, 379, 380; untitled map of Concow and Maidu territories, and "Tribal Territory with Selected Major Villages" map, in Heizer, *California*, 371, 388.

27. Kulika Opio [Gulick], "No Waiulili, Babling Waters, he Wahine Ilikini no Kaliponia," 1; Mary Kawena Pukui, E. W. Haertig, and Catherine A. Lee, *Nānā i ke Kumu (Look to the Source)* (Honolulu: Hui Hānai, 1972), 1:49.

28. Kulika Opio [Gulick], "No Waiulili, Babling Waters, he Wahine Ilikini no Kaliponia," 1; Kulika Opio [Gulick], "No Lemaine, ka Makuahine o Waiulili," 3.

29. The year 1851 marked the beginning of the treaty process that was to create a reservation near Chico. Bauer, *We Were All Like Migrant Workers Here*, 33. On the multiple purposes of reservations, see Philip J. Deloria, "From Nation to Neighborhood: Land, Culture, Policy, Colonialism, and Empire in U.S.-Indian Relations," in *The Cultural Turn in U.S. History*, ed. James W. Cook, Lawrence B. Glickman, and Michael O'Malley (Chicago: University of Chicago Press, 2008), 358.

30. Whites petitioning for removing Indians to the reservation: Indian Agent [B. S. Fairfield] to B. C. Whiting, Round Valley Reservation, February 5, 1868, and Hugh Gibson to F. A. Walker, Round Valley Reservation, September 2, 1872; need to control entry of whites onto reservation: J. L. Burchard to E. P. Smith, October 7, 1873; passes and escaping: J. L. Burchard to B. C. Whiting, October 22, 1872, Round Valley Reservation; whipping: J. L. Burchard to E. P. Smith, January 30, 1875, Round Valley Reservation: all in Correspondence of Agent/Superintendent to Superintendent, San Francisco, and to the Commis-

sioner of Indian Affairs, 1863–1873, Records of the Bureau of Indian Affairs, Record Group 75, National Archives and Records Administration-Pacific Region (San Francisco) (NARA-SF).

31. Kulika Opio [Gulick], "No Waiulili, Babling Waters, he Wahine Ilikini no Kaliponia," 1.

32. Ibid.; Kulika Opio [Gulick], "No Lemaine, ka Makuahine o Waiulili," 3.

33. Kulika Opio [Gulick], "No Lemaine, ka Makuahine o Waiulili," 3. On a grandmother's role in the raising of children after the death of their mothers, see Dixon, "The Huntington California Expedition," 245. This pattern was observed by the family of Roy Scott, a Concow whose mother died when he was two years old in 1902, after which his maternal grandmother raised him. Roy Scott interview by Dorothy Hill, n.d., "A Maidu Descendant Relates His Family History," Association for Northern California Records and Research and California State University, Chico, Oral History Program, California State University, Chico, 2.

34. Pokue [Pogue], "No na mea i ike maka ai ma Kaleponi."

35. Kenn, "Descendants of Sutter's Kanakas," 88. Henry Azbill would say a century later that the Kānaka of Verenona also dried fish to export to Hawaiʻi, a trade that I have not seen mentioned in nineteenth-century documents.

36. Hurtado, *Indian Survival on the California Frontier.*

37. Kenn, "Descendants of Sutter's Kanakas," 91.

38. Pokue [Pogue], "No na mea i ike maka ai ma Kaleponi."

39. Ibid., 91. This claim is difficult to verify. The line of descent is not readily apparent in tracing the little information about Kaʻiana's descendants in the standard published sources for chiefly genealogy: McKinzie, *Hawaiian Genealogies.*

40. In some sources, Su-My-Neh's name is given as Suwomine. "Konkow Valley Band of Maidu," http://www.maidu.com/maiduculture/bibliography/hill.html.

41. A. L. Kroeber, *Handbook of the Indians of California* (New York: Dover Publications, 1976), 399.

42. California State Census, 1852, Shasta County, 82.

43. California State Census, 1852, Contra Costa County, 7.

44. California State Census, 1852, Tuolumne County, 32.

45. U.S. Federal Census Mortality Schedules, 1860, 1, Calaveras County, California, at http://www.ancestry.com.

46. U.S. Department of the Interior, Census Office, 1870 Manuscript Census, Vernon Township, Sutter County, California, population schedule, pp. 7–8, available at http://www.ancestry.com.

47. Kini was described as Mahuka's "wahine" by 1865: Thomas B. Kamipele to Mr. L. H. Gulick, Vernon, Sutter Co., Calif., June 7, 1865, folder Kamikea-Kamipele, Hawaiian Evangelical Association Archives, Hawaiʻi Mission

Children's Society Archives, Honolulu. They were reported to be formally married by 1868: Pokue [Pogue], "No na mea i ike maka ai ma Kaleponi."

48. See Eduin Mahuka in Census Office, 1880 Manuscript Census, Vernon Township, Sutter County, California, available at http://www.ancestry.com.

49. Pokue [Pogue], "No na mea i ike maka ai ma Kaleponi."

50. A. E. Mahuka, L. H. Kapuaa, M. Nahola, et al. to Ahahui Hoopuka Nupepa Ku i ka Wa o Honolulu, December 10, 1861, Coloma, published in *Ka Hoku o ka Pakipika*, February 6, 1862, 4.

51. Barman and Watson, *Leaving Paradise*; Kenn, "Descendants of Captain Sutter's Kanakas."

52. It is generally easy to separate out the Haole born in Hawai'i by their family circumstances. For example, Samuel Emerson was a seventeen-year-old student, born in the "Sandwich Islands," living in the household of Gael Newell in Nelson, New Hampshire. He can be identified as the son of the missionary John Smith Emerson and Ursula Emerson, whose maiden name was Sewell and who was born in Nelson, New Hampshire. Manuscript Census of Free Inhabitants, 1850, New Hampshire, Cheshire County, Nelson Township, 286; "Rev. John S. Emerson" finding aid, Hawai'i Mission Children's Society Archives at the Hawaiian Mission Houses Historic Site and Archives, Honolulu. Another Hawai'i-born child of missionaries who had returned to New England, and who can clearly be differentiated from the Kānaka living in New England, is Albert F. Perkins, age five, child of the student clergyman Henry H. W. Perkins who lived in Hallowell, Maine. Manuscript Census of Free Inhabitants, 1850, Maine, Kennebec County, Hallowell Township, 194.

53. For a fine-grained social history of Nantucket and other New England whaling communities that focuses on women and gender, see Lisa Norling, *Captain Ahab Had a Wife: New England Women and the Whalefishery, 1720–1870* (Chapel Hill: University of North Carolina Press, 2000). See also Eric Jay Dolan, *Leviathan: The History of Whaling in America* (New York: W. W. Norton, 2007), 205–11.

54. Frances Ruley Kartunnen, *The Other Islanders: People Who Pulled Nantucket's Oars* (New Bedford: Spinner, 2005), 65–66, 102–4; W. Jeffrey Bolster, *Black Jacks: African American Seamen in the Age of Sail* (Cambridge, Mass.: Harvard University Press, 1977), 176–80.

55. Kartunnen, *The Other Islanders*, 64; Bolster, *Black Jacks*, 182–88; Manuscript Census of Free Inhabitants, Massachusetts, Nantucket, Nantucket County, 391–92.

56. Manuscript Census of Free Inhabitants, 1850, Massachusetts, Bristol County, New Bedford, 232. The 1860 census lists a John Swain residing with Emily. Emily (or Emma, in other sources) was listed as black on the census, and her birthplace is given as "W.I." Given usage in New England at the time, this ambiguous abbreviation might refer to the West Indies or to the Western

Islands, colonial outposts of Portugal that are today called the Azores. The fact that either birthplace is plausible is instructive: the population of the Azores included few people of African ancestry and was mostly southern European in origin. The population of the West Indies was mostly African in origin. Manuscript Census of Free Inhabitants, 1850, Massachusetts, Bristol County, New Bedford Ward 2, 47. On John and Emily Swain, see Katherine Grover, *The Fugitive's Gibraltar: Escaping Slaves and Abolitionism in New Bedford* (Amherst: University of Massachusetts Press), 53, and Amy Jenness, *On This Day in Nantucket History* (Charleston: History Press, 2014), 237; Manuscript Census of Free Inhabitants, 1850, New Hampshire, Rockingham County, Portsmouth City, 41.

57. "A Blubber Hunter's Yarn—The Sandwich Islands as Seen by an Old Whaler—Dissipated Royalty," *New York Times*, February 13, 1893, 7; Daniel R. Mandell, "Shifting Boundaries of Race and Ethnicity: Indian–Black Intermarriage in Southern New England, 1760–1880," *Journal of American History* 85, no. 2 (1998): 500. In looking at Indian people of partly African descent who identified as Indian, Mandell deploys the concept of "symbolic ethnicity," a misleading term that fails adequately to recognize that multiple and intersectional identities are real and far from just "symbolic," for indigenous people and for others. Mandell, "Shifting Boundaries of Race and Ethnicity," 485.

58. Manuscript Census of Free Inhabitants, 1850, New York, Suffolk County, Southold Township, 318.

59. James Bunker Congdon, 1863, as quoted in Grover, *The Fugitive's Gibraltar*, 50.

60. Manuscript Census of Free Inhabitants, 1850, New York, Queens County, Newtown, 95; Noel Ignatiev, *How the Irish Became White* (New York: Routledge, 1995).

61. "Sandwich Islanders—Extract of a Letter from a Gentleman Recently There," *Nantucket Enquirer*, June 12, 1833, 2.

62. "Transcript of Dorothy Hill's conversation with Henry Azbill—July 7, 1966 (from her tape)," folder 1, box 2, Dorothy Hill Collection, Manuscript 160, Special Collections, Meriam Library, California State University, Chico.

63. Queen Emma Diary, MS MC K4, Box 2, folder 13, Bishop Museum Library and Archives, Honolulu, February 8, 1884.

6. BONE OF OUR BONE

1. Iosepa Opunui, "Palapala Mai Califonia Mai," *Ka Hoku Loa*, October 1860, 13. E. O. Hall, untitled, *Ka Hoku Loa*, August 1864, 29; E. O. Hall, untitled, *Ka Hoku Loa*, September 1864, 34; Kulika Opio [Theodore Gulick], "He Palapala na ko makou elele i holo aku nei i Kapalakiko," *Ka Nupepa Kuokoa*, October 25, 1862, 4. Nuku Hiva is part of the grouping that is often referred to as the Marquesas (named for the marquis who sponsored the Spaniard who

landed there in 1595), but nineteenth-century Kānaka like Opunui often used less colonial terms: "ko Nuuhiva pae aina," meaning "the archipelago of Nuku Hiva," or simply "nā Hiwa," meaning "the Hivas." Thanks to Noenoe Silva for this point; personal communication with the author, March 6, 2015.

2. "Oihana Misionari," *Ka Hoku Loa*, August 1860, 7.

3. Ibid., 6.

4. "Bone of my bone and flesh of my flesh" appears at Genesis 2:23. As a statement of connection between two peoples, it was not limited to Pacific peoples— indeed, Tiya Miles has demonstrated the ways some Cherokee people used the phrase to proclaim their deep kinship to African Americans. Tiya Miles, *Ties that Bind: The Story of an Afro-Cherokee Family in Slavery and Freedom* (Berkeley: University of California Press, 2006), 119–22. On the meanings of bone among Kānaka, see Pukui, Haertig, and Lee, *Nānā i Ke Kumu*, 1:107, 111.

5. Cook, "Kahiki."

6. Ibid., 22–26.

7. Pukui and Elbert, *Hawaiian Dictionary*, s.v. "kapu."

8. S[amuel] M[anaiākalani] Kamakau, "Ka Moolelo o na Kamehameha, Helu 45," *Ka Nupepa Kuokoa*, October 12, 1867, 1; Kameʻeleihiwa, *Native Land and Foreign Desires*, 74; Linnekin, *Sacred Queens and Women of Consequence*, 70–72.

9. The notion that the breaking of the kapu had left a "spiritual vacuum" or "religious vacuum" has been persistent in accounts of the period. See Antony Hooper, ed., *Class and Culture in the South Pacific* (Auckland: Centre for Pacific Studies, University of Auckland; Suva, Fiji: Institute of Pacific Studies, University of the South Pacific: 1987), 159; R. Lanier Britsch, *Moramona: The Mormons in Hawaii* (Laie, Hawaiʻi: Institute of Polynesian Studies, 1989), 10; Jennifer Fish Kashay, "From Kapus to Christianity: The Disestablishment of the Hawaiian Religion and Chiefly Appropriation of Calvinist Christianity," *Western Historical Quarterly* 39 (2008): 18.

10. Pukui, Haertig, and Lee, *Nānā i ke Kumu*, 1:150.

11. For recent literature on wahi pana, their meaning, and the politics surrounding them, see Davianna McGregor, *Nā Kuaʻaina: Living Hawaiian Culture* (Honolulu: University of Hawaii Press, 2007), 5–6; Cristina Bacchilega, *Legendary Hawaii and the Politics of Place: Tradition, Translation, and Tourism* (Philadelphia: University of Pennsylvania Press, 2007), 29–59. See also Oliveira, *Ancestral Places*.

12. Translation of this and other portions of this chant are taken from "Birth Chant for Kau-i-ke-ao-uli," in *The Echo of Our Song: Chants and Poems of the Hawaiians*, ed. and trans. Mary K. Pukui and Alfons L. Korn (Honolulu: University of Hawaiʻi, 1973), 20.

13. Ibid., 14–15, 20–21. In the preceding passage, I have made the following copyedits to Pukui and Korn's text: "ke aʻa" in place of "ka aʻa" in line 2, "pū" for

"pu" in line 4, "na Kea" for "nā Kea" in line 10, "na Papa" for "nā Papa" in line 11. Mahalo to Noenoe Silva for pointing out these anomalies in the original.
14. Ibid., 15–19, 21–25.
15. Here I have deviated from Pukui and Korn's translation. They render this line as "Born of Kea was the mountain." This implies that Kea gave birth to the mountain, whereas both the grammar of the line and the rest of the mele give us every reason to understand that Papahānaumoku is the mother of the mountain and Wākea is the father. Mahalo to Marie Alohalani Brown for her thoughts on this issue.
16. "Birth Chant for Kau-i-ke-ao-uli," 17, 23.
17. Noʻeau Peralto, "'O Koholālele, He ʻĀina, He Kanaka, He Iʻa Nui Nona ka Lā," 78; Williamson B. C. Chang, "Testimony and Appendix" to the University of Hawaiʻi Board of Regents on "The Management of Mauna Kea and the Mauna Kea Science Reserve," April 16, 2015, as posted at http://scholarspace.manoa.hawaii.edu/handle/10125/35797.
18. [Woodbridge and others], *He Hoikehonua* (1836).
19. Ibid., 33–35. In the passage describing the beliefs of "pagans," I have changed a number of em-dashes into commas.
20. Woodbridge, *Rudiments of Geography on a New Plan*. On Woodbridge and his emphasis on characterizing societies, see Calhoun, "Eyes for the Jacksonian World," 1–4.
21. [Woodbridge and others], *He Hoikehonua* (1836), 32, 48, 54, 55.
22. Ibid., 40.
23. *O ka Hoikehonua no ka Palapala Hemolele* (The Geography of the Holy Scripture) (Lahainaluna: Mea Pai Palapala no ke Kula Nui, 1835); *O ka Hoikehonua no ka Palapala Hemolele* (The Geography of the Holy Scripture) (Lahainaluna: Mea Pai Palapala no ke Kula Nui, 1838). The geography book was used along with two other volumes as part of a larger religious curriculum: *O ka Hoikemanawa no ka Mooolelo Hemolele* (The Chronology of Sacred History) (Lahainaluna: Mea Pai Palapala no ke Kula Nui, 1835); and *O ke Kuhikuhi no ka Mooolelo Hemolele* (A Guide to Sacred History) (Lahainaluna: Mea Pai Palapala no ke Kula Nui, 1838). Biblical geography continued to be taught for decades: "Ka Hoike o ke Kula Kahuna Pule," in *Ka Moolelo o ka Halawai Makahiki o ka Ahahui Euanelio Hawaii, ma Honolulu, Iune 1875* (Honolulu: Heneri M. Wini, 1875), 33.
24. *O ka Hoikehonua no ka Palapala Hemolele* (1835), 5.
25. Ibid., 41.
26. Cook, "Kahiki," 24.
27. Thomas Hopu to [Edwin Welles] Dwight, September 24, 1828, "Ka Awa Loa," translated typescript, translator unknown, in folder "Hopu, Asa 1867–1869, Hopu J. K. 1896, Hopu Thomas 1829," MS H, Hawaiian Evangelical Association Archives, Hawaiʻi Mission Children's Society Archives, Honolulu;

Extracts from the Report of the Agent of the Foreign Mission School, to the American Board of Commissioners for Foreign Missions (Hartford, Conn.: Hudson and Co., 1818), 6; On Dwight, see Brumaghim, "Life and Legacy," 72.

28. Matthew 18:20, King James Version.
29. Cook, "Kahiki," 67.
30. Noelani Arista, "Forward," in Kepelino, *Kepelino's Traditions of Hawaii*, vi–vii, especially vii, n. 10.
31. Cook, "Kahiki," 68.
32. Kuykendall, *The Hawaiian Kingdom*, 1:393–94.
33. Samuel Manaiākalani Kamakau, "Ka Moʻolelo o nā Kamehameha, Helu 96," *Ke Au Okoa*, December 26, 1868, as it appears in Kamakau, *Ke Kumu Aupuni*, 105.
34. Samuel Manaiākalani Kamakau, "Ka Moʻolelo o nā Kamehameha, Helu 94," *Ka Nupepa Kuokoa*, October 24, 1868, as it appears in Kamakau, *Ke Kumu Aupuni*, 99; Kamakau, "Ka Moʻolelo o nā Kamehameha, Helu 95," *Ka Nupepa Kuokoa*, December 19, 1868, as it appears in Kamakau, *Ke Kumu Aupuni*, 102; Kamakau, "Ka Moolelo o na Kamehameha, Helu 96," as it appears in Kamakau, *Ke Kumu Aupuni*, 103; Kamakau, "Ka Moʻolelo o nā Kamehameha, Helu 120," *Ke Au Okoa*, June 24, 1869, as it appears in Kamakau, *Ke Kumu Aupuni*, 235.
35. On diplomatic tensions and relations between Hawaiʻi and the United Kingdom, see Beamer, *No Mākou ka Mana*, 99.
36. Cook, "Kahiki," 69.
37. Ibid.
38. Aikau, *Chosen People, Promised Land*, 109, 110, 124.
39. Cook, "Kahiki," 36.
40. Samuel Kauwealoha's description of the origins of the mission is invaluable here. A large portion of it is published and translated into English in Nancy J. Morris, "Hawaiian Missionaries in the Marquesas," *Hawaiian Journal of History* 13 (1979): 48–49. Chief of Oomoa: "No ka Hae Hawaii. Ko J. Kekela Palapala Aloha hope i na Ekalesia a pau ma Hawaii nei," *Ka Hae Hawaii*, March 9, 1859, 196; Puu married to Matunui's daughter: Letter of James Kekela to Rev. R. Armstrong, Oomoa, Fatuiva, January 10, 1857, as printed in *Ka Hae Hawaii*, April 8, 1857, 5. Effort to obtain guns: Dening, *Islands and Beaches*, 150–55. A number of accounts make it clear that Puu was an essential go-between between Matunui and the American missionaries and imply that he was an architect of the Hawaiian–Fatu Hivan connection that used North American missionary teachers as a resource of teachers, of travel, and so on.
41. "Iles Marquises," *Feuille Religieuse du Canton de Vaud* 29 (1854): 411–12.
42. Cook, "Kahiki," 36.
43. Ibid., 86.
44. Kekahi Mau Haumana o Ke Kulanui, *Ka Mooolelo Hawaii*.
45. Silva, *Aloha Betrayed*, 46, 55.

46. Eleanor Harmon Davis, *Abraham Fornander: A Biography* (Honolulu: University Press of Hawaiʻi, 1979), 43, 50–51.
47. Fornander, *Fornander Collection*; Fornander, *Account of the Origins of the Polynesian Race*.
48. Silva, *Aloha Betrayed*, 87–123, 104–8.
49. Kamehiro, *The Arts of Kingship*.
50. S. K. Kawailiula, "Mooolelo no Kawelo," *Ka Hoku o ka Pakipika*, September 26, 1861, 1.
51. [M. J. Kapihenui], "He Moolelo no Hiiakaikapoliopele, Helu 1," *Ka Hoku o ka Pakipika*, December 26, 1861, 1.
52. Palikoolauloa, "Na Wahi Pana o Kaliuwaa," 4.
53. For more on Kamapuaʻa in the Hawaiian-language press, see the series "He Moolelo no Kamapuaa" by S. W. Kahiolo. The thirteen-part series appeared in *Ka Hae Hawaii* every week from June 26, 1861, to September 25, 1861.
54. Palikoolauloa, "Na Wahi Pana o Kaliuwaa."

7. "WE WILL BE COMPARABLE TO THE INDIAN PEOPLES"

1. "E Hoohuiia Anei Kakou?," *Ka Makaainana*, December 23, 1895, 4.
2. "Hawaiian Recognition Plan Meets Vocal Opposition," *Honolulu Star Advertiser*, August 7, 2014; J. Kēhaulani Kauanui, "Precarious Positions: Native Hawaiians and U.S. Federal Recognition," in *Recognition, Sovereignty Struggles, and Indigenous Rights in the United States: A Sourcebook*, ed. Amy E. Den Ouden and Jean M. O'Brien (Chapel Hill: University of North Carolina Press, 2013), 311–36.
3. Chadwick Allen, *Blood Narrative* (Durham, N.C.: Duke University Press, 2002), 196.
4. Hauʻofa, "Our Sea of Islands"; Aikau, *Chosen People, Promised Land*, introduction; Cook, "Kahiki"; Somerville, *Once Were Pacific*; Ty. P. Kāwika Tengan, Tēvita O. Kaʻili, and Rochelle Tuitagavaʻa Fonoti, "Genealogies: Articulating Indigenous Anthropology in/of Oceania," *Pacific Studies* 33, nos. 2/3 (August/December 2010): 161.
5. My thinking on these topics has been shaped especially by Lyons, *X-Marks*; Round, *Removable Type*; Philip J. Deloria, *Indians in Unexpected Places* (Lawrence: University Press of Kansas, 2004); Somerville, *Once Were Pacific*; Silva, *Aloha Betrayed*.
6. Robert F. Berkhofer, *The White Man's Indian: Images of the American Indian, from Columbus to the Present* (New York: Vintage Books, 1979), 1, 4.
7. Noenoe Silva with the assistance of Iokepa Badis, "Early Hawaiian Newspapers and Kanaka Maoli Intellectual History, 1834–1855," *Hawaiian Journal of History* 42 (2008): 109; Helen G. Chapin, *Guide to Newspapers of Hawaii* (Honolulu: Hawaiian Historical Society, 2000), 65.

8. John Lee Comstock, *A Natural History of Quadrupeds, with Engravings* (Hartford, Conn.: D. F. Robinson, 1829).

9. "No ka Eleka," *Ka Lama Hawaii*, March 21, 1834, 1.

10. "No ka Bipikuapuu," *Ka Lama Hawaii*, April 4, 1834, 1.

11. "No ka Aianonanona," *Ka Lama Hawaii*, April 18, 1834, 1.

12. On forests and "savagery," see Berkhofer, *The White Man's Indian*, 13.

13. Silva and Badis, "Early Hawaiian Newspapers," 118, 124; Chapin, *Guide to Newspapers of Hawaii*, 78.

14. "No ka Mahiai," *Ka Nonanona*, January 3, 1844, 84–85.

15. "Hakina 8: No na mea i hanaia i maikai ai kanaka. No ka hoolaha ana i ka Olelo a ke Akua," *Ke Kumu Hawaii*, May 4, 1835, 67.

16. "No ka Hae Hawaii," *Ka Hae Hawaii*, April 16, 1856, 28.

17. "No ke Kalaiaina—Helu 27," *Ka Hae Hawaii*, January 27, 1858, 165.

18. "Home," *Ka Hae Hawaii*, February 8, 1860, 178.

19. Silva and Badis, "Early Hawaiian Newspapers," 112; Chapin, *Guide to Newspapers of Hawaii*, 63.

20. "He Ui Misionary. Hakina 10," *Ke Kumu Hawaii*, August 19, 1835, 131.

21. "Nu Hou Kuwaho," *Ka Nupepa Kuokoa*, October 26, 1867, 2.

22. "Kialua Chenamus," *Ka Nonanona*, October 11, 1842, 47.

23. "Kaua o na Inikinikini," *Ka Hae Hawaii*, April 2, 1856, 18; "Pepehi Kanaka," *Ka Hae Hawaii*, June 13, 1860, 4.

24. Untitled, *Ka Hae Hawaii*, February 18, 1857, 202.

25. "Pepehi Kanaka."

26. Mary Frances Morgan Armstrong and Samuel Chapman Armstrong, *America: Richard Armstrong. Hawaii* (Hampton, Va.: Normal School Steam Press, 1887), 35.

27. "Kaua o na Inikinikini," *Ka Hae Hawaii*, April 2, 1856, 18.

28. "No na Inikini," *Ka Hae Hawaii*, October 24, 1860, 124.

29. Untitled, *Ke Kumu Hawaii*, January 2, 1839, 63.

30. "No ka Emi Ana o na Kanaka," *Ke Kumu Hawaii*, April 10, 1839, 90.

31. "Ka Pepehi Hoomainoino ana ma Papu Pilo (Pillow)," *Ka Nupepa Kuokoa*, May 21, 1864, 1.

32. "No ka hanaino ia o kapoe o ka Akau elawe pio ia e koka Hema," *Ka Nupepa Kuokoa*, March 9, 1865, 2.

33. J. H. Kānepu'u, "Pau ole ke ano pouli o Hawaii nei," *Ka Hae Hawaii*, June 19, 1861, 48. On Kapohaku, see "A Blind Native Preacher," *The Missionary Herald at Home and Abroad* 65, no. 11 (November 1869): 394.

34. David A. Chang, "Borderlands in a World at Sea: Concow Indians, Native Hawaiians, and South Chinese in Indigenous, Global, and National Space, 1860s–1880s," *Journal of American History* 98, no. 2 (2011): 382–87.

35. Untitled article, *Ke Kumu Hawaii*, August 1, 1838, 10.

36. Dening, *Islands and Beaches*.

37. J. H. Kanepuu, "Kanaka Hihiu Mauka o Niu," *Ka Hae Hawaii*, November 12, 1856, 147.
38. B. L. Koko, untitled, *Ka Nupepa Kuokoa*, March 30, 1865, 4.
39. "He Mau Wai-Hooluu Ano e i Hoohuiia," *Ka Nupepa Kuokoa*, May 19, 1866, 3.
40. "Hawaii," *Ka Nupepa Ku*, February 1, 1868, 3.
41. Kenn, "Descendants of Captain Sutter's Kanakas," 94. It is possible that Imikula went on to study at the Hilo Boarding School. An Imigula appears on page 65 of the enrollment ledger, Hilo Boarding School Records, AR-7, Education Files, Enrollment Records, 65, box 4, folder 29, Lyman House Museum and Archives, Hilo, Hawaiʻi.
42. The many examples include "Palapala mai kekahi kanaka Hawaii mai Kalifonia mai," *Ka Hae Hawaii*, November 7, 1860, 132; "He Palapala na ko makou eleele i holo aku nei I Kapalakiko," *Ka Nupepa Kuokoa*, October 25, 1862, 4; "No na wahine Ilikini me na keiki, me na mea ai a lakou, a me ka nui o na keiki a na kanaka ma Keomolewa nei," *Ka Nupepa Kuokoa*, January 26, 1865, 1; John Makani, "No Ioane Makani Ilikini," *Ka Nupepa Kuokoa*, November 2, 1867, 3; "Mai Oregona Mai," *Ka Nupepa Kuokoa*, August 29, 1868, 2.
43. "Ka make weliweli ana o Hale Kalaluhi ma Bill Hill County of Cottonwood, Kaleponia," *Ka Nupepa Kuokoa*, October 20, 1866, 4; "Make Wiliama G. Kahuakaipia i pana ia e ka Ilikini ma New Years Diggings, Mariposa County, California," *Ka Nupepa Kuokoa*, February 1, 1868, 4; "Hopuia," *Ka Nupepa Kuokoa*, March 29, 1862, 2.
44. "He hakaka no ka make," *Ka Lahui Hawaii*, September 14, 1876, 3.
45. "Kaua Nui ma Sugar Creek," *Ka Hoku o ka Pakipika*, April 3, 1862, 2.
46. J. B. Nakea, "Mai Pau ke Ola," *Ka Hae Hawaii*, August 14, 1861, 80.
47. Kameʻeleihiwa, *Native Land and Foreign Desires*.
48. Hawaii Ponoi, [title missing from incomplete copy at http://www.nupepa.org], *Ke Au Okoa*, December 19, 1867, 2. On the exchange of letters and editorials referred to here, see Leilani Basham, "Ka Lāhui Hawaiʻi: He Moʻolelo, He ʻĀina, He Loina, a He Ea Kākou," *Hūlili* 6, at http://www.ksbe.edu/spi/Hulili/vol_6/.
49. Chapin, *Guide to Newspapers of Hawaii*, 62; Daniel Harrington, *Hawaiian Encyclopedia*, http://www.hawaiianencyclopedia.com/part-2-glossary-i-l.asp.
50. Untitled, *Ka Nupepa Kuokoa*, February 22, 1868, 2.
51. A. W. Wekeweke, "Na mea hou o Kohala Akau nei," *Ka Lahui Hawaii*, January 18, 1877, 3.
52. "Alapohoia ka Aina Makika," *Ka Makaainana*, June 3, 1895, 3.
53. "Na hunahuna Laulaha," *Ka Makaainana*, October 7, 1895, 7.
54. "Mamao Loa ka Laulea," *Ka Makaainana*, February 3, 1896, 4.
55. "E Hoohuiia Anei Kakou?"

56. Davianna Pomaikaʻi McGregor, "Recognizing Native Hawaiians: A Quest for Sovereignty," in *Pacific Diaspora: Island Peoples in the United States and across the Pacific*, ed. Paul Spickard, Joanne Rondilla, and Debbie Hippolite Wright (Honolulu: University of Hawaii Press, 2002), 343–44.

EPILOGUE

1. Eli Keolanui, "Immigration," and 1923 Graduation Program, Hilo Boarding School Records, Lyman House Museum and Archives, Hilo, Hawaiʻi.
2. Patrick Wolfe, "Settler Colonialism and the Elimination of the Native," *Journal of Genocide Research* 8 (2006): 387–409; Erich Steinman, "Settler Colonial Power and the American Indian Sovereignty Movement: Forms of Domination, Strategies of Transformation," *American Journal of Sociology* 117 (2012): 1073–1130; Kevin Bruyneel, *The Third Space of Sovereignty: The Postcolonial Politics of U.S.-Indigenous Relations* (Minneapolis: University of Minnesota Press, 2007).
3. Keolanui, "Immigration," 1–2.
4. Prescott F. Hall, "History of Immigration," in *Immigration and Americanization: Selected Readings*, ed. Philip Davis (Boston: Ginn, 1920), 61–68; Henry Cabot Lodge, "Immigration: A Review," in Davis, *Immigration and Americanization*, 50–60; Philip Davis, "What America Means to the Immigrant," in Davis, *Immigration and Americanization*, 661–71. Unlike Hall and Lodge, Davis aligned himself with what he perceived as European immigrants' interests and was not opposed to European immigration. Keolanui thus uses his words with the least changes and saves them for the concluding section of his essay, which argues against immigration laws that discriminate on the basis of national origin.
5. Hall, "Immigration and Its Effects," 61–62; Prescott F. Hall, "The Present and Future of Immigration," *North American Review* 213, no. 786 (May 1921): 606–7.
6. J. Kēhaulani Kauanui is elaborating this idea in her current book manuscript, "Thy Kingdom Come? The Paradox of Hawaiian Sovereignty." See also Jodi A. Byrd, *Transit of Empire: Indigenous Critiques of Colonialism* (Minneapolis: University of Minnesota Press, 2011), 147–85.
7. My suggestion here is parallel to JoAnn Conrad's discussion of folklore and temporality: she notes that although the category of "folklore" is part of an effort at "standardizing, regularizing and universalizing time," its narratives constitute "potential sites of alternate temporalities." JoAnn Conrad, "The Storied Time of Folklore," *Western Folklore* 73, nos. 2/3 (Spring 2014): 343. "Twenty-Four Hawaiian Culture Lectures to Be Held This Year," *Ka Moi*, October 4, 1935, 3.
8. "Ka Buke o Kamehameha," hand inscribed 1936, no publication information given, 29, Kamehameha Schools Library and Archives, Honolulu;

"Scenes of Samoa Are Shown to Hui o Iwi," *Ka Moi*, April 7, 1933, 1; "Hawaiian Luau Given by Hui o Iwi," *Ka Moi*, March 24, 1933, 1; "Knife Dancing to Be Taught Boys by Samoan Guests," *Ka Moi*, March 17, 1933, 3; "Hui o Iwi," *Ka Moi*, December 9, 1932, 3.

9. Mary Kauimeheiwa, "Eleile Golden Days Legend Tells Fate of Prince and Princess," *Ka Moi*, March 3, 1933; Vivian Crockett, "Awapuhi, the Blossom of Papani Valley," *Ka Moi*, April 28, 1933, 6; Ululani Weight, "The Hawaiian Coat-of-Arms," *Ka Moi*, April 28, 1933, 6; Amy Crockett, "The Return of Pele," *Ka Moi*, April 28, 1933, 5; Haunani Cooper, "The Legend of Kuula, Fish-God of Hawaii," *Ka Moi Literary Annual*, 1935, 23; David White, "The Tale of Punia and the Sharks," *Ka Moi Literary Annual*, 1935, 10–11; Eugene Burke, "Pohaku Ula," *Ka Moi Literary Annual*, 1934, 16; Frank E. Midkiff and John H. Wise, "A First Course in Hawaiian," no publication information, Kamehameha Schools Library and Archives, Honolulu; Frank E. Midkiff to Mrs. Lydia Aalaonaona Roy Akana, April 14, 1978, Kamehameha Schools Library and Archives, Honolulu. Mahalo to Stacy Naipo of the Kamehameha Schools Archives for guiding me to these records.

INDEX

Account of the History of the Polynesian Race (Fornander), 220
adoption, 167–68, 172, 176, 238
African Americans: in California settlements, 182; Kānaka categorized as, 161, 179–82, 185, 191–94; Kanaka relations with, 161–62, 186–91; racial liability of blackness, 191; textbook representation of, 118, 135, 154. *See also* United States
Ahahui Euanelio Hawaii (Hawaiian Evangelical Association; HEA), 145, 195–96, 217
Ahahui Hoopuka Nupepa Kuikawa o Honolulu, 184
Aholo, Lydia, 162
'ahu'ula (feathered cloak), 43, 49
aikāne relationship, 43–47, 90, 183. *See also* love; sexuality
'ai kapu, 199–200, 269n30
Aikau, Hokulani K., xvii, 161, 214
'āina: land distribution and rights, 34, 150–51, 243–44; term, 129, 130, 138. *See also* akua
'Āinamoana, 122, 124, 128–33, 149–50, 155, 286n50
akua: cultural record of, 221–23; gender and, 9, 152; genealogy of Hawai'i and, 19–20, 31; genealogy of Hawaiian Islands and, 8–13, 19–20, 31, 197, 201–4. *See also* 'āina; Hawaiian religion; mo'olelo; *specific gods*

Alexander 'Iolani Liholiho. *See* Kamehameha IV
ali'i: cartography and, 109; Christian denominations and, 209, 212–14; education and, 109–10, 111; history of, xiii–xiv; mana of, 200–201; on palapala, 97; qualities and skills of, 41–42; regalia of, 43, 45, 49, 74; social hierarchy and, 31. *See also specific persons*
Allen, Chadwick, 229
Aloha Aina, Ke (newspaper), 153, 242
Aloha Betrayed (Silva), xvii
American Board of Commissioners for Foreign Missions (ABCFM), 97
American Geography (Morse), 126
American Indians: book engagement of, 99; at California forts, 164–65; Civil War–era newspaper depictions of, 237–38; Coloma settlement, 168, 172, 184, 195, 222; Concow, 166–68, 171–75; Kānaka categorized as, 70–71, 179–82; Kanaka-controlled newspapers on, 242–46; land rights of, 192, 237, 245; Maidu, 157–58, 166, 171, 177–78, 241, 289n3; missionary-controlled newspapers on, 227–36; on O'ahu, 240–41; racial politics and, 161, 173–74, 188, 191–93, 236; reservation system, 167, 173, 292n29; student

305

immigration essay on, 249–53. *See also* Kanaka–American Indian relations; United States
Amiuna, Ioane, 182
Ana, Keoni, 162–63
ancestors, xiii. *See also* genealogy
Andrade, Carlos, xiii
Andrews, Lorrin, 99, 100, 230, 269n31
annexation to the United States, viii, 106, 153–55, 227–29, 246–50. *See also* sovereignty, Hawaiian; United States
Apua, 18
arioi, 93, 94, 95
Arista, Noelani, 42
Armstrong, Richard, 110, 132, 232, 235–36
atlases. *See* geography education and textbooks
Aukelenuiaiku, 18
Auna and Naiomi, 93–96
Au Okoa, Ke, (newspaper), 242, 244
'awa, 19, 23, 140, 243, 270n38
Azbill, Cora, 183
Azbill, Henry, 160, 183; on Kanaka fishing industry, 293n35; on Kanaka heritage, 178; on Keaala, 191–92; Maidu cultural preservation by, 158, 289n3
Azbill, John B., 157, 160, 180
Azbill, Mele (Mary) Kainuha Keaala, 157–60, 183, 289n2
Azores, 189, 294n56

Banks, Joseph, 190–91
Barkley, Charles T., 33
Barkley, Frances Trevor, 35, 55
Barman, Jean, 165, 184–85
Barnes' Geography (Monteith), 154, 289n86
Bauer, William J., 173–74

Beamer, B. Kamanamaikalani, xii, 109, 111
Berkhofer, Robert F., 119, 230, 233
Bible: Hawaiian translations of, 81, 96, 128–29, 206; holy places in, 206–8; verses of, 96, 195, 208, 282n52, 296n4; wahi pana and, 224. *See also* Christianity
binary thought and language, 6–7, 61–68, 70, 130–32, 179, 198, 250
birds, 15–17
Bishop, Charles Reed, 106
black Americans and blackness. *See* African Americans
boarding schools. *See specific schools*
"bone of our bone" sentiment, 195–97, 296n4
bones, 15, 197, 200, 247
Book of Mormon, 214–15
books, 97–102, 113, 129. *See also* English-language texts; geography education and textbooks; Hawaiian-language texts
Bora Bora, 1–2, 14, 221. *See also* Pacific geography
Bounty (ship), 94
Brückner, Martin, 126
Brumaghim, Wayne H., 89–90
Burchard, J. L., 173

California: African American settlements in, 182
California history: Coloma settlement, 168, 172, 184, 195, 222; Concow, 166–68, 171–75; Fort Ross and Sutter's Fort in, 163–65; gold mining, 165, 166–67, 176; Maidu, 157–58, 166, 171, 177–78, 241, 289n3; race and racialization in, 191–93; reservation system in, 167, 173, 292n29; Vernon settle-

ment, 175–78, 180, 181, 293n35. *See also* Kanaka–American Indian relations; United States
Callicum (Comekala), 58, 67
Calvinism. *See* Protestantism
Canacka Boarding House (Nantucket), 186–87, 194
Canada, 65–71
cane sugar. *See* sugar plantations
cannibalism, 69, 238
Cannon, George Q., 214–15
canoe voyaging tradition, 1–2, 4–5, 256
Cartesian graphing, 21, 270n45
cartographic agency, 114–18, 127–28, 130–32. *See also* geography education and textbooks
castas system in Spanish empire, 61–63
Catholicism: Kamakau and, 209–12; Kanaka devotion to, 209; Kepelino and, 209–10; schools and publications of, 115–20. *See also* Christianity
chants and chanting. *See* mele
Chappell, David A., 35, 272n19
Charlot, John, 42, 98, 129, 152
chiefs. *See* ali'i
Chiefs' Children's School, 109–10, 111
children: group reading, 99, 113, 129; hānai adoption, 167–68, 172, 176, 238; Nuku Hivan girl in Hawai'i, 238–39; parental death and guardianship of, 174–75, 293n33. *See also* education
China: racism toward people of, 52–53, 54; trade and, 33, 36, 52, 55, 70, 102; women's rights in, 55, 56. *See also* Macao
Christianity: on aikāne, 44; American Indians and, 174;

Biblical verses, 96, 195–97, 208, 282n52, 296n4; denominational schools and publications, 115–20; Hawaiian translation of Bible, 81, 96, 128–29, 206; holy places of, 206–8; on publication of Hawaiian mo'olelo, 222–23. *See also* missionaries and missionary schools; religion; *specific denominations*
Christian Kānaka: denominations and, 208–15; Pacific kinship and, 195–97, 205, 215–18, 224–25, 296n4; religious connection with Tahiti, 79–81, 92–97. *See also* Kānaka Maoli; *specific persons*
Church of England, 212, 213–14
Church of Hawai'i, 209, 212
Church of Jesus Christ of Latter-Day Saints, 209, 214–15
civilizational hierarchy, 135–40, 231–33. *See also* race and racial hierarchy
Civil War depictions, 237–38
Clenso, Pamela, 176, 177
clothing, 43, 45, 48–49, 74, 253
coconut, 18, 30, 64–65
cohabitation, 165–66, 168–69
Coloma settlement, California, 168, 172, 184, 195, 222
colonial power. *See* sovereignty, Hawaiian
Comekala, 58, 67
common school system, 110, 112–13, 147
complementary oppositions in Hawaiian thought, 130–32
Comstock, John Lee, 230–31
Concow, 166–68, 171–75
Congregationalism, 209–12, 215–18
Connecticut, 185, 186
Cook, James: death of, 33; on

language, 271n5; narrative on first encounter, viii, 5–6, 25–30; Nuu-chah-nulth and, 65–66
Cook, Kealani, 120, 197, 198, 207–8, 214, 217–18
Cooke, Amos Starr, 111, 212
Cooke, Juliette Montague, 111, 212
Cornwall Foreign Mission School (Connecticut), 91–92, 97
cosmology. *See* Hawaiian religion
Cox, John Henry, 52
Craig, Robert D., 269n38
Cuffee, Paul, 186

Dakota War (1862), 237–38
dance. *See* hula
Davis, Isaac, 73, 76
Davis, Philip, 251, 302n4
Daws, Gavan, 28, 83
Delano, Alonzo, 167
Demos, John, 83, 91
Dening, Greg, 5, 240
Department of Public Instruction, 133
Dibble, Sheldon, 219
dictionaries, 133, 199
diet, 146, 231–32
directional perspective. *See* perspective and perspectivalism in Hawaiian thought; "view from nowhere"
discovery. *See* exploration
Discovery (ship), 66
diseases: geography textbooks on, 118–19; introduction of, 34, 84, 102; smallpox, 174, 237; solution for, 145–46
Douglas, William, 37, 63–65, 70, 276n80
Duarte, T. Kaeo, 109
Dwight, Edwin, 84–86

eating kapu, 199–200, 269n30
economic production: capitalist and non-capitalist, 37–38; labor migration, 37–38, 162–64, 185–91, 239–40; Māhele and, 243–44; of maka'āinana, 31, 42–43; textbook representation of, 135, 138, 140–42, 149
education: central administration of, 142–44, 147–48, 153; Christian teachers and, 92–97; colonial power and, 107–8, 133–34, 253; common school system, 110, 112–13, 147; establishment of national system, 105–6; Kamehameha III on, 148; Kanaka educators and politics, 97–99, 105, 107, 142–53, 283n9; memorization and, 98–99; post-annexation American, 153–55, 249–50; select school system, 109–11, 253–55; sovereignty pedagogy and, 142–44, 283n9. *See also* geography education and textbooks; missionaries and missionary schools; *specific schools*
Eel, Aihi, 176
Elbert, Samuel, 199
El Dorado County (California), 166, 168, 172
Elementary Geography (Monteith), 154, 289n86
Elikula, John, 241
Ellis, William, 92
Emerson, Samuel, 294n52
Emma, Queen of Hawai'i, 111, 193–94
England: Church of, 212, 213–14; missionary society, 92
English-language texts: dictionaries, 133; as knowledge source, x, 147; in post-annexation education,

153–55; rhetoric of loss in, xi–xii, 264n8. *See also* language and literacy; *specific titles*
enlightenment versus ignorance. *See* naʻauao versus naʻaupō
Episcopal Church, 209, 212–14
ethnobotany, 18–19
exploration, accounts of: defining discovery, viii–ix; on Japanese, 263n4; by Kamakau, 42–43; Pele's voyage, 1–2. *See also specific authors*

Fatu Hiva. *See* Nuku Hiva
Felice (ship), 58, 63, 68
Fermantez, Kali, xiii
fishing, 175–76, 186, 293n35
folklore as category, 253–55, 302n7
food and diet, 146, 231–32
Forbes, David W., 278n1
Fornander, Abraham, 220, 267n16, 268n18
Fornander Collection of Hawaiian Antiquities, 220
Fort Pillow Massacre, 238
Fort Ross, California, 163–64
France, 210
Fugitive Slave Act (1850), 191
fur trade, x, 36, 66–67, 84, 163, 228

Garrett, John, 92
gender: California settlements and, 182; Christian missionaries and, 93–95, 109; in Hawaiian moʻolelo, 9, 152; labor migration and, 162–65, 171–72; material culture and, 47–49; opposition in Hawaiian thought on, 131; students' cultural initiatives and, 253–54. *See also* marriage practices; men; women

genealogy: of gods, 19–20, 31; land and, xiii; of Mauna Kea, 201–4; Mormonism on, 214–15; Pacific kinship and, 8–13, 195–97, 201–5, 215–18, 224–25, 296n4; preservation initiatives of, 218–25, 253–55. *See also* Kānaka Maoli; mele; moʻolelo
geography: cartographic agency, 114–18, 127–28, 130–32; defining discovery, viii–ix; genealogy and, xiii, 8–13, 195–97, 201–4, 214–18; Hawaiian versus global, xii–xiii, 2–4, 103–4; of Macao, 56–58; of mana, 197–98; mapping erasure, x–xi; perspective and perspectivalism, 21–22, 117, 121–26, 134–35. *See also* Pacific geography; *specific islands*
geography education and textbooks: in common schools, 112–18; desacralization of Hawaiian lands in, 204–8; on economic production, 135, 138, 140–42; Hawaiian cartographic agency, 114–18, 127–28, 130–32; institutional history of, 108–12; Kanaka perspective in maps, 121–26; lack of texts for, 112, 284n22; list and descriptions of, 116; racial hierarchy and colonialism in, 118–21, 126–27, 154; sovereignty and politics of, 105–8; in United States, 126. *See also specific textbook titles*
Geography Made Easy (Morse), 126
gifts: makaʻāinana and, 31; power and, 70–71, 75–76; versus trade, 67–68, 73–75
gods and deity worship. *See* akua; Hawaiian religion; *specific gods*
gold mining, 165, 166–67, 176, 241

Goodyear-Kaʻōpua, Noelani, 99, 105, 106, 143
Gospel of Matthew, 96, 208, 282n52
Gray, Francine Du Plessix, 82
group reading, 99, 113, 129
Gulick, Thomas, 167–68, 172, 174
gweilo, 55

Hae Hawaii, Ka (newspaper), 235–36, 238, 240, 244
Hakauila, Lillie, 176
Hakauila, Richard, 176
Hale Kula no nā Keiki Aliʻi, 109–10, 111
Halemaʻumaʻu Crater, Kīlauea, 222
Hale Nauā, 220–21
Hall, Lisa Kahaleole, x
Hall, Mary L., 133, 287n59. *See also specific titles*
Hall, Prescott Farnsworth, 251
Hallock, Sydney and Hannah, 189
Hāloa, xiii, 203–4
Halualani, Rona Tamiko, 266n8
Hämäläinen, Pekka, 26
hānai adoption, 167–68, 172, 176, 238
"Hānau a Hua Kalani," 201–4
Haole (category), 33, 37. *See also specific persons*
Hatohobei, Palau, 63–65, 276n80
Hauʻofa, Epeli, 117, 229
Hawaiʻi: ʻĀinamoana, 122, 124, 128–33, 149–50, 155, 286n50; cultural initiatives in, ix–xii, 218–25; discourse of isolation of, 6–7, 266n8; economic geography of, 140–42; formal political hierarchy of, 31; Pacific kinship and religion, 195–97, 205, 215–18, 224–25, 296n4; Paulet Affair, 213; population of, x, 3, 145; term for, 185; United States annexation of, viii, 106, 153–55, 227–29, 246–50; visual representation of, 121–26. *See also* geography; Pacific geography; *specific islands; specific rulers*
Hawaiʻi Island: aliʻi of, 31, 73, 89, 94; American Indians on, 241; European ships at, 27–28, 29, 33, 73; ʻŌpūkahaʻia of, 21–22; wahi pana of, 20, 201–6, 222–23. *See also* Hawaiʻi
Hawaiʻiloa, 7, 14, 267n12
Hawaiian-language texts: Bible translation, 81, 96, 128–29, 206; dictionaries, 133, 199; directional perspective in, 21–22, 117, 121–29; on Kānaka in California, 165–66, 169; kaona in, 17–18; as knowledge source, 146–47; oral to written transition, 5, 7–8, 13, 81; orthography of, xviii–xix; as primary historical sources, ix–xiii; publication of Hawaiian moʻolelo, 198, 218–25, 254; by students, 253–54. *See also* geography education and textbooks; language and literacy; newspapers; *specific titles*
Hawaiian religion, references to: by Kamakau, 210–12, 222; in newspapers, 218, 219, 222–24, 230; ʻŌpūkahaʻia and, 86, 91; in textbooks, 139, 204–5. *See also* religion
Hawaiians. *See* Kānaka Maoli
Hawaii Ponoi (pseudonym), 244
helu (literary form), 42, 101, 274n40
Hiʻiakaikapoliopele, 151, 152, 222
Hikianalia (oceangoing canoe), 256
Hill, Dorothy, 191
Hilo Boarding School: origin and instruction at, 109, 110–11, 112, 153; pupils of, 153, 249, 254, 256, 301n41
Hina, 10–11, 223
"History of Immigration" (Hall), 251

Hitchcock, Harvey Rexford, 133, 139, 287n59. See also Honua Nei, Ka Hitokane. See Waiulili
Hoikehonua, He: 79, 102, 116; colonialism in, 118–21, 131; desacralization in, 204–8; grammar and style of, 127–28; influences of, 115, 277n1, 285n29; organization of, 121–22, 127
Hoikehonua no ka Palapala Hemolele, O ka, 206–7, 297n23
Hōkūle'a (oceangoing canoe), 256
Hoku Loa (newspaper), 195, 196
Hoku o ka Pakipika, Ka (newspaper), 152, 219, 221–22, 242, 243
Holy Land, 207–8
honua, 2
Honua Nei, Ka (Hall and Hitchcock): 116, 133–35, 287n59; civilizational hierarchy in, 135–40; comparison with Kānepu'u's title, 148–49; economic geography in, 135, 138, 140–42; sovereign pedagogies of, 142–44
Honua Nei a me na Mea a Pau Maluna Iho, Ka (Kānepu'u), xii–xiii, 103–4, 148–51. See also Kānepu'u, J. H.
Hopu, Thomas, 91, 92, 208
"Huaka'i a Pele, Ka," 1–2
hui aloha 'āina, 153
Hui Oiwi, Kamehameha School, 253, 255
hula, 16–17, 220–21, 222, 224
Humehume, 97

Ignatiev, Noel, 190–91
'Ī'ī, John Papa, 210
'ike (knowledge, sight): 7, 27; hierarchy and, 30–33, 76
Ilikini/Inikini, 230. See also American Indians

Imikula, William, 241, 301n41
Immigration and Americanization (Davis), 250–51
"Immigration: A Review" (Lodge), 251
Imperial Eagle (ship), 33–38
Indians. See American Indians
indigenous identity, 227–29, 249–53. See also American Indians; Kānaka Maoli
Iphegenia (ship), 39, 58, 60, 63–65, 276n80
Irish Americans, 190–91
iwi, 15, 197, 200, 247

Judaism, 212
Judd, Gerrit P., 213

Ka'ahumanu, 200, 209
Kaamoku, 94
Kā'eo (Kā'eokūlani), 25–26, 31
Kahahana, 31
Kahekili (Kahekilinui'ahumanu), 31
Kahiki and Tahiti: Armstrong on, 232; in cultural record, 221; Kanaka religious connection with, 80–81, 92–97; as mythical versus real location, 13–14, 269n38; Sahlins on, 13; term, 13, 72, 95, 279n20. See also Pacific geography
Kahikikū, 13–14
Kahikimoe, 13–14
Kahiki'ula, 223
Kaho'olawe, 10–11. See also Hawai'i
Kahoopii, S. W., 146–47
kāhuna and 'oihana kahuna: Ka'iana and, 43, 54; knowledge and, 32–33; 'Ōpūkaha'ia, 88–89; political hierarchy and, 31; women and, 94–95. See also Hawaiian religion; specific persons

Kaʻiana (Kaʻianaʻahuʻula): clothing and, 43, 45, 48–49; descriptions of, 39–42, 72–73; genealogy of, 38, 177–78, 293n39; health and death of, 59, 76; Kamehameha and, 47, 72–73, 76; Ka Wahine and, 48, 59–60; Macao experiences of, 50–54, 61–62; Meares as aikāne of, 43–47; portrait of, 45; post-voyage political career of, 71–76; voyage of, 37, 47, 71
Kaikaina o ka Ahahui Hoopuka Nupepa Ku i ka Wa o Honolulu, 184
Kalākaua, Davida: cultural initiatives and, 106, 220–21, 244; education of, 111; genealogy of, 89; on Kamehameha, 76; motto of, 150; travels by, xiv
Kalanianaʻole, Kūhio, 162
Kalaniʻōpuʻu, 31
Kaleikuahulu, 12–13
kalo cultivation, 73–74, 140, 141, 240. *See also* makaʻāinana
kamaʻāina, 263n1. *See also* Kānaka Maoli
Kamakau, Samuel Manaiākalani: xv–xvi, 210; Catholicism of, 209–10; on Cook at Waimea, 26, 28–30; on directional perspective, 22, 271n47; image of, 211; as inspiration, xviii; on Kaʻiana, 41–42, 44–47, 72
Kamakea, G. H., Jr., 172, 174
Kamakea, G. H., Sr., 168, 172
Kamakea, Samuela, 172
Kamāmalu, Victoria, 111
Kamapuaʻa, 223
Kameʻeleihiwa, Lilikalā, 31–32, 82, 244
Kamehameha I: children of, 29; genealogy of, 12; Kaʻiana and, 47, 72–73, 76; Kūkāʻilimoku and, 89;

Pākuʻi and, 9; report of theft of bones of, 247; rise to power, 13, 28, 76; Vancouver and, 74, 75
Kamehameha II, 9, 94, 200
Kamehameha III, 110, 111–12, 148, 201–4, 209, 213
Kamehameha IV, 111, 162, 164, 193, 212–13
Kamehameha V, 111, 162, 188, 193, 213, 221
Kamehameha School, 110, 253–55
Kamehiro, Stacy L., 221
Kamoku, J. W., 148
kanaka and kānaka, xix, 35, 263n1
Kanaka–American Indian relations: Azbill family, 157–60, 178, 180, 183; at California forts, 163–65; child custody and, 174–75; hānai adoption, 167–68, 172, 176; Kanaka newspapers on, 165–66, 169, 239–42; likeness of Kānaka to American Indians, 227–28, 236–39, 247–48; mixed-heritage families in Hawaiʻi, 241–42; sugar plantation incident, 242; in Vernon settlement, 175–78, 180, 181. *See also* American Indians; California history
Kanaka Bar, California, 167, 291n21
Kānaka Maoli: cartographic agency of, 114–18, 127–28, 130–32; defined, vii, 263n1; early global engagement by, ix–x, 2–6, 25–26; early identification by Haole, 33, 37, 83; first ship voyages of, 35–36; in fur trade, x, 36, 66–67, 84, 163, 228; gender, 48, 253–54; in geography textbooks, 118–19; Hawaiian Renaissance, 255–56; indigenous identity of, 227–29, 249–50; meaning of bones to, 197; as "meek" in Sermon on the Mount,

102; as Mormonism's Chosen People, 214–15; mourning of, 59–60; Pacific kinship of, 195–97, 205, 215–18, 224–25, 296n4; population abroad in 1845, 162–63; population of, 209; racial categorization in U.S. census, 178–84. *See also* Hawaiian-language texts; *specific persons*
Kanakaʻole, Edith, 255
Kanaloa. *See* Kahoʻolawe
kāne, 9, 37, 216. *See also* men
Kāne (god), 1–2, 152
Kānepuʻu, J. H., xii–xiii, 98, 103–4, 238–39, 240. *See also Honua Nei a me na Mea a Pau Maluna Iho, Ka*
kaona, 17–18, 74
Kapena, John Makini, 244
Kapiʻolani, 200
Kapohaku, Paulo, 238
kapu: of eating, 199–200, 269n30; Judaism and, 212; overthrow of, 81, 199–200, 234, 296n9; rank and, 60; and tapu in Tahiti, 93, 95; wahi pana and, 201, 204. *See also* mana
Kapuāiwa, Lot. *See* Kamehameha V
Kapuu, Hana, 176
Kapuu, Harieka, 176
Kapuu, John, 176, 177
Kartunnen, Frances, 187
Kauaʻi: birth of, 12; Cook at, vii–viii, 5, 6, 27; wahi pana of, 20, 221–22. *See also* Hawaiʻi; Kaʻiana
Kauanamano, Dala, 241
Kauapinao (Pinao), Alanakapu, 220
Kauikeaouli. *See* Kamehameha III
Kaʻula, 12
Kaʻulawahine, 10–11
Kaumualiʻi, 97
Kauwealoha, Samuel and Louisa, 216
kava. *See* ʻawa

Kawailepolepo, 114
Kawailiula, S. K., 221–22
Kawainui, B. W., 142, 287n70
Kawelo, 221–22
Kea. *See* Wākea
Keaala, Ioane, 157–58, 177–78, 191–92
Keaala, Serrah, 170
Kealiʻiahonui, 97
Keʻāpapalani, 13–14
Keʻāpapanui, 13–14
Keauokalani, Kepelino, 8, 209–10
Keʻeaumoku, 74, 75
Kekela, James and Naomi, 216
Kekūanāoʻa, 106
Kekuaokalani, 200
Kenao, J. D., 172
Kenn, Charles W.: as historian and lua teacher, 255; on Mahuka, 184–85; on Malo's instruction, 96; on Sutter's Fort, 164, 177–78, 289n1; on Vernon settlement, 175
Keolanui, Eli, 249–53
Keōpūolani, 200, 201–2
Kepelino, 8, 209–10
kiʻi (as images of gods), 96, 210
Kīlauea, 20, 222
kinship. *See* genealogy
knowledge versus belief, 6–7
Koko, B. L., 240–41
kōlea bird, 15–17
konohiki, 150–51
Korn, Alfons L., 201
Kū, 89, 90
Kuakini, 94, 96–97
Kūkāʻilimoku, 89
Kumu Hawaii, Ke (newspaper), 234, 236–37
kupua, 19–20
kūpuna, xiii. *See also* genealogy
Kuykendall, Ralph, 28

labor migration, 37–38, 162–64, 185–91, 239–40. *See also* economic production
Lāhaināluna seminary and school (Maui): publications of, 113, 114, 115–18, 230; pupils of, 96, 133, 153; as select school, 109–11, 112
lāhui, 120, 138
Lahui Hawaii, Ka (newspaper), 103, 152, 242, 245
Lakaakaa. *See* Waiulili
Lama Hawaii, Ka (newspaper), 230–31
Lānaʻi, 10–11, 25, 31
land distribution and rights: of American Indians, 192, 236, 237, 245; of Kānaka, 34, 37–38, 150–51, 243–45
language and literacy: binary thought and language, 6–7, 130–32, 198; book market, 100–102; in California's Kanaka settlements, 173; group reading, 99, 113, 129; Kanaka voices, ix, xii; national school system and, 105–6; ka palapala, 97–99, 113; post-annexation education, 153–55; Tahitian missionaries and, 96–97; written language, 5, 7–8, 13, 81. *See also* English-language texts; Hawaiian-language texts; Malo, Davida; ʻŌpūkahaʻia
Latin America: agriculture in, 240; Kānaka in, 163, 172; newspapers on people of, 235, 246–47
Lee Boo, 276n80
Lehua, 8, 11–12, 268n18
Lemaine (Waiulili's mother), 173, 174–75
Liholiho. *See* Kamehameha II
Liliʻuokalani: education of, 111;
genealogy of, 89; government of, 153, 158; overthrow of, 106, 153, 227; religion of, 214; traveling in United States, 162
Lodge, Henry Cabot, 251
London Missionary Society, 92
Lono, 29, 89
love, 17–18, 269nn30–31. *See also* aikāne relationship; sexuality
Lunalilo, 111, 214
Lyons, Scott, 108

Macao: description of, 52, 54, 55; geography of, 56–58; illustrations of, 51, 56–57; Kaʻiana in, 40, 47, 50–53; Ka Wahine in, 36; national differences in, 50; religion in, 53–54, 71; trade in, 52, 55, 70; unnamed Kānaka Maoli in, 36, 64–65, 276n72. *See also* China
Maguindanao, Philippines, 60–63, 64, 70
Māhele, 243–44
mahiole (feathered helmet), 43, 45, 49, 74
Mahuka, A. E., 241
Mahuka, Albert, 180
Mahuka, Edward: family of, 170, 172, 174–75; historical record on, 184; racial categorization of, 179–81, 183
Mahuka, Ellen, 170, 180
Mahuka, Jenny (Kini), 169, 170, 176, 180, 293n47
Mahuka, Rebeka, 172, 173, 174–75
Maidu, 157–58, 166, 171, 177–78, 241, 289n3
Maine, 185, 186
makaʻāinana: land and, 243–44; as "meek" in Sermon on the Mount, 101–2; role of, 31, 42–43; term, 31.

See also kalo cultivation; Kānaka Maoli

Makaainana, Ka (newspaper), 227, 242, 246–48

Makunui. See Matunui

Malo, Davida: on genealogy and geography, 13; historical contributions of, 8, 210, 219; on historical memory, 267n13; missionaries and, 81; on perspective, 271n47; as school superintendent, 106; Tahitians and, 96

malo (loincloth), 43, 49, 253

mana, 200–201; geography of, 197–98; of Mauna Kea, 203–4; methods to increase, 31–32; missionaries and, 199. See also kapu; wahi pana

manaʻo, 7

Mandell, Daniel R., 295n57

Manuel, George, 229

Manuiki, 164–65, 290n13

Maoli. See Kānaka Maoli

maps and mapping. See geography education and textbooks

Maquinna, 66, 67, 68

maritime industries of New England, 185–91, 239–40

Marquesas Islands, 4, 5–6, 14, 189, 295n1. See also Nuku Hiva; Pacific geography

marriage practices: as category, 168–71; cohabitation, 165–66, 168–69; of Kānaka and American Indians, 165–66

Massachusetts, 185–91

material culture, 47–49

Matthew 18:20, 96, 208, 282n52

Matunui (Matuunui), 216–17, 298n40

Maui: aliʻi on, 25; economic geography of, 140–41; genealogy of, 10–11, 14, 20, 38, 89; missionaries on, 79, 96, 100, 102, 277n1. See also Hawaiʻi; Lāhaināluna seminary and school (Maui)

Mauna Kea, 201–4

Meares, John, 36; description of Kaʻiana by, 39–41, 73; as Kaʻianaʻs aikāne, 43–47; on Kānaka, 69–70; portrait of, 46; on racism of Nuu-chah-nulth, 69–70; trade practices in Yuquot, 67–68

mele: record of, 220–23, 253; "Hānau a Hua Kalani," 201–4; on Hawaiian geography, 6–8, 151; "Ka Huakaʻi a Pele," 1–2; "Ka Mele a Pakui," 9–14, 203; memorization and, 98–99. See also moʻolelo; wahi pana; specific mele

"Mele a Pakui, Ka," 9–14, 203

Memoirs of Henry Obookiah, 84–86, 87–88, 279n11

memorization and education, 98–99, 113

men: labor migration and, 162–65, 171–72. See also aikāne relationship; gender; women

Mexico, 172, 235, 240, 246–47

Michael, Bernardo, 114–15

Midkiff, Frank, 253, 254

Miller, David G., 39, 72

Mindanao, Philippines, 60–63, 64, 70

mining. See gold mining

missionaries and missionary schools: catechism by, 234; Catholic, 209; denominational makeup of, 115, 208–15; desacralization of Hawaiʻiʻs geography by, 198–99, 204–8; early history in Hawaiʻi, 81; first pupils of, 91; gender and, 93–95, 109; Kānaka as, 195, 215–18; on marriage, 171; on mele and

moʻolelo, 151–52; Mormon, 214; in New England, 91; on overthrow of the kapu, 200, 296n9; ka palapala and, 97–99; publications of, 115–20, 229–36; student immigration essay, 249–53; Tahitian, 92–97. *See also* Christianity; Christian Kānaka; education; geography education and textbooks; *specific school names*
Mitchell, Donald, 253
Mitchell, William Augustus, 132
moana, 8, 129–30
mōʻī, 31. *See also specific persons*
Moʻikeha, 14
Molokaʻi: agriculture on, 140, 144; aliʻi on, 31, 220; genealogy of, 10–11, 14; Kānepuʻu and, 98, 104; mele mentioning sites on, 151; Pelekunu, 145, 146
moʻo, 19–20
moʻokūʻauhau. *See* genealogy
moʻolelo: on Hawaiian geography, 6–8, 151; publication of, 198, 218–25, 254; use of, xiii. *See also* Hawaiian-language texts; mele; wahi pana; *specific moʻolelo*
Mooolelo Hawaii, Ka (Lāhaināluna Students and Dibble), 219
"Mooolelo no Kawelo" (Kawailiula), 221–22
Mormonism, 209, 214–15
Morse, Jedidiah, 126
Mowachaht Nuu-chah-nulth. *See* Nuu-chah-nulth, Yuquot
"mulatto" categorization, 179, 185, 188
mutiny of *Felice*, 68

naʻauao versus naʻaupō, 198, 207–8, 218, 224, 232. *See also* civilizational hierarchy
Nagel, Thomas, 117
Naiomi and Auna, 93–96
Nakea, J. B., 243, 244
Nantucket Enquirer, 191
Nantucket Island, Massachusetts, 185, 186–87, 194
nation and national differences, 50–53, 61–62, 138
Native Americans. *See* American Indians
Native Hawaiians. *See* Kānaka Maoli
Natural History of Quadrupeds (Comstock), 230–31
Nāwahī, Emma Aima, 153
Nāwahī, Joseph Kahoʻoluhi, 153
New Bedford, Massachusetts, 186, 187, 189
New England, Kānaka in, 185–91, 194, 239–40, 294n52
New England Calvinism. *See* missionaries and missionary schools
newspapers, Hawaiian-language: on Hawaiian sovereignty, 152, 153, 184, 192–93, 244–45; on Kanaka–American Indian relations, 165–66, 169, 239–42; Kanaka-controlled, representation of American Indians in, 227–29, 242–46; missionary-controlled, representation of American Indians in, 229–36; printing press and, 86–87, 219; publication of traditions, 218–25, 254
Newtown, Massachusetts, 186–87
Niʻihau, 10, 12, 31, 39, 268n18. *See also* Hawaiʻi
Nisenan, 168, 172
Niuolahiki, 30
noble savage discourse, 40–41, 69, 273n35
Nonanona, Ka (newspaper), 232, 235
Nootka Island, 65–71
Nuku Hiva (Nuʻuhiwa): geography

INDEX   317

of, 19, 295n1; kinship and religion in, 195–97, 205, 215–18, 224–25, 296n4; in mele, 14; students from, 91; 'ūlei from, 19, 23. *See also* Marquesas Islands; Pacific geography
Nupepa Kuokoa, Ka (newspaper), 87, 166, 238, 240–41, 245, 287n70
Nuu-chah-nulth, 36, 65–71
O'ahu, 11–12, 221–24, 240–41. *See also* Hawai'i
Obookiah, Henry. *See* 'Ōpūkaha'ia
'ōhelo 'ai, 18
Oiaio, Ka (newspaper), 242
'oihana kahuna. *See* kāhuna and 'oihana kahuna
ōiwi, xvi, 197, 253, 263n1. *See also* Kānaka Maoli
Olepau, 241
Oliveira, Katrina-Ann R. Kapā'anaokalāokeola Nākoa, xiii, 267n13
Olopana, 14
opium trade, 52, 55, 70
'Ōpūkaha'ia: biography and genealogy of, 80, 81, 84, 87–89, 185; on geographic terminology, 21; illustration of, 85; memoir by, 84–86, 87–88, 279n11; mistaken descriptions of, 83–84
Opunui, Iosepa, 195–96, 215
Oregon, missionary activity, 235
Osorio, Jonathan Kamakawiwo'ole, xii, 108
Our World, or, First Lessons in Geography, for Children (Hall), 133, 136–37, 139, 287n59. *See also Honua Nei, Ka*

Paaniani, John, 241
Pā'ao, 14
Pacific geography: connection between Tahiti and Hawai'i,

80–81, 92–97; exploration and settlement of, 4–5, 218–25; kinship and religion, 195–97, 205, 215–18, 224–25, 296n4; Mormons on, 214–15; Polynesia term, 4; trade in, 33–36, 64–65. *See also* geography; *specific islands*
pae 'āina, 4, 8
Pāku'i, 9–12, 268n16, 268n18
Palapala Aina a me na Niele e Pili Ana, He Mau, 114–16, 121–25, 127–29, 284n23
Palapala Aina a me na Niele no ka Hoikehonua, He Mau, 116
palapala, ka, 97–99, 113. *See also* language and literacy
Palau, 63–65, 276n80
Palestine, 207–8
Palikoolauloa, M. K., 223–24
Papahānaumoku, xiii, 9–12, 202–3, 297n15
Papaula, S. W., 87–90, 279n11
Pauahi, Bernice, 111
Paulet Affair, 213
Pele: bones of, 20, 270n42; geography textbook on, 205–6; "Ka Huaka'i a Pele," 1–2; mo'olelo regarding, 152, 222; 'ōhelo 'ai and, 18; tourist representations of, 1
Pelekunu, Moloka'i, 145, 146
Perkins, Albert F., 294n52
Perry, James and Mary, 187
perspective and perspectivalism in Hawaiian thought, 21–22, 117, 121–29, 134–35
Peru, 163
Philippines, 60–63, 64, 70
pigs, 73–74, 223
Pinao, Alanakapu, 220
Piper, Karen, 117, 121
plantations. *See* sugar plantations
plants, 18–19

Pogue, J. F., 169, 181–82
Polapola, 1–2, 14, 221. *See also* Pacific geography
Polynesia. *See* Pacific geography
Polynesian Voyaging Society, 256
pono, 130, 146
Poohea, J. W. Mahelona, 145–46, 149
population, of nineteenth century Hawaiʻi, x, 3, 145
Portlock, Nathaniel, 49, 52, 53–54, 275n61
priests and priesthood. *See* kāhuna and ʻoihana kahuna
print culture. *See* palapala, ka
printing press, 86, 219
prohibition. *See* kapu
prostitution, 35
Protestantism: on American Indians, 230; Kanaka adherence to, 208; schools and publications, 115–20. *See also* Christianity
Pukui, Mary Kawena, 16, 199, 201, 255
Puu, 216–17, 298n40

race and racial hierarchy: categorization in U.S. census, 178–82; and Chinese, 52–53; in geography textbooks, 107–8, 118–20, 135, 139–40, 154; in Hawaiian legislature, 244; Indian identification and rhetoric, 70–71; and Kākaka in United States, 158–61, 191–94; in Maguindanao, 61–63; marriage and, 171–72; in New England, 161, 185–91; noble savage discourse, 40–41, 69, 273n35; and Nuku Hiva child, 238–39; and Nuu-chah-nulth, 69–70. *See also* African Americans; American Indians
Ralston, Caroline, 272n19
reading. *See* language and literacy
religion: aliʻi and, 209, 212–14; belief versus knowledge, 6–7; comparison and classification of, 91, 204–5, 210–12; connection between Tahiti and Hawaiʻi, 80–81, 92–97; cosmology, 130–32; kāhuna and ʻoihana kahuna, 31–33, 43, 54, 88–89, 94–95; kiʻi, 96, 210; in Macao, 53–54, 71; in Maguindanao, 61; newspaper disagreements and, 230; Pacific kinship and, 195–97, 205, 215–18, 224–25, 296n4; of Portlock, 275n61. *See also* Hawaiian religion; kāhuna and ʻoihana kahuna; *specific gods; specific religions*
resistance. *See* sovereignty, Hawaiian
restrictions. *See* kapu
Richards, William, 96, 100
Rifkin, Mark, 107
Royal School, 109–10, 111
Rudiments of Geography on a New Plan (Woodbridge), 115–18, 120, 285n29
Russian fort, in United States, 163–64

sacred power. *See* mana
sacred sites. *See* wahi pana
Sahlins, Marshall, 13
salt production, 140
Sāmoa: Kamehameha students and, 253–54; migration and, 4, 5, 19, 25, 221; race and, 188. *See also* Pacific geography
sandalwood trade, 102, 282n56
Sandwich Islands, 185. *See also* Hawaiʻi
schools. *See* education
Scott, Roy, 293n33
seal and sea otter hunting and trade, x, 36, 66–67, 84, 163, 228, 240
Sermon on the Mount, 100–101, 282n52

Seto Levin, Stephanie, 130
sexuality: aikāne relationships, 43–47, 90, 183; increased mana and, 31–32; in moʻolelo, 152; question of male–male relations in California, 182–83; relations in New England, 187; sexual labor, 35, 164–65. *See also* love
Shoemaker, Nancy, 161
Silva, Noenoe: *Aloha Betrayed*, xvii; on English-language texts, x; Hawaiian-language texts, 87, 184, 219; on Hawaiians and missionaries, 82; on history and colonialism, xii; on on moʻolelo in newspapers, 152
smallpox, 174, 237
Smith, John, 58
Society Islands. *See* Kahiki and Tahiti
songs and singing. *See* mele
sovereignty, Hawaiian: cultural initiatives and, 220–21; education and, 153–55, 249–50, 283n9; folklore category and, 253–55, 302n7; geography and, 81, 105–8, 138–39; Hawaiian Renaissance, 255–56; hui aloha ʻāina, 153; indigenous identification and, 227–29; of Kanaka teachers, 144–53; newspapers and, 184, 192–93, 245–48; pedagogies of, 142–44; wahi pana and, xvi, xviii, 198, 201–4, 221–24. *See also* annexation to the United States
Spaulding, Ephraim, 79, 102, 277n1
Special Newspaper Publishing Association of Honolulu, 184
Starr, Augustus, 167
storied places. *See* wahi pana
stories and storytelling. *See* moʻolelo
sugar plantations: education and, 106; newspaper representation of, 245; as site of Kanaka–American Indian conflict, 242; textbook representation of, 135, 138, 140–42
Su-My-Neh, 177–78
Sutter, John, 164–65, 178
Sutter's Fort, California, 164
Swain, John and Mary, 187, 294n56

taboo. *See* kapu
Tahiti. *See* Kahiki and Tahiti
Taiana. *See* Kaʻiana
Tāne. *See* kāne
taro. *See* kalo cultivation
"Tatars," 50, 275n52
"Tawnee," 36, 64–65, 276n72
teachers. *See* education; missionaries and missionary schools
textbooks. *See* books; geography education and textbooks
Thrush, Coll, 69, 205n13
Toketa, 94, 96–97
trade: books and, 100–102; fur, x, 33–36, 64–67, 84, 163, 228, 240; versus gifts, 67–68, 73–74; opium, 52, 55, 70; power structures of, 67–68, 70–71, 73–76; sandalwood, 102, 282n56
tribute. *See* gifts: makaʻāinana
tropics descriptions, 118–19, 131, 154
Tyanna. *See* Kaʻiana

Uhumākaʻikaʻi, 221–22
ʻūlei, 19, 23
ʻulu, 19
ʻUmi, 15–16
United Nations Declaration on the Rights of Indigenous Peoples (UN-DRIP), 229
United States: annexation of Hawaiʻi, viii, 106, 153–55, 227–29, 246–50; census racial categorizations, 178–84; Fugitive

Slave Act, 191; geography textbooks in, 126; Kamehameha IV and, 164, 213–14; race and racism in, 158–62, 173–74. *See also* African Americans; American Indians; California history; *specific places*
Universal Atlas (Mitchell), 132–33
Universal Geography. *See Rudiments of Geography on a New Plan*
Upolu, Sāmoa, 19

Vahi Hoikehonua, He, 118, 119
Vaitatu, 216–17
Valeri, Valerio, 270n38
Vancouver, George, 72, 73–76
Venuti, Lawrence, 21, 270n44
Vernon (Verona) settlement, California, 175–78, 180, 181, 293n35
"view from nowhere," 117–18, 122, 127, 150, 285n32
volcano. *See* Pele

waʻa. *See* canoe voyaging tradition
waʻa kaulua, 1–2, 256
Waccanish, 68
Wahine (Hawaiian woman on *Imperial Eagle*), 33–39, 47–48, 54–55, 59, 65, 66
wahine, 33
wahi pana: cultural record of, xvi, xviii, 198, 201–4, 221–24; defined, xiii; missionary textbook on, 205–7. *See also* mana; mele; moʻolelo
"Wahi Pana o Kaliuwaa, Na," 223–24
Wailuku Female Seminary, Maui, 109, 110
Waiulili, 166–68, 172, 174, 180
Wākea, xiii, 9–12, 202–4, 297n15
Wampanoag, 186, 189
war: Civil War depictions, 237–38; increased mana and, 31–32; by Kamehameha, 76; ʻŌpūkahaʻia and, 87–88
Watson, Bruce McIntyre, 165, 184–85
Wekeweke, A. W., 245
Westerners. *See* Haole; *specific persons*
Westervelt, W. D., 269n38
whaling industry, 161, 185–91, 239–40
"What America Means to the Immigrant" (Davis), 251
Whippey, William, 186–87
Whitney, Henry, 132, 286n58
women: Christian missionaries and, 93–95; cultural initiatives by, 254; in Macao, 54–55; material culture and, 47–49; rights in China, 55, 56; term for, 33; violence against, 35, 164–65, 167; Wailuku Female Seminary, 109, 110. *See also* gender; marriage practices
Woodbridge, William Channing, 115–18, 205
World Congress of Indigenous People (WCIP), 229
"Wynee" (Walter), 34

Young, John, 73, 76
Yuquot (Nootka Island), 65–71

Zamboanga, Philippines, 60–63, 64, 70

DAVID A. CHANG is professor of history at the University of Minnesota. He is the author of *The Color of the Land: Race, Nation, and the Politics of Landownership in Oklahoma, 1832–1929.*